Environmental Economics: Profiting from Waste Minimization

A Practical Guide to Achieving Improvements in Quality, Profitability, and Competitiveness through the Prevention of Pollution

By
Darin G. Childers
3302 Moss Creek Drive
Grapevine, TX 76051

1998

Water Environment Federation
601 Wythe Street
Alexandria, Virginia 22314-1994

Water Environment Federation

Founded in 1928, the Water Environment Federation is a not-for-profit technical and educational organization. Its goal is to preserve and enhance the global water environment. Federation members number more than 42,000 water quality professionals and specialists from around the world, including engineers, scientists, government officials, utility and industrial managers and operators, academics, educators and students, equipment manufacturers and distributors, and other environmental specialists.

For information on membership, publications, and conferences, contact

Water Environment Federation
601 Wythe Street
Alexandria, VA 22314-1994 USA
(703) 684-2400

IMPORTANT NOTICE

The contents of this publication are for general information only and are not intended to be a standard of the Water Environment Federation (WEF).

No reference made in this publication to any specific method, product, process, or service constitutes or implies an endorsement, recommendation, or warranty thereof by the Federation.

The Federation makes no representation or warranty of any kind, whether expressed or implied, concerning the accuracy, product, or process discussed in this publication and assumes no liability.

Anyone using this information assumes all liability arising from such use, including but not limited to infringement of any patent or patents.

Authorized for Publication by the Water Environment Federation
Quincalee Brown, *Executive Director*

Library of Congress Cataloging-in-Publication Data

Childers, Darin G., 1962–
    Environmental economics : profiting from waste minimization / by
Darin G. Childers.
      p. cm.
    Includes bibliographical references and index.
    ISBN 1-57278-124-6
    1. Industrial management—Environment aspects. 2. Waste
minimization. 3. Environmental economics. I. Title.
    HD30.255.C52   1998
    658.5'67–DC21   97-52223
                                   CIP

Copyright © 1998 by the Water Environment Federation
Alexandria, VA 22314-1994 USA
ISBN 1-57278-124-6
Printed in the USA **1998**

To my lovely wife, Julie, for her enthusiastic support and unyielding confidence, and to my pride and joy, Jennifer—may her world be filled with kindness, where the environment is revered as life's partner and a gift to be nurtured.

I would also like to thank my parents, Barbara and Donald, and sister, Dawn, without whom I would not have the opportunities that I have today.

PREFACE

Environmental Economics: Profiting from Waste Minimization is a practical guide to the implementation of waste minimization. This book not only covers the management techniques for initiating, planning, and executing waste minimization strategies but, just as importantly, provides the tools for accomplishing these objectives.

Waste management approaches that have been conducted by adding end-of-pipe pollution controls and disposing of waste with an "out-of-site, out-of-mind" perspective are a practice of the past. Waste minimization requires a new way of looking at things, with new technological approaches to realize the benefits. Those who wish to benefit from waste minimization techniques must first be able to understand them. *Environmental Economics* provides the first step in that direction.

A significant portion of the pollution problems facing the world today, such as global climate change, ozone depletion, and nonpoint source pollution, are not totally amenable to traditional pollution control regulation and enforcement. Protection of the environment requires preventing environmental problems by reducing or eliminating the generation of pollutants in the first place. This philosophy has taken on the phrase "waste minimization" and refers to the reduction, elimination, or recycling of pollutant discharges to the air, water, or land. Waste minimization is the key to future improvements in environmental protection but, more importantly, this philosophy produces significant benefits from a business perspective. These benefits frequently result in significant reductions in manufacturing/production costs; more efficient use of valuable resources; improvements in process/product quality; and reductions in waste generation, treatment, and disposal costs. All of these benefits yield a competitive edge in the marketplace. Additional positive results include a reduction in the generator's long-term liability for the hazardous waste produced and the potential for beneficial publicity for "doing the right thing" and being proactive.

In the future, dedicated systems that include life cycle analyses that take into consideration raw materials and associated activities, energy use, product use, and waste disposal will be required to determine the risk or true cost to the environment. The products of the future will require greater information input and less energy, materials, and labor. Companies that reflect this trend in their production will be better prepared to remain competitive. Given the increasing production costs associated with managing waste, investments in the knowledge needed to reduce these costs will be an important factor in keeping a company profitable.

Although the full potential of the benefits of waste minimization is recognized by many organizations, lack of information and knowledge is frequently the barrier to implementing waste minimization programs. Industries often lack the knowledge, motivation, or capabilities to conduct the research necessary to successfully install waste minimization projects. *Environmental Economics* strives to shape the fundamental attitudes of the business community and provide the knowledge that waste minimization can frequently produce more efficient operations while reducing the costs of regulatory compliance.

ABOUT THE AUTHOR

Darin G. Childers is an Environmental Engineering Specialist with Bell Helicopter Textron Incorporated (BHTI). He received his Bachelor of Science in Chemical Engineering from the University of Southern California in 1984 and a Master of Science in Hazardous and Waste Materials Management from Southern Methodist University (SMU) in 1994. He is currently a Doctor of Philosophy candidate in Applied Science at SMU focusing on a dissertation topic entitled *The Chemical Reduction of Volatile Organic Compounds* and is a Certified Hazardous Materials Manager (CHMM) through the Institute of Hazardous Materials Management.

With nearly 15 years in industry, Mr. Childers has acquired a vast array of experience in disciplines ranging from materials and processes research and development, manufacturing engineering, and project management to those revolving around environmental, health, and safety and the implementation of International Organization Standards (ISO) 9000 and 14000. Specific environmental responsibilities have included Clean Air Act compliance, involving a wide range of issues such as air emissions monitoring, dispersion modeling, engineering controls, and permitting; the design and automation of wastewater treatment facilities and management of their daily operations; and Resource Conservation and Recovery Act (RCRA) compliance, including oversight of daily operations at a U.S. Environmental Protection Agency (U.S. EPA) Part B permitted Treatment, Storage, and Disposal Facility (TSDF). Prior to joining BHTI, he managed numerous research and development projects and their transition to production while gaining extensive experience in the fields of composites and surface finishing. As a result, Mr. Childers is dedicated to the pursuit of waste minimization practices and the belief that business goals and environmental protection need not be opposing forces, but instead, integrated to yield mutual rewards.

As an active member in the American Electroplaters and Surface Finishers Society (AESF), he has published papers relating to both managerial and technical topics, some of which include *Environmental Economics in the Aerospace Industry, "To Be or Not To Be" ISO 14000 Certified,* and *Integrating Quality and Environmental Management Systems.* Mr. Childers is a member of the Air and Waste Management Association (AWMA), where he has served as education chair, and is a member of the Water Environment Federation (WEF) and the Academy of Hazardous Materials Managers. He has also served as an instructor for the Environmental Institute for Technology Transfer through the University of Texas at Arlington covering topics relating to hazardous waste management and groundwater contamination.

ABOUT THE REVIEWERS

Edward Forest received his B.S. degree in Chemistry from Brooklyn College in New York in 1955 and his M.A. and Ph.D. degrees in Physical Chemistry from Princeton University in New Jersey in 1963.

From 1963 to 1983 Dr. Forest was employed at Xerox Corporation in both research and product development capacities. From 1974 to 1977 Dr. Forest managed the development of a high-quality, high-throughput, color-imaging process based on inorganic and organic photoconductive pigment inks and the operations of a chemical pilot plant to manufacture the photoconductive materials and inks. In 1977 he founded the Xerox Research Laboratory in Dallas, Texas, which he managed until 1983. The focus of the laboratory was decentralized, low-throughput, ink-jet printing, both color and monochrome.

Dr. Forest has been at Southern Methodist University, the School of Engineering and Applied Science, for 12 years as Visiting Professor, where he is responsible for the management and development of interdisciplinary graduate programs and collaborative sponsored research programs. In this capacity, he has developed graduate programs in Materials Science and Engineering, Hazardous and Waste Materials Management, Software Engineering, Manufacturing Systems Management, Systems Engineering, and undergraduate programs in Environmental Engineering and Science. He currently teaches materials science and environmental programs.

His current research interests are in electronic materials, chemical degradation of volatile organic compounds, and materials structure/properties. Dr. Forest has published and is the recipient of patents in the following areas: photoconductivity, microwave spectroscopy, molecular structure, weak intermolecular interactions, electrooptical imaging systems, liquid crystal displays, ink-jet printing, facsimile recording systems, and biological photodynamic effects.

Donald L. Legg is Director of Environmental, Health and Safety for Bell Helicopter Textron, Inc. His responsibilities include regulatory compliance for both U.S. and Canadian facilities, with a strong emphasis on proactive programs. Such programs include pollution prevention, industrial accident and illness prevention, and reduction of regulatory compliance costs. Mr. Legg currently serves as the Environmental Committee Chairman of Texas Association of Business and Chambers of Commerce; Chairman, Aerospace Industries Association Environmental, Health and Safety Committee; member, adjunct faculty at Southern Methodist University in the Graduate School of Engineering and Applied Science; registered lobbyist for the State of Texas; registered Qualified Environmental Professional with the Institute of Professional Environmental Practice.

He began his career as an aircraft maintenance officer with the U.S. Air Force, attaining the rank of Captain. Following this service, Mr. Legg spent 6 years at Ford Motor

Company as plant engineer, senior design engineer, and construction supervisor. During his 12 years at Texas Instruments, he served as a site safety and environmental manager, facilities manager and facilities design engineer, and corporate environmental manager. Mr. Legg received his B.S. in Mechanical Engineering from the University of Akron and his M.S. in Systems Management from the University of Southern California.

Earl W. Turns (CEF) is an environmental consultant, materials and processes engineer specializing in elimination of hazardous materials at their source. He graduated from East Texas State University with a B.S. degree in Chemistry in 1950. Examples of completed seminars include personnel, project and new business management, business and technical communications, invention marketing, and heterogenous catalysis. His company, Earl W. Turns and Associates, was registered in 1994 and has serviced clients in the aerospace, metal-finishing, and legal professions.

Mr. Turns has 45 years of varied experience in materials and processes laboratory testing and processes implementations in aerospace and metal-finishing industries. Those experiences include photoengraving, electroplating, anodizing, adhesive bonding, and electrochemical and vacuum processes. He has published 21 technical papers in journals of six technical societies. He was granted three patents and two of those are/were in significant aerospace production. Awards and/or commendations were presented to him for exemplary performance by the American Electroplaters and Surface Finishing Society (AESF), U.S. Air Force Blue II Team, Aerospace Industrial Association (AIA), and the Lockheed Martin Corporation. Professional society affiliations include the American Chemical Society, National Association of Corrosion Engineers, AESF, AIA, Texas Inventors Society, and Air Quality Advisory Committee of the North Central Texas Council of Governments.

CONTENTS

1   Waste Minimization Strategies   1

Incentives for Implementation   1
     Resource Conservation and Recovery Act   2
     Pollution Prevention Act   3
     Emergency Planning and Community Right-to-Know Act   3
     U.S. Environmental Protection Agency's Project Excellence and Leadership   4
     U.S. Environmental Protection Agency's Common Sense Initiative   4
     Business Incentives and Factors Affecting Waste Minimization   4

Source Reduction   7
     Product Changes   8
     Process Changes   9
         Input Material Changes   9
         Technology Changes   9
             Process Automation   10
             Equipment Redesign   10
             Technology Development   10
             New Process Implementation   11
                 Distillation Recovery   11
                 Adsorption   11
                 Filtration   12
                 Ultrafiltration   12
                 Reverse Osmosis   12
                 Electrolysis   12
                 Electrodialysis   12
         Operating Practices   12
             Source Segregation   12
             Housekeeping   13
             Inventory and Material Control   13
     Case Studies   14
         Printed Circuit Board Manufacturing Study   14
         Chemical Manufacturing Study   16

Recycling   17
     Use/Reuse   17
     Reclamation   18
         Resource Recovery   18
         Refrigerant Recycling   18

Opportunity Assessments: A Phased Approach   19

Waste Minimization Opportunity and Assessment Manual
    (EPA-625/7-88-003)   19
User's Guide: Strategic Waste Minimization Initiative (EPA-625/11-91-004)   19
Industrial Pollution Prevention Opportunities for the 1990s
    (EPA-600/8-91-052)   20
Program Planning and Organization   20
Assessment Phase   22
Feasibility Analysis   24
Implementation   25

References   26

Suggested Readings   27

2   Process Definition and Control Techniques   29

Chemical Engineering and the Material Balance   29
    Engineering Calculations   30
        Units and Dimensions   30
            Unit Conversions   30
            Unit Manipulation   31
            Pressure Unit Conversions   32
            Absolute and Gauge (Relative) Pressure   34
            Temperature Unit Conversions   34
        Chemical Equations and Stoichiometry   35
            Atomic Weight Calculations   35
            Molecular Weight Calculations   35
            Balancing a Chemical Equation   37
            Reaction Yields   37
        Techniques and Conventions in Problem Solving   38
            Process Optimization Using Chemical Equations   39
            Process Optimization by Unit Manipulation   39
    Material Balances   40
        Material Balance   40
            Material Balance Basics   42
            Sludge Drying   42
            Solvent Recovery   44
        Analytical Approaches to Material Balance Problems   45
            Material Balance for the Combustion of Methane   48
            Material Balance to Illustrate the Benefits of Countercurrent
               Rinsing   49
        Recycle, Bypass, and Purge Calculations   52
            Recycle of Biosolids at a Wastewater Treatment Plant   53

Statistical Process Control   55
   Statistical Process Control Overview   56
      Frequency Tables and Histograms   57
      Probability Plots and Control Charts   61
   Process Capability   67
      Variables Data Collection   68
      $\bar{X}$-R Charts   69
      Constructing an $\bar{X}$-R Chart   71
      Types of Patterns   72
      Capability Determination Methods   75
   Quality Characteristics   79
      C-Charts   80
      U-Charts   81
      NP-Charts   82
      P-Charts   82

Data Acquisition Systems   83
   Instrumentation   84
      Liquid Flow Measurements   84
      Liquid Level Measurements   87
      pH Measurement   91
         pH and Hydrogen-Ion Concentration   92
      Conductivity Measurements   95
      Oxidation-Reduction Potential   97
      Temperature   99
      Pressure   101
   Computer Control and Integration   103

Suggested Readings   106

3   Chemical Alternatives and Process Modifications   107

Metal Finishing   109
   Drag-Out Reduction   110
      Withdrawal Rates and Drain Time   110
      Racking and Drain Boards   111
   Rinsewater Management   113
      Countercurrent Rinsing Case Study   113
      Countercurrent Rinsing   113
      Static Rinsing Versus Spray Rinsing   115
      Flow Restrictors and Conductivity Cells   117
      Rinse Bath Agitation and Contact Time   117
   Solution Recovery and Recycling   119
      Ion Exchange   120
      Electrolytic Recovery   124

Membrane Processes   126
Evaporation   130
Bath Composition   131
Chemical Replacement   132
Chromium   133
Cyanide   134
Metal Stripping   134

Solvent Cleaning and Degreasing   136
Available Technologies   142
Aqueous and Semiaqueous Cleaners   143
Petroleum Hydrocarbons   147
Hydrochlorofluorocarbons   148
Supercritical Fluids   149
Miscellaneous Organic Solvents   151
Technological Developments   153
Wet Oxidation Cleaning   153
Absorbent Media Cleaning   154

Surface Coating   155
Low-Cost Emission Control Strategies   160
Spray Application Techniques   160
High-Volume, Low-Pressure Spray Guns   161
Air-Assisted Airless Processes   161
Electrostatic Spraying   162
Equipment Options   163
Automated Multicomponent Proportioners and Mixing   163
Enclosed Gun Cleaners   164
Recovery Techniques   165
Oxidation   165
Chemical Replacement   166
Solvents and Water-Based Paints   167
High-Solids Low Volatile Organic Compound Based Paints   167
Water-Based Paints   168
Calculation of Volatile Organic Chemical Content   168
Stripping Options   170
Heavy Metals   171
Plastic Media Blasting   172
Dry Ice Blasting   173

References   174

Suggested Readings   174

4   Engineering Decision Analysis and Economics   177

    Tools of Engineering Economics   177
        Interest Factors and Depreciation   177
            Straight-Line Depreciation   181
            Sum-of-Digits Depreciation   181
            Declining-Balance Depreciation   181
            Sinking-Fund Method   182
        Inflation and Taxes   183
            Inflation and Taxes Scenarios   184
        Cash Flow   185

    Methods for Project Analysis   187
        Present, Annual, and Future Worth   187
            Compound-Amount Factor   188
            Present-Worth Factor   188
            Sinking-Fund Factor   188
            Series Compound-Amount Factor   188
            Capital-Recovery Factor   189
            Series Present-Worth Factor   189
        Net Present Worth and Rate of Return   190
            Net Present Worth for a Waste Minimization Proposal   191
            Net Present Worth Evaluations Versus Assets of Unequal Lives 193
            Minimum Acceptable Rate of Return   196
            Project Comparison Using Rate of Return   197
        Benefit/Cost Ratio   199
            Benefit/Cost Analysis   199

    Sensitivity and Risk Analysis   200
        Risk Analysis of the Ion-Exchange Program   201

    Hidden Costs   203

    Suggested Readings   204

5   Standards Driving Environmental Economics   205

    ISO 9000 Standards   206
        Overview of Core Requirements   207
        Quality Systems   210
            Management Responsibility   211
            Quality System   212
            Contract Review   212
            Design Control   212
            Document and Data Control   213
            Purchasing   214
            Control of Customer-Supplied Product   214
            Product Identification and Traceability   214

Process Control   214
Inspection and Testing   215
Control of Inspection, Measuring, and Testing Equipment   215
Inspection and Testing Status   216
Control of Nonconforming Product   216
Corrective and Preventive Action   216
Handling, Storage, Packaging, Preservations, and Delivery   217
Control of Quality Records   217
Internal Quality Audits   218
Training   219
Servicing   219
Statistical Techniques   219

ISO 14000 Standards   219
Environmental Management Systems   221
Initial Review and Gap Analysis   223
Self-Declaration Versus Registration   224
Environmental Auditing   226
Life-Cycle Assessment   229
Inventory Analysis   230
Impact Assessment   231
Interpretation   232
Reporting and Critical Review   233
Inclusion of Environmental Aspects in Product Standards   233

References   239

Suggested Readings   239

Appendix A 40 CFR 261—Identification and Listing of Hazardous Waste   241

Appendix B 1990 Clean Air Act Amendments—Section 112: List of 188
           Hazardous Air Pollutants (HAPs)   261

Appendix C Discrete Compounding Interest Factors—Single Payment and
           Uniform Series   269

Index   279

LIST OF TABLES

2.1 Dimensions with multiple units.   32
2.2 Summary of effluent characteristics.   50
2.3 Factors for determining control limits.   70
2.4 Sample data.   71
2.5 Sample $\bar{X}$-R control chart parameters.   72
2.6 pH versus hydrogen- and hydroxide-ion concentration at 25°C.   91

3.1 Costs and savings of the countercurrent rinsing study.   114
3.2 Strong-acid cationic exchange affinities.   122
3.3 Ion-exchange applications.   123
3.4 Noncyanide alternatives for metal stripping.   135
3.5 Chlorinated cleaning compounds and their traditional applications.   137
3.6 Title VI—chemicals of concern.   140
3.7 Title VI—phase-out requirements.   141
3.8 Composition requirement for approved cleaning solvents—NESHAP for
     aerospace manufacturer and rework facilities.   141
3.9 Critical-point values for selected compounds.   150
3.10 Physical characteristics of several commonly used organic solvents.   152
3.11 Pigments used as colorants.   167
3.12 Coating data example.   169
3.13 Coating data example with thinner added.   169

4.1 Interest factors for discrete compounding.   187
4.2 Waste minimization proposal cost and material data.   191
4.3 Rate-of-return values.   197
4.4 Rate of return and net present value for options 1 and 2.   199
4.5 Benefit/cost analysis summary.   200
4.6 Ion-exchange proposal.   202

5.1 Generic product categories.   208
5.2 Quality system requirements for ISO 9001 and 9002.   211

LIST OF FIGURES

1.1	Source reduction methodologies.	8
1.2	Recycling methodologies.	18
1.3	Program planning and organization.	21
1.4	Assessment phase.	23
1.5	Feasibility analysis.	24
1.6	Implementation.	26
2.1	Conceptualization of pressure: force per unit area.	33
2.2	Gauge versus absolute pressure.	33
2.3	Multimedia mass balance.	41
2.4	Sludge drying process flow diagrams.	43
2.5	Solvent recovery process flow diagram.	45
2.6	Material balance, no chemical reaction.	46
2.7	Material balance, chemical equation.	47
2.8	Traditional rinsing scheme.	51
2.9	Countercurrent rinsing scheme.	51
2.10	Recycle streams.	53
2.11	Bypass streams.	53
2.12	Purge streams.	54
2.13	Activated-sludge process flow diagram.	54
2.14	Frequency table.	58
2.15	Data compilation form.	59
2.16	Normal distribution.	60
2.17	Completed frequency table.	62
2.18	Sample histogram of compiled data.	63
2.19	Probability plot of cumulative percentage data.	64
2.20	Standard deviation and mean for normal distribution.	65
2.21	$\bar{X}$-R control chart.	67
2.22	$\bar{X}$-R chart trend.	73
2.23	$\bar{X}$-R chart runs.	74
2.24	$\bar{X}$-R chart outlier.	74
2.25	$\bar{X}$-R chart mixture.	75
2.26	$\bar{X}$-R chart sudden changes in level.	75
2.27	Capability ratio values and product specification limits.	77
2.28	C-chart illustration.	81
2.29	NP-chart illustration.	83

2.30	Laminar and turbulent flow patterns.	85
2.31	Open-channel flow devices.	86
2.32	Float and magnetic proximity switches.	88
2.33	Capacitive liquid measurement.	89
2.34	Ultrasonic liquid measurement.	90
2.35	Simplified pH control system.	94
2.36	Conductivity measurement.	96
2.37	Thermocouple circuits.	99
2.38	Radiant energy principles.	101
2.39	Pressure-sensing devices.	102
2.40	Hardware configurations for data collection.	104
3.1	Conceptualization of process chemical drag-out.	111
3.2	Drain board configuration.	112
3.3	Drag-out reduction and racking.	112
3.4	Traditional single-tank rinsing configuration.	114
3.5	Countercurrent rinsing two-tank configuration.	115
3.6	Static and countercurrent rinsing in tandem.	116
3.7	Conductivity cells.	118
3.8	Ion-exchange rinsewater recovery.	121
3.9	Electrolytic recovery.	125
3.10	Electrodialysis.	127
3.11	Ultrafiltration.	128
3.12	Osmosis and reverse osmosis.	129
3.13	Reverse osmosis applications.	130
3.14	Evaporation for chrome plating bath recovery.	131
3.15	Use of chlorofluorocarbon compounds in the U.S.	138
3.16	Chemical structures and formulas of common semiaqueous cleaning compounds.	146
3.17	Phase diagram for carbon dioxide.	149
3.18	Process flow schematic for carbon dioxide supercritical fluid cleaning system.	151
3.19	Typical wet oxidation process.	153
3.20	High-volume, low-pressure spray gun application system.	161
3.21	Air-assisted airless spraying technique.	162
3.22	Electrostatic spray process.	163
3.23	Automated proportioner and mixing.	165
3.24	Plastic media blasting selection criteria.	173
4.1	Comparison of depreciation methods.	180
4.2	Cash flow diagram.	185
4.3	Cash flow diagram for existing paint-mixing process.	186
4.4	Cash flow diagram for waste minimization opportunity—multicomponent paint mixing.	186
4.5	Cash flow diagram for option 1 (20-year period).	193

4.6   Cash flow diagram for option 2 (20-year period).   194
4.7   Cash flow diagram for minimum acceptable rate of return.   197
4.8   Plot of rate of return versus net present value.   198

5.1   Quality management structure.   208
5.2   Environmental management system evolution.   220
5.3   Environmental management system continuous improvement cycle.   222
5.4   Steps toward ISO 14001 registration.   225
5.5   Stressor-effects network of nitrogen oxides emissions.   232
5.6   Product standards and their effects on the environment.   235
5.7   Conceptualization of the material balance.   237

1 Waste Minimization Strategies

Incentives for Implementation

The term *waste minimization,* or *waste reduction,* was adopted in the 1980s during a period of rapid growth of environmental concerns and ensuing formulation of environmental statutes and regulations in the U.S. Waste minimization derives its meaning from a proactive desire to avoid the generation of wastes and pollution rather than managing them after they are created. From a regulatory perspective, waste minimization refers to activities in which the volume or toxicity reduction of hazardous wastes is accomplished through either source reduction or recycling techniques. Of these two latter approaches, source reduction is often preferred to recycling because it avoids or greatly minimizes the generation of wastes from the onset.

This chapter is intended to provide a basic understanding of waste minimization and to act as a vehicle for initiating effective programs. Several industrial examples are included to add credence to the practice and energize its pursuit. By reviewing the basic concepts of project management and the sequence of activities required for effective implementation of cost-saving and environmentally friendly alternatives, other more specific tools may be introduced. In addition, the topics that follow this chapter form the staple of any and all waste minimization efforts.

Although the concept of waste minimization refers to a proactive approach to environmental issues, it is often perceived as being counter to the direction of today's ever-evolving world market. Rather than focusing on methods to reduce wastes, it is often more appropriate to focus on issues of product quality and operating costs, for example. These items have not traditionally been associated with addressing environmental topics. However, by combining manufacturing efficiencies and enhancements with quality while considering the environment, greater rewards may be realized. Indeed, business profitability and the environment are no longer on opposite ends of the spectrum. Moreover, penalties are typically incurred when environmental issues are considered as an afterthought.

This problem can be illustrated by examining the plight of a theoretical painting facility, where long-term environmental impacts were not considered in its design. For example, the facility is constructed and employs what are perceived as traditional coatings. After construction and start-up operation of the facility, however, facility operators realize

1

that the coatings currently in use contain high levels of volatile organic compounds (VOCs) and will no longer be in compliance with state and federal regulations. To continue operations, facility operators must either install expensive abatement equipment to incinerate the facility's emissions or find substitute coatings with low concentrations of VOCs to comply with regulations. If facility operators immediately install abatement equipment, the resulting scenario may be high capital and maintenance costs and perhaps an attitude that environmental issues are always in conflict with business survival. However, if a more positive approach is taken, such as a genuine desire to have the business mesh with the public and the environment, alternate coatings may be found or developed with low VOCs so that the high-dollar abatement process may be avoided altogether. Although this scenario is a simplified one, its example should be heeded. The earlier environmental issues are addressed, the less likely they are to become a problem. Moreover, when evaluated alongside everyday business decisions, environmental issues may be pivotal to business quality and profitability.

The benefits of successful waste minimization programs frequently result in significant reductions in manufacturing/production costs, more efficient use of valuable resources, improvements in process/product quality, and reductions in waste generation, treatment, and disposal costs. Additional benefits of waste minimization programs may include a reduction in a generator's long-term liability for any hazardous wastes produced, and the potential for positive publicity for being proactive and "doing the right thing."

Before elaborating on some of the strong business incentives for implementing waste minimization programs, a brief review of the federal laws that require this approach are highlighted. These statutes include the Resource Conservation and Recovery Act (RCRA), the Pollution Prevention Act, and the Emergency Planning and Community Right-to-Know Act. Also briefly discussed are the relatively new and upcoming regulatory approaches embodied under the U.S. Environmental Protection Agency's (U.S. EPA's) Project Excellence and Leadership, or Project XL, and the Common Sense Initiative (CSI). These statutes and regulations have specific sections that address waste minimization issues, indicating the importance the federal government has placed on the topic. Numerous states have also passed legislation regarding pollution prevention. Although each state's interpretation of pollution prevention may be different, waste minimization has become a requirement even though it may best serve as a powerful business tool.

RESOURCE CONSERVATION AND RECOVERY ACT

As stated by the U.S. Congress in Section 1003(b) of the 1984 Hazardous and Solid Waste Amendments (HSWA) to RCRA,

> The Congress hereby declares it to be the national policy of the United
> States that, wherever feasible, the generation of hazardous waste is to be
> reduced or eliminated as expeditiously as possible. Waste that is nevertheless
> generated should be treated, stored, or disposed of so as to minimize the
> present and future threat to human health and the environment.

In addition, HSWA requires that generators of hazardous wastes report on activities initiated to reduce the volume and toxicity of waste streams and certify that the program to minimize waste generation is in place. The following certification is found above the generator's signature on each uniform hazardous waste manifest:

> If I am a large-quantity generator, I certify that I have a program in place to reduce the volume and toxicity of waste generated to the degree I have determined to be economically practicable and that I have selected the practicable method of treatment, storage, or disposal currently available to me which minimizes the present and future threat to human health and the environment; OR, if I am a small-quantity generator, I have made a good-faith effort to minimize my waste generation and select the best waste management method that is available to me and that I can afford.

POLLUTION PREVENTION ACT

The Pollution Prevention Act of 1990 states that pollution should be prevented or reduced at the source wherever feasible, and pollution that cannot be prevented should be recycled in an environmentally safe manner. In cases where prevention or recycling opportunities are not feasible, wastes should be treated. However, disposal or other releases into the environment should be conducted only as a last resort. The following are the fundamental goals of U.S. EPA's Pollution Prevention Research Program (1991b):

- Stimulate private-sector development and use of products, technologies, and processes that result in reduced pollution;
- Expand the reusability and recyclability of wastes and products and the demand for recycled materials;
- Identify and promote the implementation of effective socioeconomic and institutional approaches to pollution prevention;
- Establish a program of research that will anticipate and address future environmental problems and pollution prevention opportunities; and
- Conduct a vigorous technology-transfer assistance program that facilitates pollution prevention strategies.

EMERGENCY PLANNING AND COMMUNITY RIGHT-TO-KNOW ACT

The Emergency Planning and Community Right-to-Know Act serves to identify wastes and processes for which pollution prevention techniques can be initiated on a large scale. Large U.S. manufacturing operations with toxic chemical production or usage levels above a certain amount are required to report annually on their air, water, or land emissions in addition to any off-site transfers to waste management facilities of more than 600 types of toxic chemicals or chemical categories. Although these manufacturing operations can claim chemical identities as trade secrets, the practice is rare. An optional section of the report allows manufacturers to list waste minimization activities and the effect these activities have had on toxic releases. This information is compiled in U.S. EPA's Toxic Release Inventory, which is available to the public. Data from the Toxic Release

Inventory—useful to U.S. EPA in determining which toxic chemicals should be reviewed and possibly regulated differently—serve as diagnostic tools to help manufacturers identify and address sources of pollution. The Toxic Release Inventory is also a primary mechanism for the public to access details on a given manufacturer's operations.

U.S. ENVIRONMENTAL PROTECTION AGENCY'S PROJECT EXCELLENCE AND LEADERSHIP

President Clinton, through the Reinventing Environmental Regulation Initiative, directed the U.S. EPA to create Project XL on March 16, 1995. The entire effort is designed to allow companies the freedom to develop new technologies and innovative strategies for achieving cleaner and cheaper environmental results than may be accomplished through the current regulatory system.

Although there are guidelines for this policy, each project under Project XL will essentially be granted regulatory flexibility by U.S. EPA in exchange for commitments to achieve better environmental results than would be accomplished through full compliance with existing regulations. U.S. EPA's goal is to implement 50 pilot projects in the four following categories: Project XL projects for facilities, various industrial sectors, government agencies, and communities.

U.S. ENVIRONMENTAL PROTECTION AGENCY'S COMMON SENSE INITIATIVE

U.S. EPA's CSI represents an entirely different approach to environmental policy in the U.S. Rather than using traditional tactics such as employing a regulatory framework that has been set up to handle issues on a media-by-media basis without regard for intrinsic differences between the wide range of industries to which they apply, U.S. EPA has developed a more wholistic method. That method operates under a philosophy that requires that whole facilities and industries be looked at for their overall impact on the environment.

Teams developing this new approach comprise individuals from industry, environmental justice and community organizations, labor, and regulatory bodies. The idea is to provide consensus-based strategies to public health and environmental protection that are flexible and innovative alternatives to the system currently in place. Industries currently participating in the strategizing include automobile manufacturing, computer and electronics, iron and steel, metal finishing, petroleum refining, and printing.

Several CSI projects include reducing duplicate reporting requirements, streamlining the permitting process, increasing community involvement in the environmental decision-making process, and creating better incentives with fewer barriers to the implementation of pollution prevention projects.

BUSINESS INCENTIVES AND FACTORS AFFECTING WASTE MINIMIZATION

What are the incentives to implementing waste minimization programs considering all the regulatory issues? According to *Hazardous Waste Minimization* (Freeman, 1990), the benefits of waste minimization often cited by U.S. EPA are minimizing quantities of regulated hazardous wastes generated, thereby reducing waste management and compliance costs; improving product yields; reducing or eliminating inventories and releases of hazardous chemicals reportable under Title III of the Superfund Amendments and Reauthorization Act; and lowering Superfund, corrective action, and toxicity tort liabilities.

Although these benefits may be somewhat limited in that they tend to focus on the term *wastes*, they may be reduced to a concept that is referred to as *process and quality control*, which is more recognizable to industry. In terms of mass balance, the more efficient the process is, the higher and more consistent the quality and, subsequently, the less the quantity of waste produced. This often leads to higher profit margins and may also be expanded to what is known as life-cycle design/engineering.

Another important business incentive to implementing waste minimization programs is positive publicity. Because the public today is more informed about environmental issues and realizes the potential effects that hazardous wastes and the release of pollutants can have on human health and the environment, companies that exhibit a strong environmental awareness and consistently work toward waste reduction may improve their public image.

Similar to any business-related action, waste reduction projects require justifications for their expense, regardless of the payback period. A frequent question that arises when considering the implementation of waste minimization projects is "If we already have an adequate disposal arrangement, why should we go to all the trouble?" Two significant reasons to do so are that resource conservation and recycling can greatly reduce raw material costs while lowering disposal costs, thus increasing profitability, and the efficiency provided by implementing waste minimization often yields tighter controls on production processes, which, in turn, provide higher product quality.

In addition, technologies exist today for industry to dramatically reduce or, in some cases, eliminate hazardous wastes discharged into the environment. Given enough time and attention, the waste reduction concept could significantly reduce the occurrence of Superfund sites, allow manufacturers to escape the liability insurance spiral, and produce a healthier and cleaner environment.

Like any new idea, however, the concept of waste minimization has its supporters and critics. Those who are skeptical of the effectiveness of waste reduction have cited such issues as unsupportive regulations, unreliable machinery, and high-cost roadblocks. Nevertheless, for the majority of those companies that have implemented waste reduction programs, the results have been consistently positive and, more times than not, profitable. Moreover, those same businesses often report that pitfalls do exist but that the methodology and concept of waste minimization are sound and make sense economically and environmentally.

In some cases, waste reduction solutions may require considerable capital investment. In fact, some companies are implementing waste reduction programs to meet evolving state pollution control regulations and to avoid problems associated with the rising costs of landfilling and the dwindling availability of disposal options. However, most companies have found that capital investments can pay for themselves in a relatively short period of time—usually less than 2 years and sometimes only a few weeks. Traditionally, the substitution of raw materials can turn out to be cheaper and the implementation of new equipment more energy efficient while providing a company with product quality improvements.

The various stages of a company's life may also play a significant role in the success with which waste minimization programs are implemented. For example, companies in their formative stages are typically undergoing development and expansion, whereas those that have been in operation for a number of years may have settled into somewhat of a stabilized existence. The latter type of business often encounters more difficulty in initiating a waste minimization program because of its established production patterns. However, factors that differentiate established companies may pertain to the type of overall business. Compare, for example, industries with manufacturing processes that rarely change (for example, the aircraft industry) versus those that are constantly evolving (for example, the electronics industry). In general, it is easier to implement changes when processes must be redesigned as a business necessity rather than for the reason of "doing the right thing."

In addition to these factors, a company's size and financial status can be critical to implementing waste reduction strategies. Business resources such as research facilities and environmental/engineering expertise often depend on the size of the company. Large facilities or companies with multiple production facilities can support these specialized areas more easily than their smaller counterparts. Smaller facilities, in many cases, may not have the time or personnel to learn about waste reduction opportunities. Similarly, financial status is an important factor. In general, the larger the company is, the easier it is to allocate funds for process improvements and proactive measures, which are the crux of waste minimization.

Regardless of a company's size and financial status, management structure dominates the success or failure of a comprehensive waste minimization program. Therefore, it is important that waste reduction efforts be initiated and promoted from top management and disseminated to all management levels within a company. In addition, employee involvement, such as the formation of waste reduction teams, can yield innovative solutions and a sense of accomplishment that may serve to strengthen the entire effort.

Strategies for marketing waste minimization are essentially the same as those for marketing anything else. For example, general marketing goals for most companies are to attract customers and to retain their interest over time; marketing goals for the concept of waste reduction are no different. In addition, securing the attention of company employees and keeping them focused and motivated are some common goals. One way to attract their attention is to establish a competition, or goal, in a particular area of production that will create a sense of accomplishment. Finally, ideal areas to focus marketing strategies for waste minimization on are tied to economic incentives, regulatory compliance, and corporate objectives, rather than on the prospective technologies that are required to implement waste minimization strategies.

Determining the target audience for a waste minimization program is also critical. For most businesses, the primary target audience consists of productions managers and engineers, for example. This audience may be expanded to operators, suppliers, and customers as the program develops.

As the target audience is identified, it is important to establish a plan for getting participants to buy into the project. From a company's perspective, these plans may differ with the employee's point of view. While upper management may be interested in overall company benefits such as profit margins and product quality, for example, shop floor operators may be interested in how a program may streamline production processes and affect their jobs. Thus, the marketing plans for these two groups will be quite different. Similarly, where awards and recognition may be a suitable incentive for employees, educational efforts and a heightened awareness of the technical and business ramifications from waste reduction may be a more viable approach for management.

The next phase in effectively marketing a waste reduction program is to establish a long-term commitment to the program and to instill independent thinking and attitudes to get project participants to "spread the word" about it on their own. If the foundation of the program is properly laid out from the onset by identifying the appropriate target audiences and establishing a framework for its acceptance, achieving long-term commitment to a waste reduction program may be as simple as holding regular team meetings to keep the concept of waste minimization fresh in the employees' minds. Moreover, if the implementation of ideas is handled well and success stories can be accumulated, advocates of a program will be generated and waste minimization will naturally progress to an everyday effort.

Finally, it is important that once the concept of waste reduction has been accepted, interest in it does not fade. Methods that ensure that management and employees continue to contribute to waste reduction are critical to the long-term success of established and future programs. A common approach for companies in the past was to buy pollution control equipment to avoid regulatory compliance issues and enforcement. This approach, however, was frequently costly and provided little financial benefit to companies.

Today, most companies look for more progressive means to achieve regulatory compliance, and they buy into waste minimization for multiple purposes (although regulatory compliance may still be the primary objective). The additional benefits from today's approaches have expanded to include areas such as process efficiencies, decreased operating costs, and improved public image. Eventually, the concept of waste minimization may be perceived as a sound business practice, rather than a tool for environmental or regulatory compliance.

Source Reduction

There are many technologies, techniques, and management approaches involved in waste reduction. This section describes some of the most commonly used methods for implementing these projects. Source reduction is the reduction or elimination of hazardous wastes at the source and typically occurs within a process. The concept, as illustrated in Figure 1.1, includes changes to the product itself and source control measures such as input material changes, technology changes, and improved operating practices. It is considered the highest in priority of all the waste minimization approaches because it focuses on the elimination/reduction of wastes from the onset.

The following are a number of examples in which source reduction techniques have been found to be successful. These examples demonstrate the potential environmental and economic benefits of waste minimization and may help to clarify the definition of source reduction in contrast to recycling/reuse methodologies. Many of the examples are success stories that have been extracted from U.S. EPA documents. Furthermore, they are representative of the ongoing efforts between the regulatory community and businesses to develop open channels of communication and of the pursuit of mutually beneficial solutions to environmental problems.

PRODUCT CHANGES

Product changes primarily pertain to compositional changes and modifications made to the final product, or the method of its use, to minimize environmental impacts. The following example is indicative of this process and serves as an introduction to measures

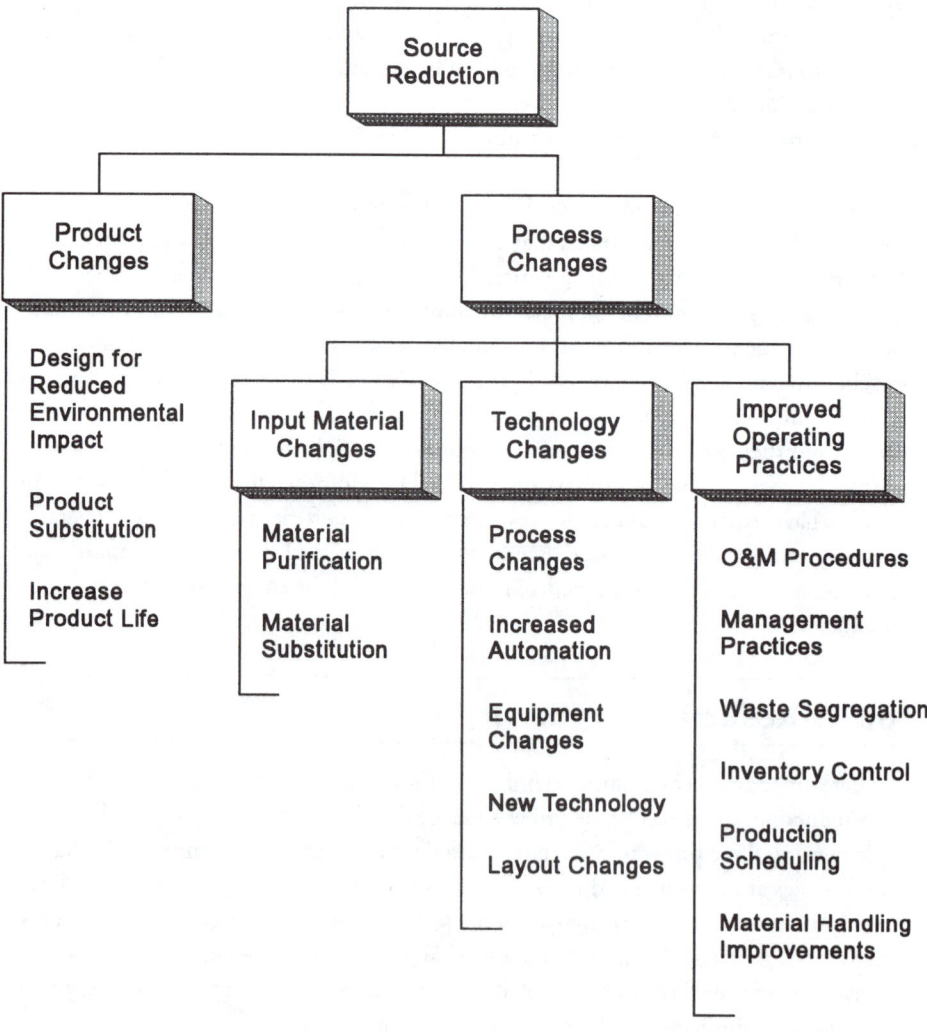

Figure 1.1   Source reduction methodologies (O&M = operations and maintenance).

such as surface coatings, chemical alternatives, and process modifications discussed in Chapter 3.

> Riker Laboratories of California previously used an organic-based solvent as a carrier for coating its pharmaceutical tablets. Faced with stricter air pollution regulations, the firm first investigated the possibility of installing pollution control equipment. The initial cost of such equipment would have been $180,000, and in addition, $30,000 would have been required to operate the equipment for an entire year. As an alternative to such a high expense, Riker decided to explore the waste reduction approach. By making a $60,000 investment in equipment modifications, the company was able to switch from an organic solvent-based coating to a water-based coating. This eliminated the pollution problem completely and the need for pollution control equipment. The water-based material worked just as well in the tablet coating process and resulted in a $15,000/yr savings in raw material costs (Institute for Local Self-Reliance, 1986).

PROCESS CHANGES

Process changes focus on how a product is produced rather than on changing the effects a finished product may have on the environment. These efforts revolve around the process itself and traditionally are performed more expeditiously than product changes.

Input Material Changes

This approach to source reduction, which may be the most logical and comprehensive way to eliminate waste, involves replacing the polluting ingredients of a product with less toxic ones. However, the material substitution approach may be applicable only in certain situations. The most economical approach is to consider these effects when a product is being developed for the first time. With existing products, on the other hand, material substitution can be expensive because new chemicals may frequently require additional investments in new equipment.

The U.S. Department of Defense, for example, developed a process in which small plastic beads are air blasted at the surface of an airplane to remove paint. This eliminates the need for hazardous solvents to remove the paint. The U.S. Department of Defense estimates that this process has decreased the amount of hazardous wastes from 4 500 kg (10 000 lb) of wet sludge to 145 kg (320 lb) of dry paint chips and disintegrated plastic material per aircraft. In addition, the amount of work required per aircraft to remove the paint by air blasting is eight times less than by traditional methods (U.S. EPA, 1991).

Technology Changes

Technology changes are perhaps the last course of action to pursue in a source reduction program and are typically the most costly. Process automation, equipment enhancement or replacement, and the use of new technology fall under this category. Although not all options are discussed herein, the following overview should provide the framework for future thought.

PROCESS AUTOMATION. Bell Helicopter Textron, Inc., Fort Worth, Texas, in an effort to reduce the amount of rinsewater generated at its various chemical process lines, has implemented, where feasible, countercurrent rinsing. Countercurrent rinsing is essentially a method by which parts removed from a chemical bath are passed through a series (two or three) of rinse tanks. The water in the rinse tanks flows counter to the direction of the parts. In this way, the final rinse is maintained at the desired purity, although the other rinse tanks incur sequentially higher contaminant levels. Instrumentation and automation to control the flow of fresh water on an as-needed basis and thus control the quality of the final rinse was accomplished as well. Measurements at one of the countercurrent rinsing stations indicated weekly cost savings of $4 300. The implementation cost was approximately $13 150. Therefore, the project paid for itself in approximately 3 weeks (Childers, 1997).

EQUIPMENT REDESIGN. An example of equipment redesign achieving successful results is the case of used chlorinated solvent cleaners, such as tetrachloroethylene, which are byproducts of degreasing metal parts before painting or assembly. In most cases, a company's first step in reducing the production of these waste solvents is to examine the degreasing equipment to see if any modifications can be made to prevent solvent evaporation. Three such modifications include raising the freeboard height to 75% of the tank width instead of the conventional 50% tank width; placing covers on open-tank systems whenever the system is not in use; and installing condensing coils, known as chillers, near the top of vapor degreasers.

Federal and state regulations stipulate that some of these configurational parameters be in place to achieve compliance. It is an approach to force reductions in the emission of hazardous air pollutants, but also a coerced method of waste minimization. Nevertheless, companies that have modified or redesigned their degreasers typically have found that the improvements easily pay for themselves in a relatively short period of time.

TECHNOLOGY DEVELOPMENT. Sometimes it is not a product that causes pollution, but the process used in manufacturing the product. The solution is similar to material substitution except that the replacement of the old equipment with a new technological approach is pursued to avoid producing harmful byproducts. However, in many situations, technology must be used to recycle valuable materials and actually takes the form of add-on processes rather than replacement.

One such example of new technology development, in contrast to technology implementation, is depicted by the establishment of the National Defense Center for Environmental Excellence (NDCEE). The NDCEE, which is operated by Concurrent Technologies Corporation, has established the Demonstration Factory to showcase new environmentally acceptable technologies and processes in an actual factory setting for U.S. Department of Defense and industrial clients.

The Demonstration Factory set up by NDCEE was designed to be a low-cost, low-risk alternative to on-line, in-house evaluations of new material and equipment technologies in the areas of parts cleaning, coating removal, organic coating, and inorganic coating. The organic finishing line, for example, at NDCEE's Demonstration Factory can evaluate, demonstrate, and validate metal pretreatment and organic finishing operations with state-of-the-art, environmentally safe, and transfer-efficient manufacturing processes that include process discharge recovery and recycling. Designed to accommodate full-scale demonstration and validation of approximately 80% of manufactured products, it is capable of handling parts measuring up to approximately 0.9 m wide by 1.2 m high by 1.2 m long (3 ft wide by 4 ft high by 4 ft long) and weighing up to 110 kg (250 lb). The facility can accommodate at least two clients each week. The entire system is automatically controlled by programmable logic controllers/personal computers with graphical monitoring and control screens to archive critical process data (Docherty *et al.*, 1995).

NEW PROCESS IMPLEMENTATION. Not all companies are fortunate enough to be able to participate in a venture of this sort. Rather, most rely on the implementation of existing, reliable, and proven technologies to achieve waste reduction goals. Therefore, most of the following discussions address new process implementation and are focused on closed-loop technologies because they are more broadly applied and less product-specific. These technologies are by no means all of the solutions available, but they reflect an obvious trend that not all technologies must be new, innovative, or risky. These processes represent intelligent solutions based on economic decision analyses. Most of the topics that follow are merely introductory in nature. The discussion of chemical alternatives and process modifications in Chapter 3 may serve as a vehicle for more lengthy discussions.

Distillation Recovery. Distillation is mainly used to recover organic solvents. Because of their low boiling points, most degreasing solvents can be distilled easily, yielding a recovery rate of nearly 85 to 90% of the original solvent. The debate over whether this is technically a recycling method or closed-loop approach is dependent on the specific application. When a company faces the decision of whether to invest in its own distillation unit or to contract the services of an off-site recycler, there are advantages and disadvantages to both choices. An off-site recovery arrangement eliminates the capital expenditures needed for distillation equipment, as well as the maintenance and operating costs associated with such equipment. However, the generator must assume liability for wastes transported to the off-site facility. Off-site recycling may also occasionally result in contamination of the recovered solvent with a substance from another user's process. Regardless of the choice, distillation recovery may prove to be quite profitable.

Adsorption. Adsorption systems, which recover solvents from solvent-laden air, are well known in the chemical industry. The technology has been around for at least 40 years. Small companies, in particular, have been reluctant to use adsorption because of the capital investment required. This may change as solvent prices continue to increase and as RCRA forces companies to look at waste reduction technologies. Adsorption systems, which can achieve recovery efficiencies of more than 90%, consist of three stages:

1. A bed of activated carbon or a resin to adsorb solvent vapors from passing air. Many new types of resins are now commercially available, and a specific resin can be purchased to selectively adsorb certain materials.
2. The saturated carbon or resin is then stripped of the solvent temporarily attached

to it. This is known as *desorption*. Desorption can be accomplished by several methods such as low-pressure steam, a solvent that removes the adsorbate, an acidic or caustic wash followed by thermal activation processes, and indirect heating via a nitrogen purge. Steam is most frequently used because it is inexpensive, can be easily condensed back to a liquid, and does not present a disposal problem.

3. The solvent in solution is recovered, either through simple decantation if the solvent is immiscible in water, or otherwise through distillation, which was described in the previous section.

Filtration. Waste streams containing particulate matter can be passed through filters, which capture the particles for reuse. The porosity of the filter will determine which particles are recovered. Some industries add lime to their waste stream before passing it through the filter. Through this action, heavy metals are precipitated and more easily removed.

Ultrafiltration. Ultrafiltration separates colloids or macromolecules from lower-molecular-weight solutes. The most popular membrane components include organic compounds such as cellulose nitrate, cellulose acetate, polysulfone, aramids, polyvinylidene fluoride, and acrylonitrile polymers. Ceramic membranes are also in use and demonstrate their usefulness through their ability to be cleaned by more aggressive chemicals such as acids and bases.

Reverse Osmosis. Reverse osmosis separates macromolecular or ionic components from a solution by using a semipermeable membrane. Pressure reverses the normal osmotic flow of water across the membrane, thus concentrating and purifying the solution.

Electrolysis. Electrolysis is an electrically charged cell that attracts ions and removes them from the waste stream. The electroplated metal can then be removed from the cathode and sold as scrap. Many metals can be removed in this way, including copper, manganese, nickel, chromium, and silver. The recovery rate is excellent, ranging from 90 to 99.9%.

Electrodialysis. Electrodialysis uses an applied electrical field to draw ions through a selectively permeable membrane. These physically rugged membranes are often composed of a polystyrene matrix crosslinked with divinylbenzene and rendered either anion or cation selective.

Operating Practices

SOURCE SEGREGATION. Source segregation is one of the most elementary approaches to waste reduction. This method not only allows the nonhazardous constituents to be separated from those that are hazardous, thereby minimizing the quantity of hazardous wastes, but also creates the opportunity for uncontaminated (hazardous or nonhazardous) streams to be used for use/reuse applications or recycling activities.

One such example is the segregation of solvent waste streams into individual solvent components. This may mean walking through the manufacturing process to see where waste solvents are currently combined and then ensuring that individual solvents are kept separate. This separation will make recovery easier and less expensive. Where solvents cannot be separated by management efforts, fractional distillation, as previously discussed, may be used to separate combined solvents into their individual components.

Another benefit of the segregation of waste components is the fact that the resulting wastes may be marketed for sale, rather than paying the costs associated with disposal options. A waste exchange is a service in which used materials may be sold for use in other industries for a variety of production processes. Materials that one industry views as wastes may be a valuable resource to another industry. Waste exchanges are informational networks that operate similar to newspapers. Industrial firms fill out forms describing the wastes they produce or the raw materials they need. This information is then printed in a publication that is distributed to all industries on the mailing list. The search is complete when a firm is contacted by someone that needs and is able to recycle the raw materials present in the firm's wastes. Both parties benefit, either from the sale and avoided disposal costs of previously unmarketable wastes or from receiving raw materials at an inexpensive price.

HOUSEKEEPING. In addition to representing a common-sense approach to waste reduction, improving housekeeping at a facility is often the most effective first step toward achieving waste reduction, particularly for smaller companies. Consider a waste audit that has identified a number of problems relating to poor housekeeping at an industrial facility. The first problem, storing hazardous wastes outside a facility and improperly labeling them, increases the risk of spillage that can result in contamination of surrounding surface water. The second problem, unauthorized dumping of chemicals into the on-site wastewater treatment system, can overload treatment capacity and result in the violation of effluent guidelines. Correcting these problems can eliminate a company's violations of the environmental regulations and reduce its risk to long-term liability. Educating employees on the proper disposal of chemical wastes will also produce a safer work environment.

INVENTORY AND MATERIAL CONTROL. Another important waste reduction technique accomplished through management initiatives is the control of process raw materials, intermediate products, final products, and any associated waste streams. Inventories that result in out-of-date, off-specification, contaminated, or unnecessary raw materials; spill residues; or damaged in-process or final products can result in significant costs, not only in the purchase of these materials or in operations/labor, but in waste disposal as well.

The two primary methods of inventory management are inventory control and material control. Inventory control is associated with the types and quantities of materials used within a company. Material control pertains to the actual handling of raw materials and the associated finished products and resulting waste streams.

A common practice adopted within the last decade to control inventory is a method referred to as "Just-in-Time." This approach requires more involvement and oversight than the option of having large inventories, but it can drastically reduce errors that result in out-of-date or off-specification raw materials as a result of extended shelf life. Misconceptions, such as buying in bulk to reduce purchasing costs, are a frequent mistake. Just-in-Time is a management technique that requires efficiency and constant monitoring to achieve its ultimate potential. Computer tracking and automation often help streamline any difficulties. Where instrumentation and programmable logic controllers are the tools for efficiency in chemical and mechanical processes, people and computer databases are the requirements for optimal Just-in-Time operation.

Material control is most effectively accomplished by properly maintained equipment, personnel training, and instrumentation methods that regulate the use of raw materials. Eliminating equipment and piping leaks, metering chemicals, and outlining process specifications not only eliminate waste, but create rather beneficial side effects—quality enhancements and increased profits.

CASE STUDIES

In many cases, individual elements of source reduction are not employed, but several are used in an integrated approach. To illustrate the benefits of these initiatives, two U.S. EPA case studies are presented.

Printed Circuit Board Manufacturing Study

U.S. EPA's *Waste Minimization Assessment for a Manufacturer of Printed Circuit Boards* (U.S. EPA, 1992c) was conducted as part of the agency's University-Based Assessments Program. The study was based on a plant that annually manufactures 4×10^5 m^2 (4.3×10^6 sq ft) of printed circuit boards. Both the screens used to transfer ink patterns to the circuit boards and the circuit boards themselves are produced at this facility.

Raw materials for producing silk screens include photographic film sheets, developer, glue, methyl ethyl ketone, polyester mesh fabric, screen emulsion, and tape. After a laser printer produces the pattern on film sheets, the film is developed and rinsed. The mesh fabric is stretched and glued onto metal frames, and an emulsion is spread onto this screen. The film is taped to the screen and exposed to ultraviolet (UV) light to transfer the pattern to the emulsion coating. A water rinse removes exposed emulsion and leaves the inverse of the circuit pattern on the screen. The frames are heat cured.

Raw materials for producing the circuit boards include presized zinc/copper-coated fiber-glass panels and screen printing inks. Initial reference holes (6 to 10) are punched on the panels that are then scrubbed in a mechanical wet scrubbing operation to remove the protective zinc layer from the copper coating. Rinsewater is filtered before being pumped to the plant's wastewater treatment system. Wastewater (87 ML/a [23 mil. gal/yr]) from all the various processes undergoes treatment at $56 750/yr. Sludge ($1.7 \times 10^4$ kg/a [3.8×10^4 lb/yr]) is landfilled at $14 160/yr. Paper filters (1 200 kg/a [2 700 lb/yr]) containing copper, zinc, and brush particles are landfilled at $1 730/yr.

After the panels are printed with etch-resistant ink and inspected, they are etched. This etching process removes unnecessary copper coating and leaves the circuit pattern on the panel. Necessary raw materials include hydrochloric acid, chlorine gas, caustic soda, and microetch solution. Panels are first conveyed through etch tanks of hydrochloric acid, dissolved gaseous chlorine, and recirculated rinsewater. A four-stage cascade rinse and a stripping etch-resist bath follow. Finally, panels are microetched to remove oxides, rinsed in a two-stage countercurrent rinse, and dried. When the solution is saturated with copper (725 m^3/a [1.92×10^5 gal/yr]), it is bled to a storage tank and recycled at $110 530/yr. Etch-tank fumes are diverted to a continuously operating fume scrubber. Water from this bath containing caustic soda beads is filtered to remove etch-resistant ink particles. Sludge (0.45 m^3/a [120 gal/yr]) and paper filters (5 440 m/a [17 860 ft/yr]) are landfilled at $1 730/yr and $1 900/yr, respectively. Fumes go to the scrubber, and spent microetch solution goes to the plant's water treatment system.

Panels are then screen-printed with a solder mask ink, which is UV-cured at 150°C (300°F); screen-printed in white to identify board type; screen-printed in black on the underside to identify circuit components; and, again, UV-cured at 150°C (300°F). Waste ink from all four screen-printing operations is scraped from machines and reused. Screen and machine-cleaning rags (4.8×10^4/yr) containing ink and xylene/propylene solvent are recycled off site at $21 890/yr. Broken screen mesh (7 280 m/a [2.388×10^4 ft/yr]) containing emulsion is landfilled at $1 730/yr.

In the first stage of the punching operation, 1 500 to 2 000 holes are punched into panels for component wiring. In the second, compound-punching stage, the final wiring holes are punched and the panels are cut into boards. Panel webbing and slugs (1.961×10^5 m/a [6.435×10^5 ft/yr]) are landfilled at $35 600/yr.

Ninety percent of the products are sealed with an epoxy-based protective coating. This coating, the function of which is similar to that of the solder coat, acts as a base for component attachment and is applied to punch boards. The necessary raw materials include sulfuric acid, hydrogen peroxide, the epoxy-based coating, the coating's thinner, and isopropyl alcohol (used to clean the protective coating tank). After the boards are microetched to remove oxides, they receive a high-pressure spray rinse, a four-stage cascade rinse, and a roll coating of the epoxy. The tank is cleaned twice a year and the bottom drained to the plant's waste treatment system. Water goes to the waste treatment system. Water is reused in the high-pressure spray rinse. Spent coating solution (0.625 m^3/a [165 gal/yr]) is disposed off site as hazardous waste at $7 330/yr; the coating thinner (1.041 m^3/a [275 gal/yr]) evaporates. The boards are then inspected. Approximately 5% (7.54×10^4 m/a [2.475×10^5 ft/yr]) of the panel area is rejected and landfilled at $2 080/yr.

The following waste minimization opportunites were determined:

- A closed-loop, chilled-water system using recirculated water to cool the UV ovens and etch tanks could reduce wastewater 60% at an annual savings of $40 000. The payback period for the $76 640 implementation cost would be 1.9 years.
- A steam generation system to heat spent etch solution and drive off a portion of the water would reduce the volume of spent etchant 35% at an annual savings of $33 510. The payback period for the $28 180 implementation cost would be 0.8 year.
- Substituting reusable polymer membrane filters for the copper- and zinc-containing paper filters in the mechanical wet scrubbing operation would reduce filter waste 96% and save $3 100 annually. The payback period for the $7 700 implementation cost would be 2.5 years.

Chemical Manufacturing Study

A plant that annually produces approximately 136 million kg (300 mil. lb) of acrylic emulsions, low-molecular-weight (LMW) resins, and herbicides and other specialty chemicals was evaluated in U.S. EPA's *Pollution Prevention Case Studies Compendium* (U.S. EPA, 1992a and 1992b) as part of the agency's University-Based Assessments Program.

The raw materials used to produce acrylic emulsions and LMW resins include monomers, additives, activators, and catalysts. Monomers are pumped from tanker trucks to storage, then to holding/premixing tanks, and sometimes to additive, activator, or catalyst holding tanks for mixing. From the premixing tanks, raw materials are mixed in one of three steam-heated, pressure-regulated reactors where the polymers are formed. The resulting acrylic emulsion polymers or LMW resin products are pumped to blend tanks where other ingredients are added (for example, formaldehyde as a preservative). At this point, acrylic emulsion polymers are pumped through tightly woven cloth filters to separate unwanted clumps of product. The emulsion or LMW resin product is then pumped to storage tanks or directly to drums.

These two production processes generate several wastes on an annual basis. Burnable liquids are generated as a result of bad mixtures or bad reactions. They are also generated from incorrect temperatures or batch weight of solution. Approximately 6 985 kg (1.54 $\times$ 10^4 lb) of this material is disposed of annually as hazardous waste at a cost of $77 110. Off-grade methylolacrylamide and acrylamide are generated in several ways. In some cases, bad batches of commercial product are received. The material has a short shelf life, and off-grade material can also result from operator or equipment error. From the combination of these circumstances, approximately 2 313 kg (5 100 lb) of this material is disposed of as hazardous waste at a cost of $40 760. In addition, the acrylic emulsion process produces used filters and trapped product from the filtering process. Approximately 2 000 kg (4 400 lb) of this material is landfilled off site at a price of $33 080. The LMW resin process produces unsalable LMW resins in the amount of 9 470 kg (2.088 $\times$ 10^4 lb), disposed of as hazardous wastes for $116 200.

The herbicide and specialty chemicals produced are mixed in a pressure- and temperature-regulated reactor where a specified reaction occurs. The product is then pumped to a blend tank where substances to reduce the viscosity are added. From this blend tank, the products are loaded onto railcars and shipped. The highly acidic propionic acid generated by reaction is recycled to reactors for further use. The low-acidic-content acid is used in the wastewater treatment system to neutralize caustic wastewater. Approximately 2 700 kg (6 000 lb) of propionic acid waste is generated from spills while loading and unloading material; $13 510 is spent to dispose of this material as hazardous. Annual cleaning of the reactor and blend tanks results in 450 kg (1 000 lb) of herbicide residues that is disposed of as hazardous waste at a cost of $24 150.

In handling general plant wastes, a pollution abatement system was installed mainly to remove vapors with irritating odors (a small amount of the vapors is organic). Vapors (99.97%) from the monomer storage area, backfire preventers, and the lower explosive limit monitors enter a natural-gas-fired thermal oxidizer at 760°C (1 400°F). Stack gases

(1.116×10^4 m^3 [3.942×10^5 cu ft]) pass directly to the outside atmosphere. Recovered heat is used to heat boiler water. An on-site wastewater treatment facility treats wastewater from the resin line (including the LMW dispersant process), laboratory, air compressor, cooling water, and herbicide-reactor and blend-tank cleanings. The 1.36×10^5 kg (3×10^5 lb) of wastewater sludge is landfilled at an annual cost of \$456 800. The 477 ML (126 mil. gal) of treated wastewater is sewered at an annual cost of \$2 121 700.

The following waste minimization opportunities were determined:

- To reduce the amount of off-specification products, upgrading the redundant sensing and control devices on the reactor raw material lines would reduce burnable liquids 75%, composited absorbed monomers 19%, off-grade methylolacrylamide/acrylamide 71%, and unsalable products 15%. The annual savings would be \$139 810, and the payback period for the \$365 080 implementation cost would be 2.6 years.
- The installation of a gas-fired dry-off oven in the wastewater treatment system would reduce the volume of sludge hauled off site. Annual savings of \$92 730 could be realized as a result of a \$70 230 investment.

Recycling

Recycling refers to the use or reuse of wastes as an effective substitute for a commercial product or an ingredient or raw material in an industrial process. Recycling, however, also includes the concept of reclamation. *Reclamation* is the recapturing of useful constituent fractions within a waste material or the removal of contaminants from wastes to allow the primary material to be reused. Figure 1.2 illustrates these concepts.

Technologies for achieving closed-loop recycling were introduced in the previous section and references were made to chemical alternatives and process modifications (see Chapter 3). As such, they are not reiterated here. However, several industrial examples that demonstrate the potential environmental and economic benefits of waste minimization via recycling methodologies are provided. Because recycling, in general, deals with the handling of wastes once they have been produced, it is not surprising that source reduction is often the approach of choice.

USE/REUSE

For years, Bell Helicopter Textron, Inc., has used plastic media blast as a paint stripping material to replace methylene chloride. However, in cooperation with U.S. Technologies Inc., a supplier of the plastic media blast (PMB), an arrangement was made to lease the virgin PMB and return the spent contaminated material for use in a thermoforming operation to produce pen holders, bathroom sinks, and other useful items. This project, approved by both federal and state regulatory communities, eliminated the generation of hazardous wastes (spent methylene chloride) and produced an annual cost savings of nearly \$45 000 (Childers, 1997).

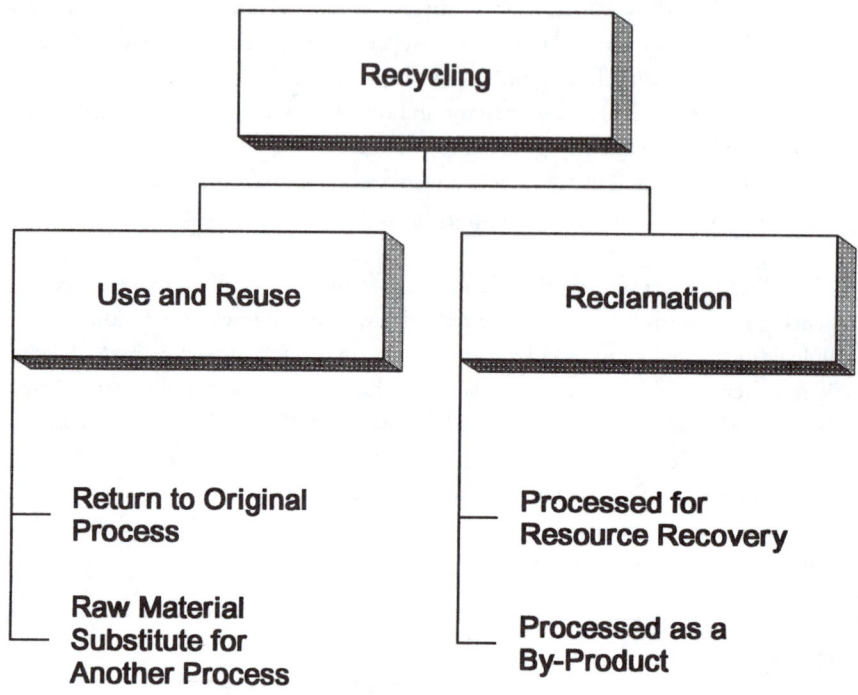

Figure 1.2   Recycling methodologies.

RECLAMATION

Resource Recovery

An assessment of a steel-making facility showed that calcium fluoride (fluorspar) in the sludge generated during neutralization of the pickling-line wastewater could be recovered. By recycling the fluorspar, the company saved the money it would have spent to buy it and also reduced the volume of sludge requiring disposal by 30% (U.S. EPA, 1991b).

Refrigerant Recycling

Automotive air conditioner refrigerant is the largest source of ozone-depleting chlorofluorocarbons in the U.S. U.S. EPA sampled and analyzed refrigerant from more than 200 automobile air conditioners to develop a standard of purity for recycled refrigerant. In 1988, a committee that included representatives of the Motor Vehicle Manufacturers Association (MVMA), the Mobile Air Conditioner Society, and the Society of Automotive Engineers reviewed and approved this standard and recommended it to MVMA. A program for certification of refrigerant recovery devices has subsequently been initiated. When fully implemented, this research could represent an important milestone in reducing the effects of automotive air conditioner refrigerant in depleting the earth's stratospheric ozone layer (U.S. EPA, 1991b).

Opportunity Assessments: A Phased Approach

A large portion of businesses generating hazardous wastes can implement waste reduction programs at a reasonably low cost with low risk, and achieve a prompt return on their investment. Surveys conducted by the National Academy of Sciences characterize most of American industry as being in the early, low-cost phase of waste reduction. Although not every company will be able to achieve the same levels of success, with the proper management commitment numerous low-cost opportunities for waste reduction will present themselves.

The approach as defined by many within the regulatory community, primarily U.S. EPA, breaks down the process of waste minimization into the following four phases:

1. Program planning and organization,
2. Assessment,
3. Feasibility analysis, and
4. Implementation.

Each of these phases is broken out into its own subsection to provide a more detailed schematic of the individual processes that should take place. For additional tools and references, three pivotal publications released by U.S. EPA can provide a host of materials to assist in waste minimization activities. One of these publications includes a fairly extensive software program available free through U.S. EPA's Technology Transfer Network bulletin board or Internet home page. The contents of these U.S. EPA documents are summarized briefly below.

WASTE MINIMIZATION OPPORTUNITY ASSESSMENT MANUAL (EPA-625/7-88-003)

This manual (U.S. EPA, 1988) explains how to conduct a waste minimization assessment and develop options for reducing hazardous wastes generation at a facility. It presents the management strategies needed to incorporate waste minimization into company policies and structure, and shows how to establish a company-wide waste minimization program, conduct assessments, implement options, and make the program an ongoing one. Included in the appendices are worksheets useful in carrying out assessments, sample assessments of common industrial processes, and an example of an economic feasibility analysis.

USER'S GUIDE: STRATEGIC WASTE MINIMIZATION INITIATIVE (EPA-625/11-91-004)

This user's guide (U.S. EPA, 1991c) describes software that was developed as a tool for demonstrating concepts of process analysis after a waste minimization audit. The software requires user-supplied information for process definition, and material inputs and products for each unit operation and for outputs associated with waste streams. U.S. EPA's user's guide also provides a scheme for identifying and prioritizing (on a cost or volume basis) waste reduction opportunities in process units and treatment operations, provides mass-balance calculations and process flow diagrams, and directs the selection of candidate waste minimization strategies. This software system was developed in conjunction with the U.S. EPA's *Waste Minimization Opportunity Assessment Manual* (U.S. EPA, 1988). The latter manual can provide the information necessary for input to this software.

INDUSTRIAL POLLUTION PREVENTION OPPORTUNITIES FOR THE 1990s (EPA-600/8-91-052)

In this document (U.S. EPA, 1991a), a set of criteria was developed for the purpose of subjectively prioritizing industry segments for their pollution prevention potential and their opportunity for improvement. Using this set of criteria, high-priority industries were selected from a list based on the Standard Industry Classification for investigation of the need or opportunity for waste reduction through source reduction and/or material recycling. Information concerning the opportunities in particular industries was then solicited through associations, companies, researchers, and government agencies. A final list of 17 industries was compiled. In addition, the investigation identified a number of generic research or technological needs for which industry contacts believed research could lead to waste minimization applicable to more than one industry.

It should be noted that almost all of these sources provide management overviews of the waste minimization process and give examples of successful projects. Virtually none contain analytical tools or technical essays on the process for developing waste minimization as a stand-alone process. As a result, the concepts that relate waste minimization, quality, process control, and technological advancements are provided within this publication as a single source for addressing both environmental and business concerns.

PROGRAM PLANNING AND ORGANIZATION

The beginning of a waste minimization program must strongly incorporate the backing of upper management and strive to achieve a change in the thought processes of all personnel. Views that categorize wastes as something that must be dealt with at the end of a process must be changed. Attitudes need to be molded that emphasize the "big picture." The system or process must be considered as a whole, from the inflow of materials, through their handling and processing, to final product output and waste disposal.

Planning and organization of a waste minimization program involves getting management commitment for the program, defining a policy for the company, stating overall waste minimization goals, and building an assessment task force. Figure 1.3 outlines these objectives in relation to the entire opportunity assessment effort.

For a waste minimization program to be successful it should have top management support. Managers must educate themselves on the concepts involved by attending seminars sponsored by universities or trade organizations or by consulting articles and books. Additional efforts may include networking with other industries. The sharing of ideas and information across industrial boundaries may not have been a frequent process in the past, but in relation to waste reduction, it is an excellent mechanism for fostering regulatory and community support. The results from education provide a strong foundation of

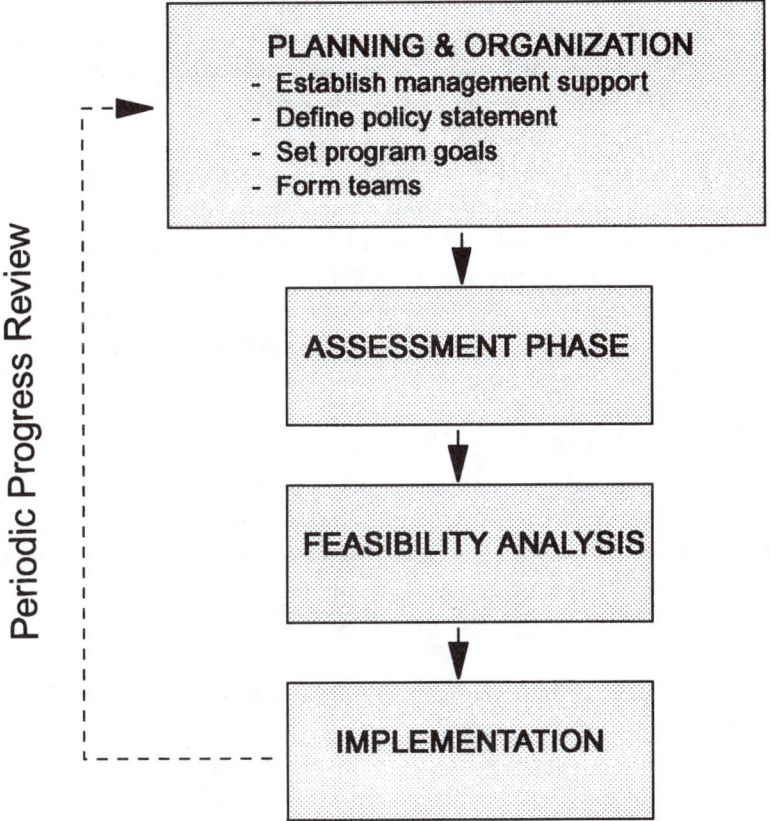

Figure 1.3 Program planning and organization.

support for waste reduction programs and should be followed by the documentation of a company waste minimization policy statement. This sets the tone for the company's attitude and the direction of the overall program.

To ensure that progress is made with the program, it is imperative that middle managers and their employees be involved, as these people can be pivotal to a program's efficiency and overall success. Frequently, the most effective ideas are those generated from employees who are actually running the processes in question. To provide an energized flow of ideas and encourage serious employee participation, many companies have provided additional motivation via monetary incentives.

Cross-functional departments and organizations need to be involved in the waste minimization process as well. Therefore, it is important to assemble a waste minimization project team comprising a number of different members. The team of individuals selected on any given project should include those with direct responsibility and knowledge of the particular waste stream or area in which the process is operated, and a few that have absolutely no involvement in the process at all. This latter group of individuals may bring ideas to the project team that are not colored by exposure to the process everyday.

Moreover, these individuals may be selected from within the company or hired as outside consultants. Other excellent vehicles for waste minimization projects include establishing process action teams and quality groups. These groups were initially formed to improve process efficiencies and product quality and to reduce scrap; they are a resource that should be tapped whenever possible.

As with any project or program, it is always a good idea to set goals to provide direction. Although most of the goals set for a waste minimization project should be realistic, a few ambitious ones should be selected as well. Especially challenging goals tend to promote entirely new ways of thinking because they usually cannot be solved by quick or traditional means. In any event, it is best if the goals are measurable and quantifiable rather than qualitative. This allows measurements to be taken that provide indicators of progress.

ASSESSMENT PHASE

The assessment phase involves a number of steps, as shown in Figure 1.4.

Conducting a mass balance for any identified processes is usually the best place to start. This step should include an assessment of all materials entering and leaving a process. In cases in which specific projects have not been identified and overall facility practices are being analyzed, a review of the following may be in order:

- Identification and characterization of raw material inventories,
- Listing of waste streams generated,
- Resource Conservation and Recovery Act hazardous wastes manifests,
- National Pollutant Discharge Elimination System reports,
- Emergency Planning and Community Right-to-Know Act Form R reports, and
- Process flow diagrams and equipment inventories.

Developing a basic understanding of the processes that generate wastes at a facility is essential. Where appropriate, flow diagrams should be prepared to identify the quantity, types, and rates of the processes selected.

All waste streams in a facility should be evaluated for potential waste minimization opportunities. However, because resources are typically limited, the need to concentrate waste minimization efforts in a specific area may be more appropriate. Considerations of waste quantities, hazardous properties of the waste, applicable regulations, employee safety, economics, and other parameters should be evaluated in selecting a process for optimization. In situations where hazardous chemical properties and employee safety are of paramount concern, investigations should be based on the toxicity of the chemicals used and focused directly on their elimination as a priority. Projects that require little investment and have the potential for significant savings should be selected first. These projects will also serve as motivation for additional projects in the future.

After facility information is compiled, data should be reviewed to educate individuals on the processes they are going to investigate. The ensuing inspection should begin where raw materials are received then continue with the targeted processes to the points at

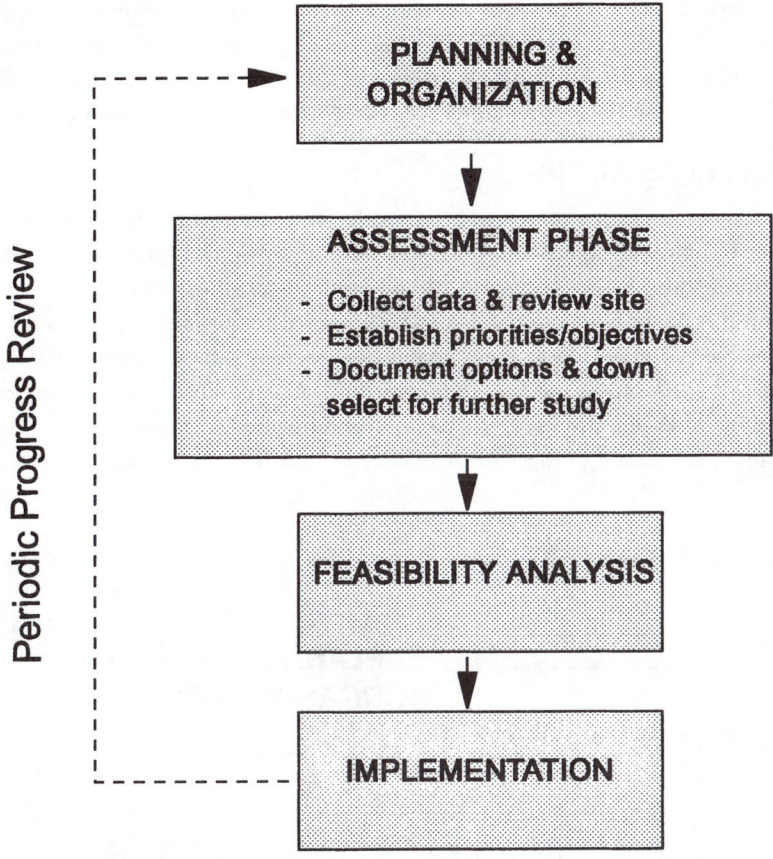

Figure 1.4   Assessment phase.

which products and wastes leave. Attention should be given to areas where wastes are generated. A "subprocess" for each main process should be also be considered. Typical subprocesses include maintenance operations, storage for raw materials, finished product handling and work areas, and "work in progress" that may be floating between operations.

After pertinent data are gathered, a list of potential projects that will encompass waste minimization options for further consideration should be generated. Technical and economic concerns are considered in the following feasibility stage. No options should be ruled out at this time. Information from the site inspection, as well as trade associations, government agencies, technical and trade reports, equipment vendors, consultants, and plant engineers/operators, may serve as sources for ideas.

A screening process to select the most promising options for complete technical and economic feasibility studies should then begin. Either an informal review or a quantitative decision-making process may be used to identify marginal or impractical options that may be eliminated from further consideration. Regardless of the method used, it is important to determine the benefits that may be gained. Whether the technology exists and has a good track record or needs to be developed, overall costs and the timetable required for implementation should be included on a "decision tree." Again, the first items selected should be those that require little investment but have the potential for significant savings. There are typically a wide number of these low-risk options to choose from.

FEASIBILITY ANALYSIS

For a waste minimization option to be seriously considered for funding, it must be shown to be technically and economically feasible for use. Figure 1.5 represents a flow chart illustrating this phase of the waste minimization opportunity assessment.

A technical evaluation determines whether a proposed option will work for a specific application. Therefore, process and equipment changes should be assessed for their effects on waste quantity, workplace safety, product quality, labor, compatibility with existing manufacturing flows/procedures/rates, operation and maintenance, and utility costs. All of these parameters should fit together in an effective manner and be carefully consid-

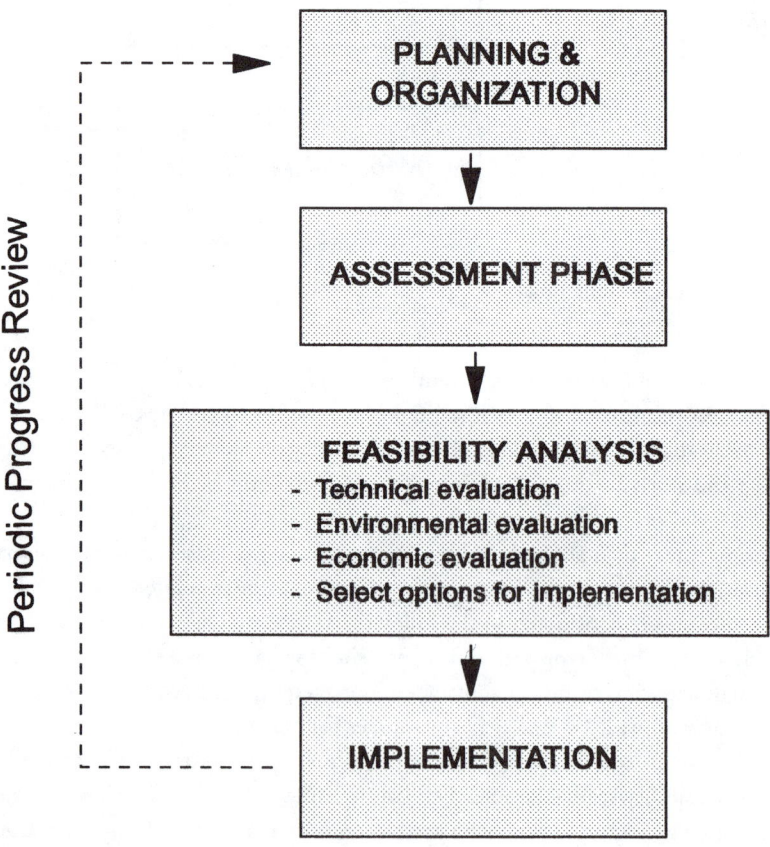

Figure 1.5   Feasibility analysis.

ered. To determine the best approach, overall benefit, and cost worthiness of a proposal, an economic evaluation of the alternatives should also be conducted.

Economic analyses are executed using typical engineering economics approaches that measure profitability via payback period, return on investment, and net present value. The costs of a waste minimization project are typically separated between capital costs and operating costs. Also taken into account should be any savings or changes in revenue. Chapter 4 provides a detailed discussion of economic analyses that are illustrated by numerous industrial examples.

Process modifications, especially those involving the purchase of equipment, should be reviewed by either a committee set up for that purpose or, in some cases, the chief executive officer or president of the company. For approvals to be granted, decision makers should be given sound information on the capital funds required and the potential benefits. Unfortunately, this is typically the weakest link in a waste minimization opportunity assessment. Conducting waste minimization because "it is the right thing to do" is not enough. Waste minimization should be used as a financial tool, not as a crutch for compliance. Factors such as profitability, cost avoidance, quality, enhanced productivity, competitiveness, and decreased liability risks should be integrated into every analysis. It is important to remember that many waste reduction opportunities are low cost and/or offer a quick payback.

IMPLEMENTATION

A waste minimization project that has shown promise through both technical and economic feasibility reviews is ready to be implemented. The steps of this phase of the waste minimization opportunity assessment are illustrated in Figure 1.6.

Even after a budget has been procured for a waste reduction project, resistance to the project may exist. Resistance generally derives from employee distrust for something new. This may be conceptualized by considering an example of a manufacturing director that needs to solve a water pollution problem. One solution for a director is to install an "end-of-pipe" treatment process. The other solution is to eliminate the source of the pollutant by changing the process or product. The director must decide between something that is familiar and proven (that is, end-of-pipe treatment) and an option that could erase the problem entirely, but is unproven. This scenario is not uncommon. Indeed, changing a process or product is often perceived as a higher risk alternative. With systematic investigation and implementation, however, greater process control and economic benefits are often the results.

There are better ways to decrease resistance among staff than to simply dismiss all employees who refuse to become educated or participate in the process. One approach is to appoint resistant individuals to positions on the planning and implementation committees. Over the course of this process, these resistant individuals may become so drawn into the program that they become the most avid supporters.

After a waste minimization project has been implemented, continued management support, ongoing efforts to track wastes, and identification of new opportunities are essential. A company should also monitor the extent to which pollution is being reduced and how much money is being saved in terms of energy, raw materials, and disposal

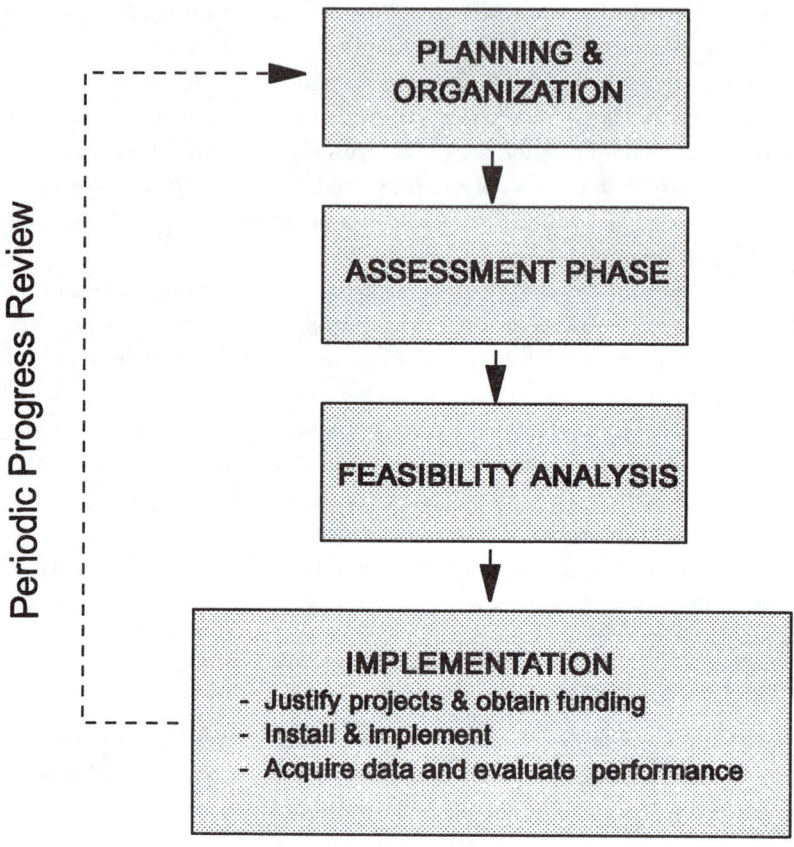

Figure 1.6   Implementation.

costs. This information is not only important when calculating investment paybacks, but is essential when the company files its manifest and biennial reports on waste reduction efforts under the 1984 RCRA amendments. There is no reason to ease up just because one project or area has been addressed. Success stories should be used to foster new ideas. Indeed, waste minimization is an ongoing process subject to continual changes and requires monitoring and reassessment. In fact, many project managers will evaluate the initial results and feed the information back to the beginning of the opportunity assessment process. This allows for fine-tuning and continuous improvement of the project.

REFERENCES

Childers, D.G. (1997) Environmental Economics in the Aerospace Industry. *18th Annu. Am. Electroplaters Surf. Finishers Soc./Environ. Protection Agency Pollut. Prevention Control Conf. Proc.,* Orlando, Fla.

Docherty, M., *et al.* (1995) Pollution Prevention in Organic Finishing Processes. *Proc. 88th Annu. Meeting Exhibition Air Waste Manage. Assoc.,* San Antonio, Tex.

Freeman, H.M. (1990) *Hazardous Waste Minimization.* McGraw-Hill, Inc., New York, N.Y.

Institute for Local Self-Reliance (1986) Toward Pollution-Free Manufacturing. Am. Management Assoc. Briefing., Washington, D.C.

U.S. Environmental Protection Agency (1988) *Waste Minimization Opportunity Assessment Manual.* EPA-625/7-88-003, Office Res. Dev., Washington, D.C.

U.S. Environmental Protection Agency (1991a) *Industrial Pollution Prevention Opportunities for the 1990s.* EPA-600/8-91-052, Office Res. Dev., Washington, D.C.

U.S. Environmental Protection Agency (1991b) *Pollution Prevention.* EPA-600/M-91- 036, Office Res. Dev., Washington, D.C.

U.S. Environmental Protection Agency (1991c) *Users Guide: Strategic Waste Minimization Initiative.* EPA-625/11-91-004, Office Res. Dev., Washington, D.C.

U.S. Environmental Protection Agency (1992a) Evaluations of Waste Minimization Technologies at the General Dynamics Pomona Division. In *Pollution Prevention Case Studies Compendium.* EPA-600/R-92-046, Risk Reduction Eng. Lab., Office Res. Dev., Washington, D.C.

U.S. Environmental Protection Agency (1992b) Waste Minimization Assessment for a Chemicals Manufacturer. In *Pollution Prevention Case Studies Compendium.* EPA-600/R-92-046, Risk Reduction Eng. Lab., Office Res. Dev., Washington, D.C.

U.S. Environmental Protection Agency (1992c) *Waste Minimization Assessment for a Manufacturer of Printed Circuit Boards.* EPA-600/S-92/008, Office Res. Dev., Washington, D.C.

SUGGESTED READINGS

U.S. Environmental Protection Agency (1992) *Facility Pollution Prevention Guide.* EPA-600/R-92-088, Office Res. Dev., Washington, D.C.

U.S. Environmental Protection Agency (1993) *Measuring Pollution Prevention Progress Proceedings.* EPA-600/R-93-151, Office Res. Dev., Washington, D.C.

Wentz, C.A. (1989) *Hazardous Waste Management.* McGraw-Hill, Inc., New York, N.Y.

2 Process Definition and Control Techniques

In addition to a basic understanding of waste minimization strategies, it is important to gain a thorough knowledge of the analytical techniques that allow proper initial and ongoing assessments of manufacturing processes, as well as the methods for measuring and controlling process efficiency. This chapter provides an overview of the concepts that are pivotal to achieving these goals.

The most basic tool in evaluating a process involves a chemical engineering principle referred to as *material balance*, or *mass balance*. The material balance is simply an extension of the laws governing the conservation of matter; that is, mass (or energy) is neither created nor destroyed and is constant for all isolated systems. Simply stated, the mass into a process equals the mass out of a process, even though the form of the materials may have been changed along the way. In the case of an automobile, for example, the gasoline and air fed into the cylinder of the engine leave the cylinder after combustion in the form of carbon dioxide, water, excess air, and various compounds of incomplete combustion. The combined mass of the gasoline and air entering the process must equal the sum of the mass of the reaction products leaving the cylinder.

After a process has been evaluated in this manner, techniques for increasing its efficiency may be used. In relation to waste minimization, the ramifications are easily realized because of the reduction in raw material usage, lower scrap rate, and other indicators. These techniques are referred to as statistical process control and, to achieve desired results, frequently require the implementation of in-process measurements and automation using various types of instrumentation. Statistical process control (SPC) and instrumentation concepts are elaborated on later in this chapter, after a strong foundation has been established using the material balance.

Chemical Engineering and the Material Balance

Before addressing the concepts of the material balance, it is helpful to review some general background information. The topics of background information discussed in this review are tied to the basic elements of chemistry and physics. By providing this review, the principles of the material balance may be easier to understand and, more important, easier to apply.

ENGINEERING CALCULATIONS

This section begins with a discussion of units and dimensions, provides examples of unit and dimension calculations, then expands to issues dealing with chemical equations and stoichiometry. The latter concepts provide a basis for defining the techniques and conventions commonly used in solving mass balance problems.

Units and Dimensions

Because they are a fundamental part of acquiring accurate calculations, units and dimensions represent one of the most important discussions covered in this chapter. Dimensions are the basic structures of measurement and include time, temperature, length, mass, pressure, and so on. Units are the means for defining dimensions. For example, where the dimension is length, the unit of metres (feet) may be used. For the dimension of temperature, the units may be Celsius (Fahrenheit). Unit expressions should always be attached to values and numbers.

UNIT CONVERSIONS.   Consider the following conversion problem:

$$10 \text{ lb} = x \text{ g} \ (1 \text{ lb} = 453.6 \text{ g})$$

$$55 \text{ gal} = x \text{ L} \ (1 \text{ gal} = 3.785 \text{ L})$$

To convert pounds to grams, the equation should be set up so that the pound units cancel out, leaving only gram units, as follows:

$$(10 \text{ lb})(453.6 \text{ g}/1 \text{ lb}) = 4\,536 \text{ g}$$

Note that because 1 lb equals 453.6 g, dividing one by the other is equivalent to 1. Therefore, multiplying 10 lb by the ratio of 453.6 g/1 lb does not increase or decrease the mass of the material in question. This exercise can be repeated for converting gallons to litres, as follows:

$$(55 \text{ gal})(3.785 \text{ L}/1 \text{ gal}) = 208 \text{ L}$$

Density is the ratio of mass per unit volume and is typically expressed as grams per cubic centimetre (pounds per cubic foot). Determining the density of a substance is a relatively simple process. In the case of a liquid (for example, water [H_2O]), a beaker may be weighed with a known volume of liquid. By subtracting the mass of the beaker, the mass of the water can be obtained. Taking the mass of the water and dividing it by the known volume yields a value referred to as density. Solids may be evaluated in a similar fashion by using a beaker filled with a compatible liquid (that is, one that does not cause any reactions) and determining the volume of the liquid that is displaced. The mass of the solids is the difference between the total mass of the beaker, liquid, and solids minus that of the beaker and liquid. Again, dividing the mass of the solids by the volume of displaced liquid provides a density value for the solids. Although these procedures for determining density are simplified, the overall concept is valid.

Another term often used to refer to the density of a substance is *specific gravity* (sp gr). Although traditionally thought of as a dimensionless number, specific gravity is actually a ratio of densities. The ratio for most liquids and solids is determined by taking the density of the liquid or solids in question and dividing it by a reference substance, typically the density of water at 4°C (1 in the metric system). Therefore,

$$\text{Specific gravity} = \frac{\text{(lb/cu ft) of substance A}}{\text{(lb/cu ft) of H}_2\text{O at 4°C}} = \frac{\text{(g/cc) of substance A}}{\text{(g/cc) of H}_2\text{O at 4°C}} \qquad (2.1)$$

A simple way to convert specific gravities to densities when using the metric system is to remember that a specific gravity value is essentially equal to that value expressed in grams per cubic centimetre. This is not the case when attempting to use the English system. For example, when substance A exhibits 1.9 sp gr,

$$1.9 = (\text{Density of substance A in g/cc})/(1 \text{ g/cc H}_2\text{O at 4°C})$$
$$1.9 \text{ g/cc} = \text{Density of substance A in g/cc}$$

However,

$$1.9 = (\text{Density of substance A in lb/cu ft})/(1 \text{ g/cc H}_2\text{O at 4°C})$$
$$1.9 \text{ (1 g/cc H}_2\text{O})[(62.43\text{lb/cu ft})/(1 \text{ g/cc})] = \text{Density of substance A in lb/cu ft}$$
$$119 \text{ lb/cu ft} = \text{Density of substance A in lb/cu ft}$$

UNIT MANIPULATION. Consider a degreasing operation that uses perchloroethylene (C_2Cl_4) with 1.623 sp gr in a tank that holds 50 gal. In the degreasing operation, 5 gal of perchloroethylene is needed to top off the tank each week because of evaporative losses. How many pounds of perchloroethylene are emitted to the air each week?

First, the specific gravity of perchloroethylene must be converted to a density value with the units of pounds per gallon as follows:

$$1.623 = (x \text{ g/cc C}_2\text{Cl}_4)/(1 \text{ g/cc H}_2\text{O})$$
$$1.623 \text{ g/cc} = \text{Density of C}_2\text{Cl}_4$$

Second, a unit conversion must be performed to yield a perchloroethylene density value in pounds per gallon as follows:

$$(1.623 \text{ g/cc C}_2\text{Cl}_4)(1 \text{ lb/453.6 g})(1 \text{ cc/1 L})(3.785 \text{ L/1 gal}) = 13.54 \text{ lb/gal}$$

Finally, multiplying by the volume of perchloroethylene used each week gives the number of pounds emitted to the air. For example,

$$(13.54 \text{ lb/gal})(5 \text{ gal}) = 67.70 \text{ lb}$$

Thus, perchloroethylene emissions to the atmosphere each week are 67.70 lb.

Not all unit manipulation problems involve quantities that are expressed in units as simple as those found in the above problem. For example, units depicting pressure, force, and energy are actually a representation of multiple dimensions. Table 2.1 provides some typical examples of dimensions with multiple units. Unit manipulation can be an in-

Table 2.1   Dimensions with multiple units.

Units	Dimension
Pressure	atm, mm Hg, lb/sq in., kg/m^2
Force	N, kg, m/sec^2, lb$_f$, lb$_m$ft/sec^2, dynes, J/m
Energy	Btu, ft-lb, J, erg, g/cal, hp-hr

volved process. There is not only the complication of differing units for the same dimension (for example, joules versus grams per calories, millimitres mercury versus pounds per square inch), but the complication of those resulting from the use of the English system and Systeme International (SI). More commonly known as the metric system, SI is the standard measurement system in most countries, the U.S. being an exception.

Units of the SI system were derived by relating the basic laws of physics and observed physical phenomena to units of mass and length. In addition, a simple decimal relationship was developed. Where there are only three primary defined units in the metric system, there are four in the English system. This fourth unit in the English system is used in calculations dealing with force and can create confusion. In the metric system, force is denoted by newtons, which is a representation of kilogram-metres per second. In the English system, force is expressed as pounds (lb$_f$) and represents a pound of mass (lb$_m$) influenced by gravity (g$_c$). For purposes of discussion of waste minimization, which occurs on or near the earth's surface, a pound of force will always equal a pound from a mass standpoint (that is, lb$_f$ equals lb$_m$).

Although mass, volume, and time are relatively straightforward dimensional concepts, temperature and pressure deserve additional discussion. Pressure is defined as force per unit area and may be expressed as either absolute or relative. Figure 2.1 provides an example of this concept.

In Figure 2.1, force is applied in all directions and distributed over the entire surface area of the sphere. Here, *absolute pressure* is defined as the measurement of pressure in reference to a complete vacuum (zero pressure). Relative pressure is expressed as a quantity measured above or below a reference point, which is typically atmospheric pressure. Relative pressure is also referred to as *gauge pressure*:

$$\text{Atmospheric pressure} + \text{Gauge (relative) pressure} = \text{Absolute pressure} \qquad (2.2)$$

Using an illustration similar to Figure 2.1, these concepts (specifically, gauge versus absolute pressure) are also demonstrated in Figure 2.2.

PRESSURE UNIT CONVERSIONS. If a scuba diver descends to a depth of 100 ft, and the pressure at sea level is 1 atm, what is the pressure exerted on the diver's body as measured in atmospheres?

The primary conversion factor needed for this problem is

$$1 \text{ atmosphere} = 406.8 \text{ in. H}_2\text{O (at } 4°\text{C)}$$

Therefore,

Figure 2.1   Conceptualization of pressure: force per unit area.

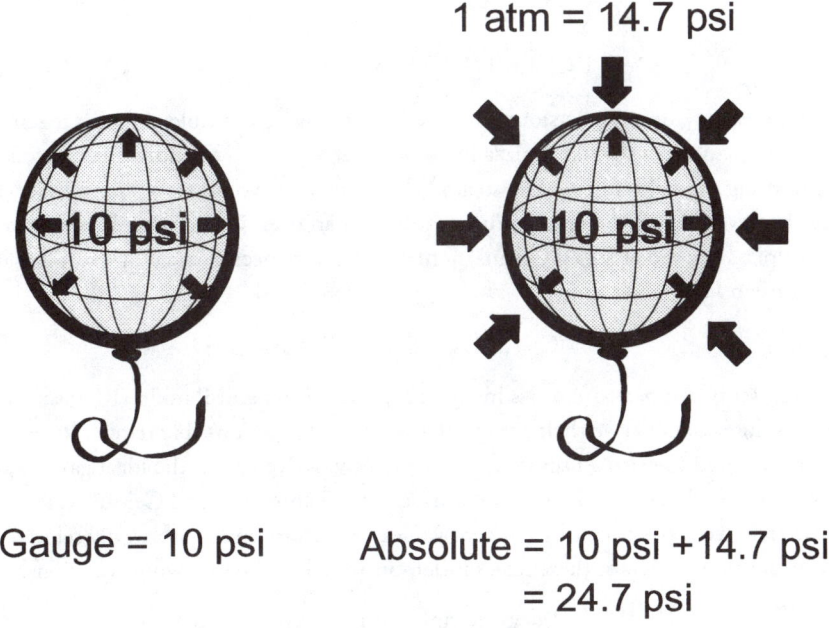

Gauge = 10 psi   Absolute = 10 psi +14.7 psi
                                  = 24.7 psi

Figure 2.2   Gauge versus absolute pressure (psi × 6 895 = Pa).

$$100 \text{ ft } (12 \text{ in.}/1 \text{ ft})(1 \text{ atm}/406.8 \text{ in. } H_2O) = 2.95 \text{ atm}$$

The scuba diver is experiencing a pressure almost three times as great as that on the water's surface.

ABSOLUTE AND GAUGE (RELATIVE) PRESSURE. Before the scuba diver begins the dive, the diver's tank must be filled to a pressure (as registered on the tank's gauge) of 3 000 psi. What is the true pressure in the tank?

An empty tank is in equilibrium with its surroundings. Therefore, the pressure in the tank equals 14.7 psi when empty, and the gauge will register 0 psi. The true pressure when the tank has been filled is as follows:

$$3\,000 \text{ psig (gauge)} + 14.7 \text{ psi (atmospheric)} = 3\,014.7 \text{ psia (absolute)}$$

TEMPERATURE UNIT CONVERSIONS. The concepts of temperature are not that dissimilar to pressure in that different scales exist and each depends on a reference point by which the scale is defined. The Celsius scale is based on scientific parameters in which 0°C represents the value corresponding to the freezing point of water and 100°C represents the boiling point of water. In contrast, these points are seemingly more arbitrary on the Fahrenheit scale: 32°F corresponds to freezing water and 212°F to boiling water. It is important to understand that these two temperature scales do not change with equal increments. That is, an increase in 1°C equals an increase in 1.8°F. The following equations relate the Celsius and Fahrenheit scales:

$$(°F - 32)0.555\,6 = °C \qquad\qquad\qquad (2.3)$$

$$(°C)(1.8°F/1°C) + 32 = °F \qquad\qquad\qquad (2.4)$$

For temperature unit conversions, consider the following example. A U.S. citizen traveling in Europe decides to bake a pizza in an oven set at 325°. The individual is accustomed to using the Fahrenheit scale, so a 325° temperature would seem appropriate. In Europe, however, where the temperature reading for an oven would most likely be in Celsius, a pizza cooked at 325° (Celsius) actually will have been cooked at the following temperature in Fahrenheit units:

$$(325°C)(1.8°F/1°C) + 32°F = 617°F$$

Another set of temperature scales involves the Kelvin (K) and Rankin (R) absolute scales. Absolute scales derive their zero values from the lowest possible temperature that is known or believed to exist. Absolute zero is primarily derived from the ideal gas law and the law of thermodynamics. This is in contrast to the Fahrenheit and Celsius scales, which are relative scales because their zero values stem from points arbitrarily selected. The relationship for each of these scales is demonstrated in the following equations:

$$°C + 273.15 = K \text{ (the degree symbol is not used with Kelvin)} \qquad (2.5)$$

$$°F + 459.58 = °R \qquad\qquad\qquad (2.6)$$

These two equations indicate that 0 K corresponds to −273.15°C and 0°R is equal to −459.58°F.

Chemical Equations and Stoichiometry

Although some of the topics covered in this section may be an extension of the section titled Units and Dimensions, they have been included in this discussion of chemical equations and stoichiometry.

The first concept that should be understood is the mole. *Mole* is a term typically used in chemistry to denote a specific number of atoms, electrons, or other similar particles. This quantity is defined as the amount of a substance that contains as many particles as there are in 12 g of carbon-12. This quantity or number is called *Avogadro's number*. Again, there are differences between the SI system and the English system. A gram-mole has 6.023×10^{23} molecules, whereas a pound-mole has $(6.023 \times 10^{23}) \times 453.6$ molecules. This is not surprising when units are tracked properly. For example,

$$1 \text{ mol} = 6.023 \times 10^{23} \text{ molecules/g (by definition)}$$

Or

$$1 \text{ g-mol} = 6.023 \times 10^{23} \text{ molecules}$$

Converting from grams to pounds yields the following:

$$(6.023 \times 10^{23} \text{ molecules/1 g-mol})(453.6 \text{ g/1 lb})$$

Or

$$1 \text{ lb-mol} = (6.023 \times 10^{23} \text{ molecules})(453.6)$$

Before elaborating on the practical use of molar quantities, it is important to know how to determine the number of moles that are present in a given situation. To determine the number of moles of a substance, the atomic or molecular weight (mass) of the substance must be known. Atomic weights are found on the periodic table for each element. For example, the atomic weight of an individual oxygen atom is 16. What does the value 16 mean? This value represents 16 g for every mole of atoms (6.023×10^{23} atoms). However, because oxygen atoms typically exist in pairs, the molecular weight is actually 32 g for every mole of oxygen molecules (O_2). Because an understanding of these issues is important, several examples are provided.

ATOMIC WEIGHT CALCULATIONS. For atomic weight calculations, consider the following example: How many gram-moles are present in a pipe of solid elemental copper (Cu) weighing 5 lb if the atomic weight of copper is 63.55 g/g-mol?

$$(5 \text{ lb Cu})(453.6 \text{ g/1 lb}) = 2\,268 \text{ g Cu}$$

Then

$$(2\,268 \text{ g Cu})(1 \text{ g-mol/63.55 g}) = 35.69 \text{ g-mol Cu}$$

MOLECULAR WEIGHT CALCULATIONS. As calculated in a previous example, if 30.7 kg (67.70 lb) of perchloroethylene is to be emitted to the atmosphere each week, determine the corresponding number of gram-moles.

The molecular weight of perchloroethylene is found by adding the atomic weights of each element, where the atomic weight of carbon is 12.01 g/g-mol, and the atomic weight of chlorine is 35.45 g/g-mol. Then, the molecular weight of $C_2Cl_4 = (2)(12.01) + (4)(35.45)$, and the molecular weight of $C_2Cl_4 = 165.82$ g/g-mol. Once the molecular weight is known, the number of moles of perchloroethylene corresponding to 30.70 kg (67.70 lb) can be calculated. For example,

$$(67.70 \text{ lb})(453.6 \text{ g}/1 \text{ lb})/(165.82 \text{ g/g-mol}) = 185 \text{ g-mol } C_2Cl_4$$

With the concepts of molar quantities in hand, chemical equations can now be introduced. A chemical equation is a method by which reactions are illustrated or summarized. Methane (CH_4), for example, when burned on stoves or in furnaces, may be represented by the following chemical equation:

$$CH_4 \text{ (g)} + 2O_2 \text{ (g)} \rightarrow CO_2 \text{ (g)} + 2H_2O \text{ (l)} \qquad (2.7)$$

The combustion of methane results in a reaction with oxygen molecules to yield carbon dioxide (CO_2) and water vapor. The numbers or coefficients preceding each type of molecule indicate the number of molecules necessary for the chemical reaction to take place as written. For one molecule of methane to completely combust, at least two molecules of oxygen must be present. Similarly, for each molecule of methane completely combusted, one molecule of carbon dioxide is formed along with two molecules of water.

The concept of combining elemental weights and using the ratios of molecules, as determined by the coefficients of a balanced chemical equation, to calculate the moles of one substance as related to the moles of another substance is referred to as *stoichiometry*. The notations (g) and (l) denote gas and liquid, respectively. One of the main reasons chemical equations are useful is that they provide the means to determine the amount of molecules or compounds necessary to drive the process to completion, and also indicate the quantity of products and byproducts.

To be effective, it is important that all chemical equations be balanced. If the equation for the combustion of methane did not depict two molecules of oxygen as a reactant and two molecules of water as a product, then it would appear as though an oxygen atom was generated in the process and two hydrogen atoms disappeared. For example,

$$CH_4 \text{ (g)} + O_2 \text{ (g)} \rightarrow CO_2 \text{ (g)} + H_2O \text{ (l)} \qquad (2.8)$$

$$\text{Reactant atoms} \qquad \text{Product atoms}$$

$$\overbrace{1C + 4H + 2O} \neq \overbrace{1C + 2H + 3O}$$

This reaction is not physically possible because mass must be conserved. In addition, there must be equal numbers of atoms of each type on both sides of the equation. Therefore, the proper form of the equation for the combustion of methane is

$$CH_4(g) + 2O_2(g) \rightarrow CO_2(g) + 2H_2O(l) \qquad (2.9)$$

$$\text{Reactant atoms} \qquad \text{Product atoms}$$

$$\overbrace{1C + 4H + 4O} = \overbrace{1C + 4H + 4O}$$

Several examples in this chapter will illustrate the significance and usefulness of the chemical equation as a tool in waste minimization.

BALANCING A CHEMICAL EQUATION. Assume an industrial wastewater treatment process treats hexavalent chromium to trivalent chromium and then proceeds to precipitate the trivalent chromium (Cr^{3+}) as chromium hydroxide (Cr[OH]3) using a sodium hydroxide (NaOH) solution. How would the chemical equation for the precipitation of trivalent chromium be written, assuming that the pH of the solution is such that the chromium will readily precipitate?

First, the basic structure of the equation should be written as follows:

$$Cr^{3+} + NaOH \rightarrow Cr\,(OH)_3 + Na \tag{2.10}$$

Next, the equation should be balanced using the following formula:

$$Cr^{3+} + 3NaOH \rightarrow Cr\,(OH)_3 + 3Na^+ \tag{2.11}$$

Note that the sum of all ionic charges present on the left side of the equation must also equal the charges present on the right side.

REACTION YIELDS. A typical use of chemical equations is to calculate the yield of a reaction (the amount of product formed) and to determine whether any of the reactants will remain at the completion of the process. How many moles of hydrogen chloride (HCl) are necessary to dissolve 10 g of calcium carbonate ($CaCO_3$ [limestone]), and how much calcium chloride ($CaCl_2$) is formed?

The chemical equation for this reactions is as follows:

$$CaCO_3 + 2HCl \rightarrow CaCl_2 + CO_2 + H_2O \tag{2.12}$$

$$(10 \text{ g } CaCO_3)(1 \text{ g-mol } CaCO_3/100.1 \text{ g } CaCO_3) = 0.100 \text{ g-mol } CaCO_3$$

One gram-mole of calcium carbonate reacts with 2 mol of hydrogen chloride to form 1 g-mol of calcium chloride. Hence,

$$(0.100 \text{ g-mol } CaCO_3)(2 \text{ g-mol } HCl/1 \text{ g-mol } CaCO_3) = 0.200 \text{ g-mol } HCl$$

$$(0.100 \text{ g-mol } CaCO_3)(1 \text{ g-mol } CaCl_2/1 \text{ g-mol } CaCO_3) = 0.100 \text{ g-mol } CaCl_2$$

$$(0.100 \text{ g-mol } CaCl_2)(111 \text{ g } CaCl_2/1 \text{ g-mol } CaCl_2) = 11.1 \text{ g } CaCl_2$$

Again, the molecular weight for each compound is determined by adding individual atomic weights, which may be found on a periodic table.

The previous example, using Equation 2.12, demonstrates the need to cover methods of expressing solutions in terms of concentration. The three most important terminologies are *molarity*, *molality*, and *normality*. Of these, the most commonly used terminology is molarity.

Molarity refers to the number of moles of a solute dissolved in a litre of solution. A solution is the combined mixture of solute and solvent. A solvent is typically the liquid component, whereas the solute is the substance dissolved. If 2 mol of salt (NaCl) were dissolved in 1 L of water, the resulting molarity would be equal to 2. The expression for this concentration would be $2M$ salt, where the units for M are in moles per litre. Molarity is the most commonly used expression for concentration units.

Molality is the number of moles of a solute dissolved in 1 kg of solvent. It is based on the resulting volume of solution. Hence, 2 mol of salt dissolved in 1 kg of water is a 2 molal solution ($2m$ NaCl). It is interesting to note that since 1 kg of water occupies a volume of 1 L, this particular solution is also a 2-molar solution ($2M$). As long as water is used as the solvent, this simple conversion can be used.

The third unit of concentration is normality. This expression is typically used in acid–base titration exercises. The normality of a solution is denoted by N and corresponds to the number of equivalents of solute per litre of solution. An equivalent is essentially the number of hydrogen ions (H^+) or hydroxide ions (OH^-) present in a given molecule. In the case of a $1.0M$ solution of sulfuric acid (H_2SO_4), the normality would be $2N$. The value of $2N$ is derived from the fact that a solution containing 1 mol of sulfuric acid actually yields 2 mol of H^+ ions. Similarly, a $0.1M$ solution of $Ca(OH)_2$ is $0.2N$, and a $3M$ phosphoric acid solution (H_3PO_4) is $9N$.

Techniques and Conventions in Problem Solving

Solving problems that involve engineering calculations need not be an onerous task. By following a set of guidelines, the process can be streamlined while providing much information. The following step-by-step guide may be used to analyze and solve problems:

1. Determine exactly what the problem is before getting started. Review all pertinent information and understand what is governing the situation to be investigated.
2. Sketch a picture of the problem. This is a useful tool that forces the problem to be organized in a logical fashion. Details that are often inadvertantly omitted may be uncovered by this simple procedure. A typical illustration scenario is to denote boxes as processes or equipment and lines to indicate the flows of various streams.
3. Research all of the available information and data and select only that which is required for performing the calculation. This step also includes the identification of any chemical equations that are involved. All equations must be balanced or the results could prove to be erroneous.
4. Identify the mathematical equations or formulas that govern the questions being asked. Set up these equations and make sure all of the data needed to obtain the answer are available.
5. Check all of the data/information used in the calculations to be sure that all units and dimensions are compatible. Perform all necessary conversions before executing the engineering calculations (this step is the one in which most errors typically occur).
6. Although not always possible, it is helpful to have an idea of what an answer to a problem will be before performing these problem-solving procedures. In many cases, doing so can prove to be a check on any bad calculations made or conversion factors forgotten.

To further illustrate the benefits of this step-by-step procedure, two example equations that may be used in waste minimization investigations are provided.

PROCESS OPTIMIZATION USING CHEMICAL EQUATIONS. In Equation 2.11, which deals with the precipitation of trivalent chromium to chromium hydroxide, the chemical reduction reaction of Cr^{6+} to trivalent chromium occurs at pH 2.5. However, to precipitate chromium hydroxide efficiently, the pH of the solution must be increased to a value of 9.0. If the wastewater treatment process is accomplished through batch treatment in a 1 000-gal tank, how many litres of a $10M$ solution of sodium hydroxide would need to be added to the tank after the chemical reduction of chromium is complete to bring the solution up to a pH of 9.0? Consider the trivalent chromium concentration to be negligible in comparison to the hydrogen-ion concentration (pH = $-\log[H^+]$).

What is required to change the solution from pH 2.5 to 9.0? At pH 2.5, the hydrogen-ion concentration is determined as follows: for the solution at pH 2.5,

$$-\log[H^+] = 2.5 \tag{2.13}$$

where $[H^+]$ denotes the hydrogen-ion concentration.

$$[H^+] = 3.16 \times 10^{-3} \text{ mol/L}$$

For the solution at pH 9,

$$-\log[H^+] = 9$$

$$[H^+] = 1.0 \times 10^{-9} \text{ mol/L}$$

To determine the amount of sodium hydroxide solution required to alter the pH from 2.5 to 9, an accounting of the moles of hydrogen ions in the solution must be performed. For each mole of hydroxide ions introduced into the solution, an equal number of moles of hydrogen ions is neutralized. Therefore, the moles of hydroxide ions in the following equation are subtracted from the moles of hydrogen ions present:

$$\frac{[(\text{Moles of } H^+ \text{ ions present in original solution}) - (\text{Moles of } OH^- \text{ ions introduced})]}{(\text{Total volume of resulting solution})}$$

$$= 1 \times 10^{-9} \text{ moles/L} \tag{2.14}$$

The value, 1×10^{-9} mol/L, corresponds to the hydrogen-ion concentration in the final solution (pH 9). Where x is the volume in litres of sodium hydroxide solution added, substitute the following in the equation:

$$\frac{(3.16 \times 10^{-3} M\, H^+)(1\,000 \text{ gal})(3.785 \text{ L/gal}) - (10M\, OH^-)\,(x)}{(1\,000 \text{ gal})(3.785 \text{ L/gal}) - (x)} = 1 \times 10^{-9} M^+$$

$$x = 0.119\,6 \text{ L } OH^- \text{ required} = 0.119\,6 \text{ L NaOH solution required}$$
$$x = (0.119\,6 \text{ L})(1 \text{ gal}/3.785 \text{ L}) = 0.031\,6 \text{ gal NaOH solution}$$

PROCESS OPTIMIZATION BY UNIT MANIPULATION. The effluent from the batch treatment process, depicted by Equation 2.11, is to be pumped to a clarifier at a rate of 10 gpm. The precipitated chromium hydroxide particles are so small that their settling rate is driving the speed at which the entire wastewater treatment facility can process water. To facilitate clarification, a flocculent aid consisting of a synthetic polymer at a diluted concentration of 5 mg/L is introduced. The flocculent aid assists in the formation of larger aggregate particles of the precipitated chromium hydroxide, thus greatly reducing the amount of settling time required.

If the concentration of chromium hydroxide entering the clarifier is approximately 40 mg/L, and the mixing ratio of flocculent to precipitant on a mass basis is 1:50, how would the flow rate of flocculent aid (in gallons per minute) required to optimize the clarification process be determined?

This calculation is fairly simple in that all that is needed is a mass flow rate equation that takes into account the ratio of polymer to precipitant feed stream.

$$\text{Mass flow rate of polymer:Mass flow rate of precipitant}$$

$$1:50$$

$$(5 \text{ mg/L polymer})(3.785 \text{ L/gal})(x \text{ gal/min})(1) =$$

$$[40 \text{ mg/L Cr(OH)}_3](3.785 \text{ L/gal})(10 \text{ gal/min})(1/50)$$

Hence, x is equal to 1.6 gpm polymer.

MATERIAL BALANCES

This section is divided into three areas: the concept of the material balance; approaches to solving material balance problems; and advanced topics that integrate recycle, bypass, and purge calculations. Examples are provided to illustrate the importance of these issues in relation to establishing a strong foundation for an effective waste minimization program.

Material Balance

Now that a basic review of the fundamentals in engineering calculations has been discussed, the material balance can be expounded. The material balance, for purposes of this discussion, is a method for illustrating the concept of "the conservation of mass." As with chemical equations and stoichiometry discussed earlier in this chapter, the sum of the masses of substances entering a process must equal the sum of the masses of the products of the reaction. Note that a material balance is not an illustration of the conservation of volume or the conservation of moles; rather, stated simply, it is "mass in" equals "mass out" for any system under consideration.

This concept is essential to the waste minimization evaluation process because potential or resulting improvements cannot be quantified without determining a baseline. A material balance is demonstrated in Figure 2.3.

To perform an effective analysis of a process, it is important to consider the types and characteristics of the media entering and leaving a system. Measuring the quantity of gasoline entering and leaving an automobile engine, for example, is meaningless. In this case, the amount of air entering the system (which contains numerous gaseous con-

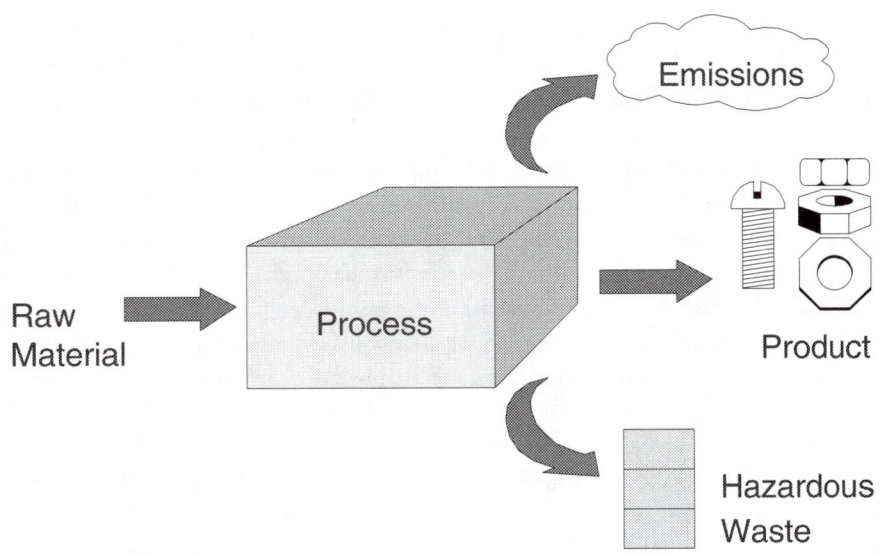

Figure 2.3   Multimedia mass balance.

stituents) and the reaction products of both complete and incomplete combustion are required.

Accumulation of mass within a system is a parameter that is frequently overlooked. For a material balance to serve its purpose, it must be accounted for. In the case of the gasoline engine, mass accumulation may not appear to be an issue because the process chamber (that is, the cylinder) is traditionally exhausted of any uncombusted raw materials and reaction products before repeating the cycle. However, what about the carbon deposits left behind? Although this question may not appear to be significant, over time it may be an important parameter in determining the overall performance of the vehicle.

Another example of accumulation can be demonstrated in the operation of a warehouse. By defining the process "vessel" as the warehouse building, the differences of deliveries into and out of the building indicate changes in inventory that are directly related to accumulation. A final example of accumulation that is perhaps the easiest to understand is that of a bank account. Ideally, the process of withdrawing and depositing monies into a bank account would be one in which the flow of cash into the account greatly exceeds the flow of cash out of the account, thereby yielding a large accumulation of monies (that is, a bankroll). In most cases, however, the flow of money into an account closely matches the flow of monies out.

The concept of steady state is equally important to the material balance. The term *steady state* refers to the constant amount of materials flowing into and out of the process over time with no buildup or depletion of material taking place within the process or system. The rate of water flowing through the nozzle of a fountain in a park, for example, can be considered steady state; however, the collection of rainwater in a bucket cannot be considered steady state. In some cases, steady state may have to be an assumed condition. In the case of the gasoline engine, the process may be considered steady state if the engine is running in a laboratory at constant revolutions per minute. However, if the engine is operating on the road, inconsistencies in speed, terrain, and individual driving habits would change the flow of fuel into the reaction chambers, thereby not yielding a steady-state process.

The first and most important step in defining a material balance problem is selecting a system for the basis of the calculation. A system defines the boundaries by which the process under investigation is governed. For the warehouse, the system was the building, and in the case of the water fountain, the system was the nozzle. These are fairly straight-forward examples. However, in the example of the gasoline engine, the system could be defined as the entire vehicle, the engine, or an individual cylinder within the engine. Therefore, selecting the proper system for a material balance can simplify the required calculations.

In many industrial processes, the assumption of steady state may be made as the flow of parts is maximized by minimizing the frequency of "starts" and "stops." This is typi-cally referred to as *continuous-flow manufacturing*. A batch process, one in which materials are fed to a system and then stopped during the reaction, may also be considered steady state over time if key assumptions are made (that is, the reaction process occurs a consis-tent number of times per shift).

To illustrate the usefulness of the material balance in waste minimization evalua-tions—and before moving on to more involved concepts and permutations—several ex-amples are provided in the following sections.

MATERIAL BALANCE BASICS. A rectangular swimming pool with the dimensions 15 ft by 25 ft is losing water at a rate of 6×10^{-5} in./min. The owner has set up an auto-matic system to add makeup water at the end of every day that is triggered by a timer set to go off every 24 hours. How much water should be metered into the pool to bring it back to its original level each time the automatic system is activated?

This problem can be solved by using a volumetric balance because the substance lost (evaporating water) is the same mass as the makeup substance (tap water). Therefore, all that is needed is to calculate the amount of water evaporated every 24 hours.

$$\text{Surface area of pool} = (15 \text{ ft})(25 \text{ ft}) = 375 \text{ sq ft}$$

$$\text{Rate of water loss} = (6 \times 10^{-5} \text{ in./min})(1 \text{ ft/12 in})(60 \text{ min/1 hr})(24 \text{ hr/d})$$

$$(375 \text{ sq ft})(7.48 \text{ gal/cu ft}) = 20.2 \text{ gpd} \qquad (2.15)$$

The makeup water added each day is approximately 20 gpd.

SLUDGE DRYING. A wastewater treatment sludge containing nearly 98% water as it is pulled out of the bottom of a clarifier is pumped at a rate of 100 gph to an aging vacuum filter for dewatering prior to disposal. The vacuum filter is able to produce a filter cake containing 15% solids. Management is proposing to replace this old piece of equipment with a new filter press that can generate filter cake containing 35% solids. If the new press costs $25 000 and sludge disposal costs are at a flat rate of $1/lb, how long will it be before the increased efficiency in water removal from the sludge pays for the new press?

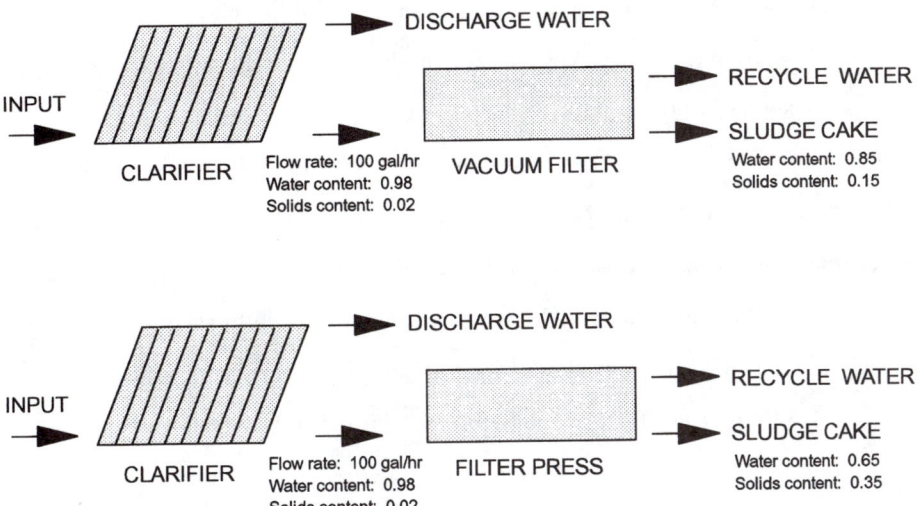

Figure 2.4 Sludge drying process flow diagrams (gal/hr $\times$ [1.051 $\times$ 10^{-6}] = m^3/s).

Both units process sludge at the same rate and will operate only 8 hours per day. The composite density of the solids is approximately 100 lb/cu ft (the time value of money should not be considered).

A good starting point is to sketch a diagram of the process, as illustrated in Figure 2.4. Because the amount of solids entering and leaving each type of filter will be the same, the only constituent that will yield savings in terms of disposal costs will be the reduction in water contained in the sludge cake. Therefore, the difference in the amount of water contained in the sludge produced from the vacuum filter minus the amount of water contained in the sludge produced from the filter press will be the vehicle by which the new filter press will pay for itself.

Mass of constituents entering the vacuum filter:

$$(62.5 \text{ lb/cu ft } H_2O)(1 \text{ cu ft}/7.48 \text{ gal})(0.98)(100 \text{ gph}) =$$
$$\text{Mass flow rate of water to vacuum filter} \quad (2.16)$$

$$819 \text{ lb/hr} = \text{Mass flow rate of water to vacuum filter}$$

$$(100 \text{ lb/cu ft solids})(1 \text{ cu ft}/7.48 \text{ gal})(0.02)(100 \text{ gph}) =$$
$$\text{Mass flow rate of solids to vacuum filter} \quad (2.17)$$

$$26.7 \text{ lb/hr} = \text{Mass flow rate of solids to vacuum filter}$$

With the mass of solids contained in the resulting filter cake known, along with its corresponding weight percentage, a mass balance of water around the system may be performed:

$$\text{Mass of solids in} = \text{Mass of solids out} = 26.7 \text{ lb/hr solids} \qquad (2.18)$$

$$\text{Mass of water in} = \text{Mass of water in sludge cake} + \text{Recycle water}$$

$$819 \text{ lb/hr } H_2O = (26.7 \text{ lb/hr})[(1 - 0.15)/0.15] + \text{Recycle water}$$

$$151 \text{ lb/hr } H_2O = \text{Mass flow rate of water in sludge cake}$$

$$668 \text{ lb/hr } H_2O = \text{Mass flow rate of recycle water}$$

Repeat the mass balance calculations for the new filter press as follows:

$$\text{Mass of solids in} = \text{Mass of solids out} = 26.7 \text{ lb/hr solids}$$

$$\text{Mass of water in} = \text{Mass of water in sludge cake} + \text{Recycle water}$$

$$819 \text{ lb/hr } H_2O = (26.7 \text{ lb/hr})[(1 - 0.35)/0.35] + \text{Recycle water}$$

$$50 \text{ lb/hr } H_2O = \text{Mass flow rate of water in sludge cake}$$

$$769 \text{ lb/hr } H_2O = \text{Mass flow rate of recycle water}$$

The amount of time required for payback to be achieved on the new filter press may now be determined as follows:

$$(\text{Mass}_{H20} \text{ in vacuum filter sludge} - \text{Mass}_{H2O} \text{ in filter press sludge})(\text{Disposal rate})(8 \text{ hr/d})$$
$$= \text{Savings rate} \qquad (2.19)$$

$$(151 \text{ lb/hr} - 50 \text{ lb/hr})(\$1.00/\text{lb})(8 \text{ hr/d}) = \text{Savings rate}$$
$$\$808/d = \text{Savings rate}$$

$$(\text{Cost of new filter press})/(\text{Savings rate}) = \text{Payback period} \qquad (2.20)$$
$$(\$25\,000)/(\$808/d) = \text{Payback period}$$
$$31 \text{ days} = \text{Payback period}$$

The purchase of the new filter press is well worth it. After approximately 1 month of service, cost savings will be incurred each month.

SOLVENT RECOVERY. An industrial process yields a waste stream of spent solvent (for example, acetone) that is contaminated with water. Using the process flow diagram shown in Figure 2.5, calculate the quantity of distillate *(D)* recovered and the amount of wastewater (*W*) produced based on the influent flow rate. All percentages given are based on weight.

$$\text{Mass of acetone in} = \text{Mass of acetone in distillate} + \text{Mass of acetone in waste} \qquad (2.21)$$
$$(5 \text{ lb/min})(0.30) = D(0.97) + W(0.05)$$

$$\text{Mass of water in} = \text{Mass of water in distillate} + \text{Mass of water in waste}$$

$$(5 \text{ lb/min})(0.70) = D(0.03) + W(0.95)$$

With two equations and two unknowns, the mass flow rates for *D* and *W* may be calculated as follows:

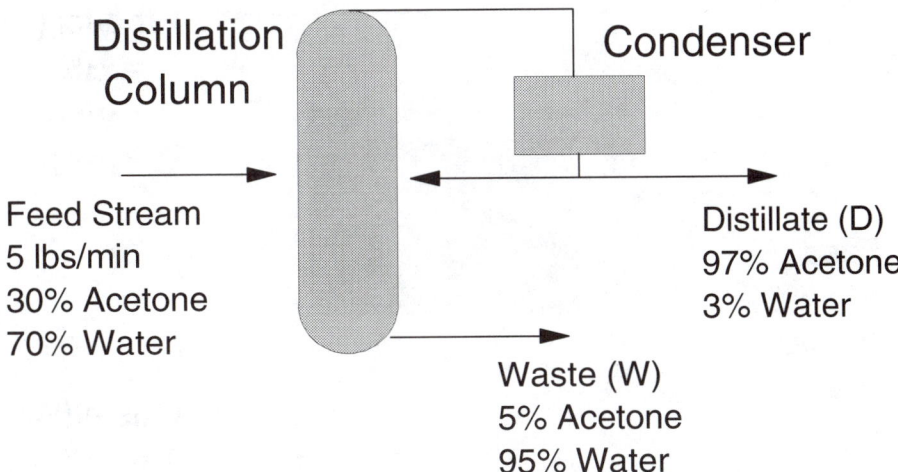

Figure 2.5 Solvent recovery process flow diagram (lb × 0.453 6 = kg).

$$1.5 = D(0.97) + W(0.05)$$

$$3.5 = D(0.03) + W(0.95)$$

Use the first equation to solve for D:

$$D = 1.55 - W(0.52)$$

Substitute this value of D into the second equation:

$$3.5 = [1.55 - W(0.52)](0.03) + W(0.95)$$

$$W = 3.7 \text{ lb/min}$$

Therefore,

$$D = 5 \text{ lb/min} - 3.7 \text{ lb/min} = 1.3 \text{ lb/min}$$

Analytical Approaches to Material Balance Problems

While previous examples in this section provide a demonstration of the material balance in action with little or no guidance on the methodology to approach the problems, this subsection attempts to illustrate a common methodology that may be used for all material balances. Although each situation may appear to have its own particular characteristics (that is, drying, precipitation, distillation, combustion), the steps for setting up a material balance for a system are identical.

The most important pieces of information to compile are the mass flow rates and compositions of all streams entering and leaving a system. Details on what is occurring within the system are also important, especially if a chemical reaction is taking place. In many situations, these data may not appear to be readily available. However, by knowing that this is the fundamental starting point, key pieces of information may be assimilated to establish this foundation.

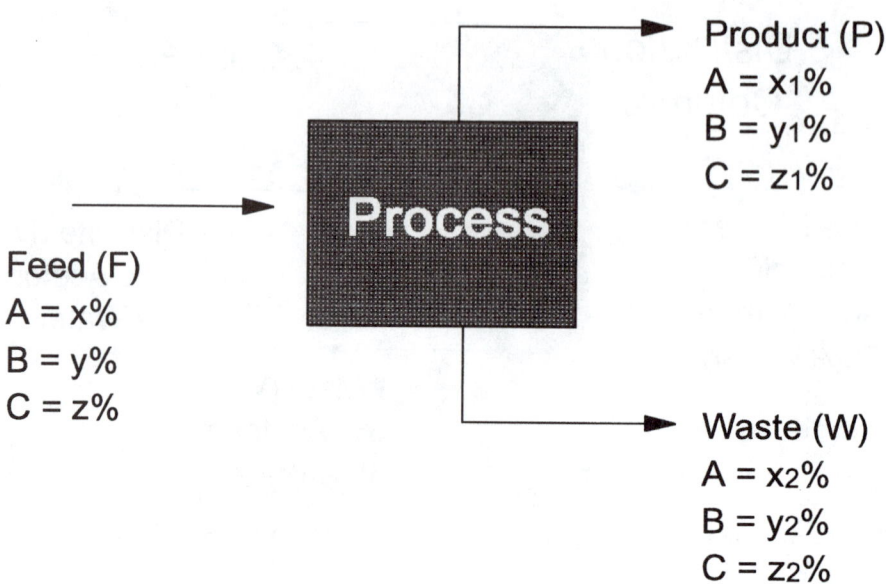

Figure 2.6   Material balance, no chemical reaction.

In Figure 2.6, the "system," which may be any of a host of processes, is denoted by a box. The lines entering and leaving the box are the various streams. Each stream has a flow rate and is broken down into its constituents on a mass-percentage basis. A total mass balance is conducted by setting the "mass in" equal to the "mass out." In Figure 2.6, no accumulation or chemical reactions are occurring within the system. A total mass flow rate balance is depicted by the following equation:

$$F \text{ (Raw materials)} = P \text{ (Product)} + W \text{ (Byproducts or waste)} \qquad (2.22)$$

To conduct a constituent balance, the total flow rate of each stream is multiplied by the applicable constituent weight percentage to yield an individual mass flow rate. A balance is then conducted around the entire system.

$$\text{In} = \text{Out} \qquad (2.23)$$

$$A\,(F)(x) = (P)(x_1) + (W)(x_2)$$

$$B\,(F)(y) = (P)(y_1) + (W)(y_2)$$

$$C\,(F)(z) = (P)(z_1) + (W)(z_2)$$

where F, P, and W depict the flow rates of each stream (for example, gpd or L/min) and the various values for x, y, and z are weight percentages for each stream constituent A, B, and C, respectively.

Each equation may be solved individually to provide the weight percentage of each constituent in the waste stream. This particular example is merely an algebraic exercise, but it should be noted that as less information is available, the more difficult the problem becomes.

Although the overall approach to the problem does not change in situations that involve chemical reactions, the sequence of calculations may become more complex.

Compared to Figure 2.6, a considerable amount of information does not appear to be present in Figure 2.7. Because a chemical reaction is involved, the first priority is to set up the following chemical equation:

$$CH_4 + 2O_2 \rightarrow CO_2 + 2H_2O$$

Because oxygen is not supplied in a pure form, but rather through air, the equation should be adapted to reflect true conditions. Nitrogen (N) and oxygen (O) are the major components in the atmosphere; typically, all other constituents may be neglected. The composition of air as depicted in Figure 2.7 is 21% oxygen and 79% nitrogen. Therefore, for each mole of oxygen consumed, 3.76 mol of nitrogen is involved (0.79/0.21 equals 3.76).

$$CH_4 + 2O_2 + 7.52N_2 \rightarrow CO_2 + 2H_2O + 7.52N_2 \qquad (2.24)$$

It is now possible to calculate the amount of air required for the complete combustion of methane. First, the mass flow rate of methane is converted to moles. Second, as discussed in the chemical equation subsection, the number of moles of oxygen and nitrogen may be found. From this, the mass flow rates of the air constituents are calculated, and a total mass flow rate for the system may be solved. Once the total mass balance is performed, a material balance at the constituent level is conducted to determine the concen-

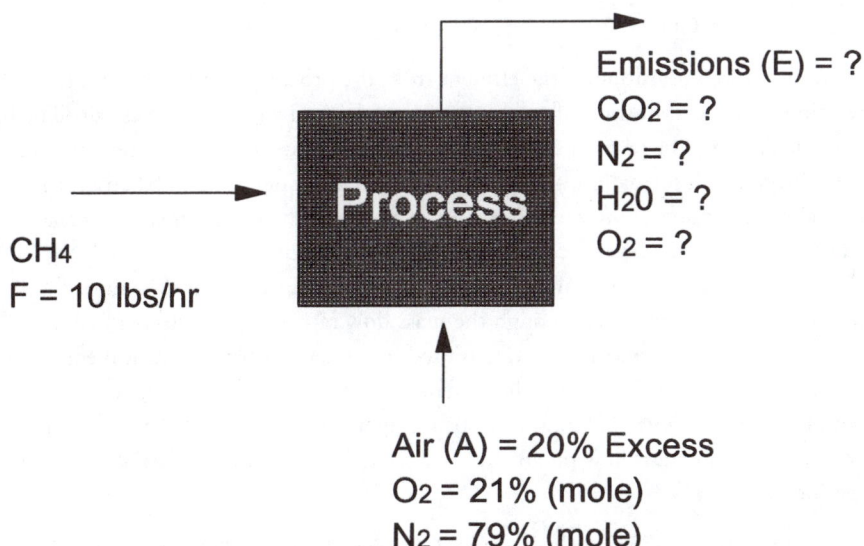

Figure 2.7   Material balance, chemical equation (lb/hr × 0.126 = kg/s).

trations of the effluent as demonstrated in Figure 2.6. These calculations are illustrated in the subsection titled Material Balance for the Combustion of Methane.

The approach to solving material balance problems is not unlike the step-by-step guidelines discussed in performing engineering calculations. The following actions outline a systematic approach to addressing material balance problems and can greatly simplify the path to a sound conclusion:

1. Always sketch a diagram illustrating the process or system under investigation.
2. Label each item in the diagram and include all available information (that is, constituent concentrations, flow rates, chemical reactions, and so on).
3. Where possible, perform calculations to determine the constituent concentrations of streams not directly identified. Use this information to find the mass of each component.
4. With the aforementioned data in place, set up equations to perform a total material balance around the system.
5. Perform a material balance at the constituent level. This step is typically more labor intensive than the overall or total material balance because of the number of equations that must be set up.

Now that some of the basic rules have been discussed, several material balance problems will be illustrated. The first is a detailed account of the process identified in Figure 2.7.

MATERIAL BALANCE FOR THE COMBUSTION OF METHANE. The combustion of methane is represented by a chemical equation that takes into account the nitrogen content found in air:

$$CH_4 + 2O_2 + 7.52N_2 \rightarrow CO_2 + 2H_2O + 7.52N_2$$

How would the composition of the effluent from the process identified in Figure 2.7 be determined if the combustion of methane were performed with 20% excess air? The basis of the calculation to be performed revolves around the mass flow rate of methane into the system. From this information, the quantity of air may be determined based on the chemical equation governing the process. After all of the incoming streams are characterized, the effluent products and concentrations may be identified.

To quantify the amount of air required for the process, the molar flow rate of methane must be calculated. Although the mass flow rate is required to perform the overall material balance, the molar flow rate is used in tandem with the chemical equation to find the molar flow rate of air into the system. The stoichiometric information from the chemical equation allows molar ratios to be set up as shown below. After these parameters are determined, the mass flow rate of air is calculated and the total material balance may be performed:

(10 lb CH_4/hr)(453.6 g/1lb)(1 g-mol CH_4/16 g CH_4) = 283.5 g-mol CH_4/hr
(283.5 g-mol of CH_4/hr)(2 g-mol O_2/1 g-mol CH_4)(1/1-.20 excess air)

= 708.8 g-mol O_2/hr

(283.5 g-mol of CH_4/hr)(7.52 g-mol N_2/1 g-mol CH_4)(1/1-.20 excess air)

$$= 2\ 664\ \text{g-mol } N_2/\text{hr}$$

Molar flow rates of air constituents can be converted to mass flow rates as follows:

$$(708.8\ \text{g-mol } O_2/\text{hr})(32\ \text{g } O_2/1\ \text{g-mol } O_2) = 22\ 682\ \text{g } O_2/\text{hr} = 50\ \text{lb } O_2/\text{hr}$$

$$(2\ 664\ \text{g-mol } N_2/\text{hr})(28\ \text{g } N_2/1\ \text{g-mol } N_2) = 74\ 592\ \text{g } N_2/\text{hr} = 164\ \text{lb } N_2/\text{hr}$$

Now, a total mass balance around the system can be performed as follows:

$$In = Out$$

$$F + A = E \tag{2.25}$$

$$10\ \text{lb } CH_4/\text{hr} + 50\ \text{lb } O_2/\text{hr} + 164\ \text{lb } N_2/\text{hr} = 224\ \text{lb effluent/hr}$$

The mass balance of the constituents is now conducted. The balance of nitrogen is simple because it is not a reactive species. The mass flow rate of nitrogen in (164 lb/hr) equals the mass flow rate out (164 lb/hr). To determine the other species, however, the chemical equation must be used again.

For every gram-mole of methane in, 1 g-mol of carbon dioxide and 2 g-mol of water are generated. Therefore, the mass flow rates for these two effluent species are as follows:

(283.5 g-mol CH_4/hr)(1 g-mol CO_2/1 g-mol CH_4)(44 g CO_2/1 g-mol CO_2)

$$= 12\ 474\ \text{g } CO_2/\text{hr} = 27.5\ \text{lb } CO_2/\text{hr}$$

(283.5 g-mol CH_4/hr)(2 g-mol H_2O/1 g-mol CH_4)(18 g H_2O/1 g-mol H_2O)

$$= 10\ 206\ \text{g } H_2O/\text{hr} = 22.5\ \text{lb } H_2O/\text{hr}$$

It is important to remember that this particular process uses 20% excess air. Therefore, molecular oxygen will also be contained in the effluent. The amount of oxygen used in the actual combustion of methane is

(283.5 g-mol CH_4/hr)(2 g-mol O_2/1 g-mol CH_4)(32 g O_2/1 g-mol O_2)

$$= 18\ 144\ \text{g } O_2/\text{hr} = 40\ \text{lb } O_2/\text{hr}$$

Therefore,

$$50\ \text{lb } O_2/\text{hr (in)} - 40\ \text{lb } O_2/\text{hr (combusted)} = 10\ \text{lb } O_2/\text{hr (effluent)}$$

A sum of the mass flow rates of the effluent components must equal the total mass effluent flow rate determined earlier:

$$27.5\ \text{lb } CO_2/\text{hr} + 22.5\ \text{lb } H_2O/\text{hr} + 10\ \text{lb } O_2/\text{hr} + 164\ \text{lb } N_2/\text{hr} = 224\ \text{lb/hr effluent}$$

The calculation is confirmed. The effluent characteristics are summarized in Table 2.2.

MATERIAL BALANCE TO ILLUSTRATE THE BENEFITS OF COUNTERCURRENT RINSING. Countercurrent rinsing is a waste minimization process by which

Table 2.2   Summary of effluent characteristics.

Effluent component	Weight %	Mass flow rate, lb/hr[a]
Carbon dioxide	12.3	27.5
Water	10.0	22.5
Oxygen	4.5	10.0
Nitrogen	73.2	164
Total	100	224

[a]lb/hr × 0.453 6 = kg/h.

rinsewaters used in chemical process lines may be greatly reduced—by as much as 97% in many cases. This procedure is an application of the material balance concept.

In traditional, single-bath rinses, water must flow through the bath at a rate that will typically maintain a contaminant concentration equal to 0.1% of the concentration of the chemical bath preceding it. In countercurrent rinsing, this procedure is maintained but is accomplished through a cascade of rinse baths with water flowing in a direction opposite to the flow of parts. Illustrations of single-bath and countercurrent rinsing techniques are provided in Figure 2.8 and 2.9, respectively. In both rinsing schemes, the drag-out from the concentrated chemical baths preceding the rinsing tanks is the same. However, as will be shown, considerably less water will be used for the latter of the two methods (assuming 1 gph of drag-out is experienced).

For the traditional approach, the following amount of water will be needed to dilute 1 gph of concentrated solution entering the rinse bath:

$$C_0(1 \text{ gph}) + C_{H2O}(x \text{ gph}) = C_f(x + 1 \text{ gph}) \tag{2.26}$$

Where

$$
\begin{aligned}
C_0 &= 1 \text{ (concentration value of 100\% influent or drag-out);} \\
C_f &= 0.001 \text{ (final rinse contaminant concentration, effluent);} \\
C_{H2O} &= 0.000 \text{ (pure water, no contaminants); and} \\
x &= \text{inflow of fresh rinsewater, gph.}
\end{aligned}
$$

Substituting in known concentration values yields

$$(1)(1 \text{ gph}) + (0)(x \text{ gph}) = (0.001)(x + 1 \text{ gph})$$

$$1\,000 \text{ gph} = (x + 1 \text{ gph})$$

$$999 \text{ gph} = x = \text{Volumetric flow rate of water}$$

The approach and calculations for the countercurrent rinsing scenario are considerably more involved. A material balance around the entire system and material balances around each individual rinse tank are required to solve the volumetric flow rate of water entering the systems and the concentration of the effluent rinsewater leaving the system and destined for wastewater treatment (note that the drag-out rate from tank to tank is a constant 1 gph).

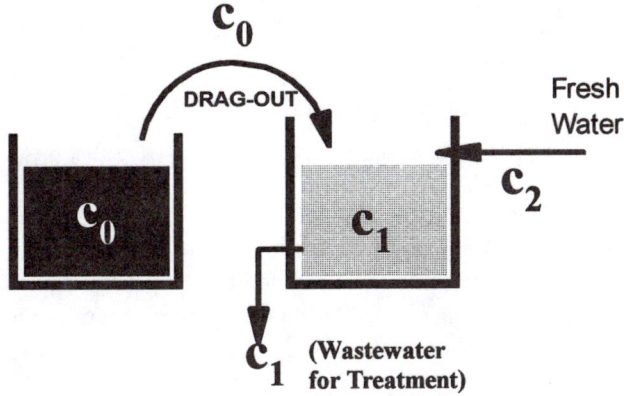

Figure 2.8   Traditional rinsing scheme.

First, a material balance is performed around tank 1 as follows:

$$C_0(1 \text{ gph}) + C_2(x \text{ gph}) = C_1(x) + C_1(1 \text{ gph}) \qquad (2.27)$$

Where

C_0 = 1 (concentration value of 100%, influent or drag-out);
C_1 = concentration of contaminants from process in rinse tank 1;
C_2 = C_f = 0.001 (final rinse concentration, effluent);
C_{H2O} = 0.000 (pure water, no contaminants); and
x = inflow of fresh rinsewater (gph).

Substituting in for know concentration values yields

$$(1)(1 \text{ gph}) + (0.001)(x \text{ gph}) = C_1(x + 1 \text{ gph})$$

$$1 + 0.001x = C_1(x + 1)$$

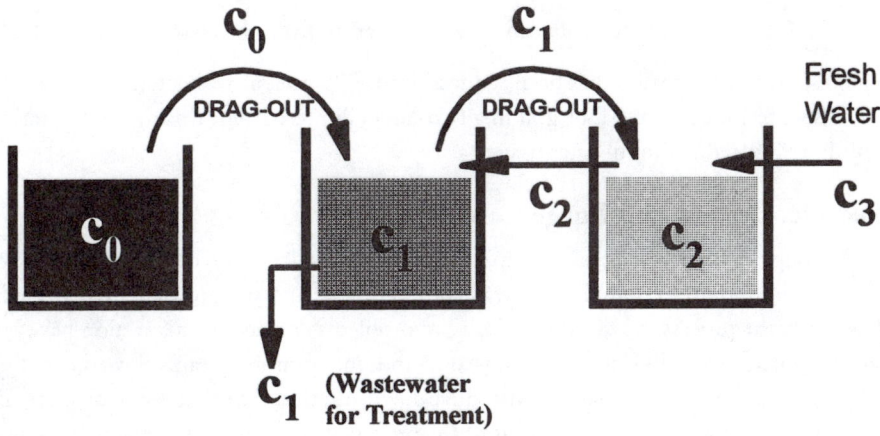

Figure 2.9   Countercurrent rinsing sheme.

Now, a material balance may be performed around tank 2 as follows:

$$C_1(1 \text{ gph}) + C_{H2O}(x \text{ gph}) = C_2(x) + C_2(1 \text{ gph})$$

Substituting in for known concentration values yields

$$C_1(1 \text{ gph}) + (0.000)(x \text{ gph}) = (0.001)(x + 1 \text{ gph})$$

$$C_1 = (0.001)(x + 1)$$

By substituting the resulting equation for the material balance around tank 2 into the resulting equation for the material balance around tank 1, a single equation with only one unknown is found. Solving for x will yield the required amount of incoming water necessary for maintaining a final rinse concentration of 0.1%:

$$1 + 0.001x = (0.001)(x + 1)(x + 1)$$

Dividing through by 0.001 yields

$$1\,000 + x = (x + 1)(x + 1)$$

$$1\,000 + x = x^2 + 2x + 1$$

$$0 = x^2 + x - 999$$

Solving by use of the quadratic equation, $[-b \pm (b^2 - 4ac)^{1/2}]/2a$,

$$x = -32.11 \text{ or } 31.11$$

Because x cannot be a negative value, the flow rate for incoming fresh water must be 31.11 gph. At this point, the concentration of the effluent from tank 1 going to a wastewater treatment process may be determined by plugging in the flow rate of water to either of the previous material balance equations. The equation for the balance around tank 2 is used as follows:

$$C_1 = C_2(x + 1)$$

$$C_1 = (0.001)(31.11 + 1)$$

$$C_1 = 0.032\,1 = 3.21\% \text{ contaminant concentration from the chemical process bath}$$

The resulting flow of water has been reduced from 999 gph, as was determined for typical rinsing techniques, to 31.11 gph in a two-tank countercurrent rinsing configuration. This is a 97% reduction in water usage.

Recycle, Bypass, and Purge Calculations

Although no new concepts concerning the material balance are discussed in this section, it is important to note the different configurations and streams that are managed in relation to the process. So far, the majority of problems covered in this section have entailed a process wherein one stream flowed in and one or more streams flowed out. Recycle, bypass, and purge calculations introduce a permutation that routes one or more of the outgoing streams back to the front of the process. Setting up a material balance is no different in these cases, although it is slightly more complex.

Recycle streams are streams in which part of the effluent of a process is rerouted back

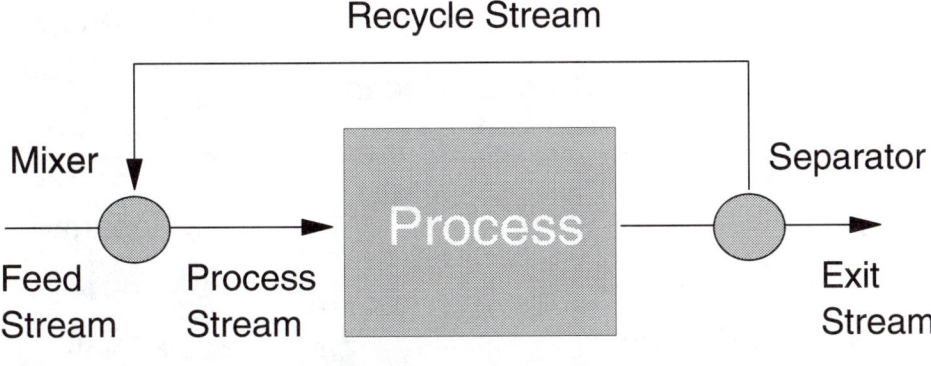

Figure 2.10   Recycle streams.

to the beginning and combined with one or more influent streams before passing through the process once again. Bypass streams are streams that pass around or skip a process entirely, the opposite of recycle streams. Purge streams are bled off of a recycle stream to reduce or remove the amount of inert compounds or undesirable materials that would otherwise be rerouted back through the process. This serves to control the buildup of high levels of contaminants. Illustrations of recycle, bypass, and purge streams are provided in Figures 2.10, 2.11, and 2.12, respectively.

RECYCLE OF BIOSOLIDS AT A WASTEWATER TREATMENT PLANT. A municipal wastewater treatment plant is using an activated-sludge process to produce a high-quality effluent. The activated-sludge process is accomplished through the use of bacteria—the main constituent of the activated sludge. These bacteria oxidize the biologically degradable material in the influent wastewater. The mixture of treated water and biosolids, known as *mixed liquor*, then flows to a clarifier where the activated sludge is settled out yielding a clear effluent. A portion of the mixed liquor is recycled back to the activated-sludge process, where it is combined with influent wastewater as seed material. The remainder of the sludge is excess solids and is disposed of as process waste.

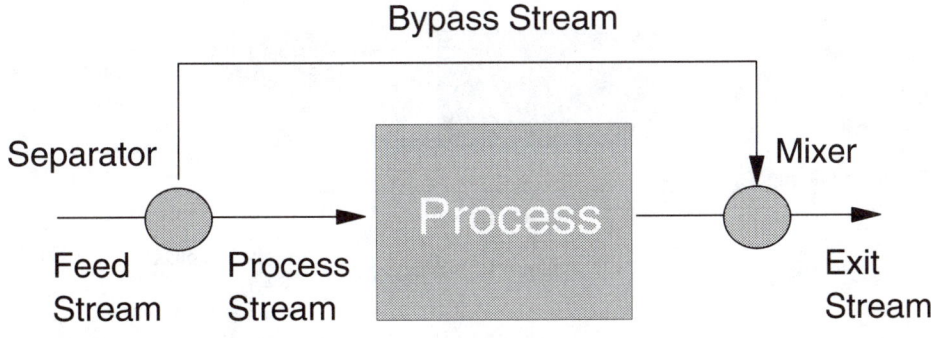

Figure 2.11   Bypass streams.

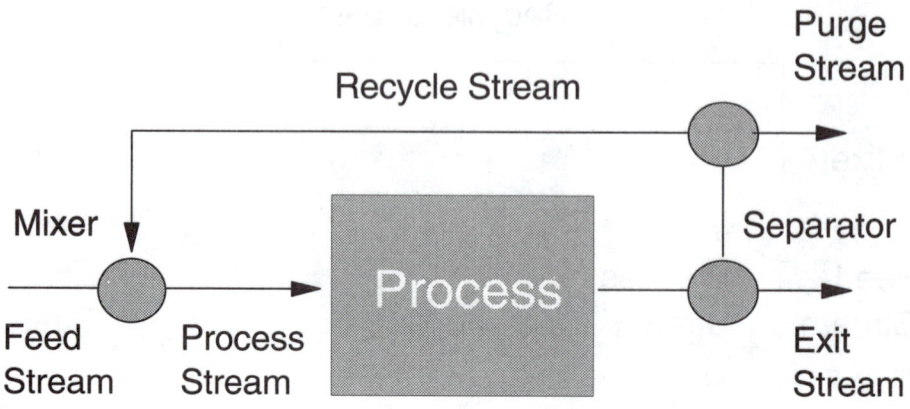

Figure 2.12   Purge streams.

The influent flow rate of wastewater to the activated-sludge process is 10 mgd, and the flow comprises 20% recycled biosolids by volume. The concentration of biosolids within the activated-sludge process tank is 3 000 mg/L. The concentration of biochemical oxygen demand (BOD) material entering the system is 550 mg/L, and the effluent is required to contain no more the 5 mg/L BOD. The concentration of biosolids being pulled off the bottom of the clarifier is 1.0×10^4 mg/L.

How would the effluent flow rate and the quantity of biosolid wastes generated each day be calculated (neglecting all growth factors, decay rates, and reaction kinetics for the bacteria cells)? A process flow diagram is illustrated in Figure 2.13. Several assumptions can be made immediately. The concentration of biosolids in the effluent is nearly zero because the biosolids are allowed adequate time to settle out in the clarifier. The concentration of biosolids in the recycle stream is equal to biosolids in the waste stream. Simi-

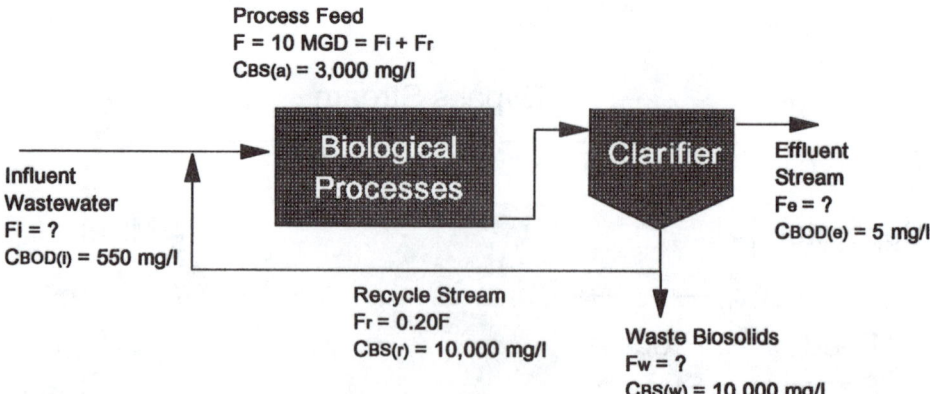

Figure 2.13   Activated-sludge process flow diagram (mgd $\times$ [3.785 $\times$ 10^3] = m^3/d).

larly, the BOD concentration in the influent and effluent streams around the clarification process is the same—5 mg/L. Therefore,

$$C_{BS(e)} = C_{BS(i)} = 0 \text{ mg/L} \tag{2.28}$$

$$C_{BS(w)} = C_{BS(r)} = 1.0 \times 10^4 \text{ mg/L}$$

$$C_{BOD(e)} = C_{BOD(w)} = C_{BOD(r)} = 5 \text{ mg/L}$$

The influent flow rate of wastewater to the entire system is determined as follows:

$$F_r = (0.20)(10 \text{ mgd}) = 2 \text{ mgd} \tag{2.29}$$

$$F_i + F_r = 10 \text{ mgd}$$

$$F_i = 8 \text{ mgd}$$

Conducting a material balance around the clarifier for the biosolids yields the volumetric flow rates for the wastes and effluent streams as follows:

$$\text{Biosolids in} = \text{Biosolids out} \tag{2.30}$$

$$[C_{BS(a)}](F_r + F) = [C_{BS(e)}](F_e) + [C_{BS(r)}](F_r) + [C_{BS(w)}](F_w)$$

Because $C_{BS(e)} = 0$ mg/L and $C_{BS(r)} = C_{BS(w)}$, the equation may be rewritten as follows:

$$[C_{BS(a)}](F_r + F) = [C_{BS(w)}](F_r + F_w)$$

$$(3\ 000 \text{ mg/L})(10 \text{ mgd}) = (1.0 \times 10^4 \text{ mg/L})(2 \text{ mgd} + F_w)$$

$$1 \text{ mgd} = F_w$$

Hence, the mass of biosolids sent for disposal each day is as follows:

$$(F_w)[C_{BS(w)}] = [(1 \text{ mgd})(1 \times 10^6 \text{ gal}/1 \text{ mgd})] \times$$

$$[(1.0 \times 10^4 \text{ mg/L})(1 \text{ g}/1\ 000 \text{ mg})(1 \text{ lb}/453.6 \text{ g})(3.78 \text{ L}/1 \text{ gal})]$$

$$(F_w)[C_{BS(w)}] = 833.33 \times 10^2 \text{ lb/d} = 41.7 \text{ tons/d}$$

Conducting a mass balance around the entire system yields the following effluent water flow rate:

$$F = F_w + F_e \tag{2.31}$$
$$8 \text{ mgd} = 1 \text{ mgd} + F_e$$
$$7 \text{ mgd} = F_e$$

Statistical Process Control

Statistical process control is an important part of waste minimization programs. Although it is frequently thought of as a quality tool, SPC is useful in making the most of a process to yield more consistent parts, thereby producing less scrap. When a part is scrapped, not only is the part itself waste, but so are all of the raw materials, energy, and manpower used to create the scrapped part. In addition to being a business/manufacturing tool, SPC can be an environmental weapon. The more consistent a process, the more efficient is the use of raw materials, energy, and manpower. This translates to higher quality products and increased profitability.

STATISTICAL PROCESS CONTROL OVERVIEW

Statistical process control is a concept and method for controlling or monitoring a process to improve overall product quality and, indirectly, product yield. Statistics is the primary vehicle by which these improvements may be made in the areas of manufacturing operations, inspection, engineering, and other business segments. Typical quality checks in the past have been conducted at the end of a process to certify product acceptance. This final inspection, however, does nothing to control the process. Rather, it channels parts to the shipping department or the scrap bin. Statistical process control requires that in-process monitoring be performed so that the actual process may be continuously evaluated and optimized. By conducting in-process monitoring, the resulting product quality may be more accurately controlled.

Although SPC is traditionally thought of as a manufacturing tool, it is useful in many situations. Statistical process control is a process control technique. It may be applied to any process where variables can be measured and tracked such as chemical process baths maintenance, painting operations, machining operations, computer or electronics assembly, wastewater treatment, computer data entry, food preparation, and heating and cooling.

The first step in using SPC is to define the process being evaluated. This may entail the study of one or many cause-and-effect relationships. The study can become more complex as the number of relationships increases. Process variation is intrinsic to any process. No two products produced in a process are exactly identical. It is the goal of SPC to minimize this variation and to define an acceptable range within which variation may occur. Variations that occur may result from short-term causes, long-term causes, "part-to-part" causes, or intermittent process upsets. These types of variations may be categorized as common causes or assignable causes.

Common causes, which are traditionally small, are caused by normal process variations. These are virtually inherent in any process and may be minimized, but generally not eliminated. Variation because of common causes produces a process that will typically be stable and predictable. Quality is tailored within a range of common causes and is manifested by the definition of process and product specifications. Specifications are formulated around a wide range of parameters such as performance, appearance, cost, customer criteria, safety, and others. Within specifications, quality is defined by attributes and variables. *Variables* are measured values that may consist of product weight, length, and composition. *Attributes* are not specifically measured values; they are more commonly associated as pass–fail criteria. An example of a product attribute occurs in the evaluation a knife blade's sharpness by using a piece of paper. If the piece of paper can be cut, the blade passes the test and is considered acceptable even though no specific determinable value was measured for blade sharpness.

Assignable causes are somewhat self-explanatory. These types of process variations are specific in nature and produce anomalies within a process. Examples of assignable causes are the use of off-specification raw materials, equipment failures, and the use of insufficiently trained personnel. Assignable causes, in contrast to common causes, produce variations in a process that lead to instabilities and unpredictable results. It is impossible to restrict a window of process variation if assignable causes have not been eliminated.

The final general area of SPC is the formation and study of *distributions*. Distributions are a visual representation of process variation. The concepts that define distributions are population and sample. A population is a grouping of products selected for evaluation. Populations may be narrow in scope (that is, products produced by an operator on a specific piece of equipment on a single shift) or wide ranging (that is, products manufactured over the period of several months by numerous operators using a number of different pieces of equipment). In either case, a sample is that portion of the products produced within a defined population that are selected for measurement or evaluation. Samples are typically selected at random, and the information gathered from these products defines a pattern of variation that is manifested as a distribution. Selecting parameters for measurement within a sample population is key to performing an analysis on process variation.

Three types of tools useful in distribution analysis are frequency tables, histograms, and probability plots. These tools are simple to construct and provide valuable information. The following subsections discuss the basics of these tools and illustrate their importance in an SPC program.

Frequency Tables and Histograms

A *frequency table*, which is the most simplistic tool for distribution analysis, consists of a tabular representation of the number of times a measurement value occurs within a sample population. The frequency at which a measured value occurs may be visualized quickly using a frequency table. The table consists of a list of measured values, typically presented vertically to the left of the table, with the frequency of these numbers illustrated horizontally increasing from left to right. An example of a frequency table is shown in Figure 2.14. A frequency table allows the spread between high and low values to be readily observed. When constructing a frequency table, it is important that the measured values are obtained randomly from a sample group. The steps required to complete a frequency table are itemized below.

1. Determine the range of allowable measurement as per a design specification (for the data plotted in Figure 2.14, specification limits of 20 and 60 will be used).
2. Construct a data compilation form that allows measurements to be recorded as they are gathered for future use in the frequency table. An example of this type of form is provided in Figure 2.15, which contains data plotted in Figure 2.14. It is important to note that the measurement instruments selected must be compatible with the range of measurements to be recorded. A device that could accurately measure values to the second decimal place, 0.05 in. (1.27 mm) for example, would not be acceptable for measurements requiring four significant figures, 0.000 5 in. (0.012 7 mm).
3. Select the measurement values or ranges of measurement values to be used on the

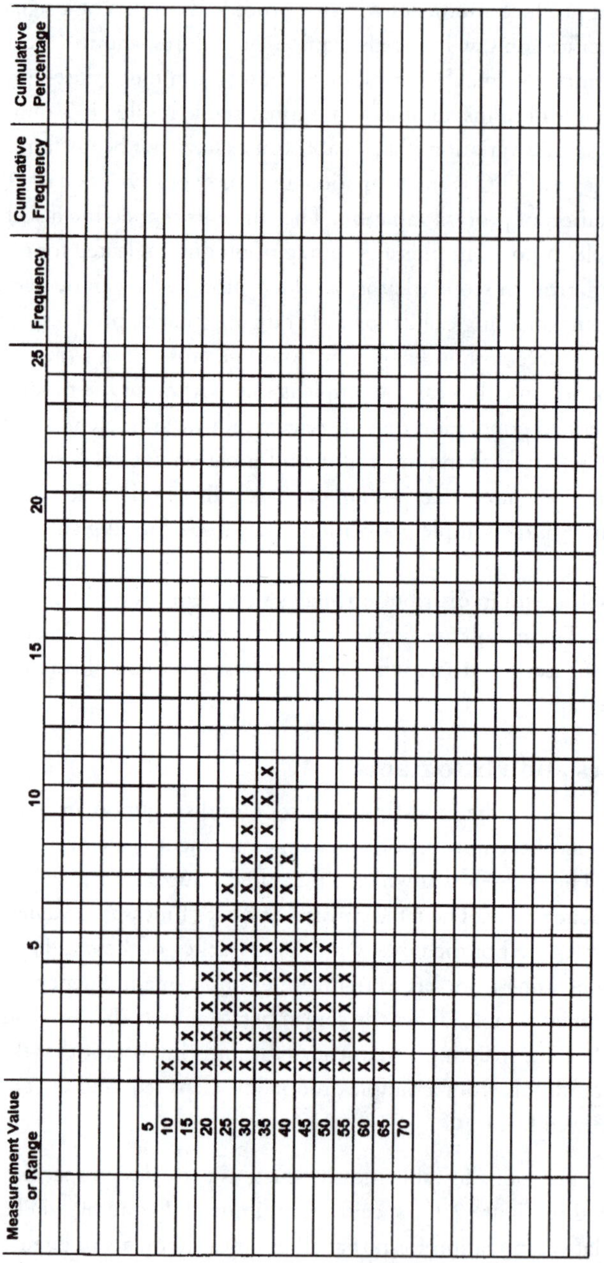

Figure 2.14   Frequency table.

Measurement Event	Measurement Value	Measurement Event	Measurement Value	Measurement Event	Measurement Value
1	10	26	30	51	30
2	15	27	25	52	35
3	30	28	35	53	25
4	20	29	40	54	40
5	35	30	15	55	35
6	30	31	35	56	20
7	25	32	45	57	45
8	35	33	30	58	50
9	30	34	20	59	55
10	20	35	50	60	50
11	40	36	35	61	55
12	45	37	40	62	30
13	35	38	25	63	
14	30	39	35	64	
15	25	40	30	65	
16	50	41	55	66	
17	40	42	40	67	
18	30	43	50	68	
19	55	44	25	69	
20	25	45	45	70	
21	40	46	65	71	
22	35	47	60	72	
23	60	48	30	73	
24	45	49	40	74	
25	35	50	45	75	

Figure 2.15   Data compilation form.

frequency table. These values are entered vertically in the far right column. Horizontal lines should be interjected in this vertical listing to indicate specification limits.

4. Plot the measurement values on the frequency table that were listed on the data compilation form. The resulting distribution that is typically found is referred to as a *normal distribution* and resembles a bell-shaped curve as shown in Figure 2.16. Patterns of this type indicate symmetry in that equal numbers of measurements fall above and below the center line of the curve.

5. Enter the number a times a given measurement value appeared on the table in the applicable area in the column labeled *frequency*.

6. Calculate the *cumulative frequency* of measurements by adding the frequency values in the order of their appearance. For example, if the frequency value for the first row of recorded measurements were 1, and the frequency value for the second row of measurements were 2, the cumulative frequency for rows one and two would be 1 and 3 (that is, 1 plus 2), respectively.

7. The last exercise to perform to complete the frequency table is to calculate the *cumulative percentage*. Cumulative percentage is a method for determining the percentage of product measurements that fall above and/or below a given value. Cumulative percentage is found by using the following equation:

Cumulative percentage $= [(\text{cumulative frequency})/(n + 1)] \times 100\%$ $\qquad$ (2.32)

where n is the total number of measurements recorded.

Regarding the calculation of cumulative frequency in step 6, the cumulative percentages for the first two rows are as follow:

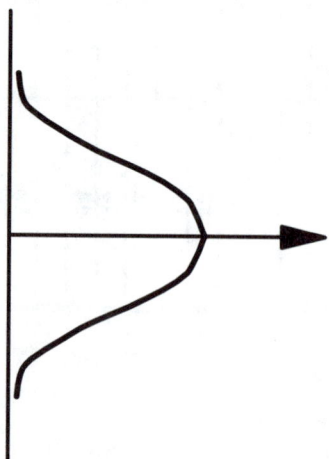

Figure 2.16 Normal distribution.

Row 1: cumulative frequency = 1:

Cumulative percentage = $[(1)/(62 + 1)] \times 100\% = 1.59\%$

Row 2: cumulative frequency = 3:

Cumulative percentage = $[(3)/(62 + 1)] \times 100\% = 4.76\%$

Steps 1 through 7 are summarized in Figure 2.17. The frequency table categorizes measured data so it may be understood. It shows the number of times a value occurs within a sample population and indicates the cumulative frequency and cumulative percentage of the measured data. As will be discussed, the frequency table is a simplification of a histogram and is only one of several tools for evaluating measured process parameters. The most important information gathered by the construction of a frequency diagram is the determination of whether a given set of measurements indicates a normal distribution for the process and the cumulative percentage. The former is useful for interpreting process stability, whereas the latter may be used to describe a population of parts between a set of parameters.

A *histogram* is a graphical representation of measured data in the form of a bar chart. It is a formalized version of a frequency table in that the shape or distribution of the data is the same. In a histogram, however, the data are plotted from left to right using vertical bars. A sample histogram of the data compiled in Figure 2.17 is shown in Figure 2.18. The axis labels in Figure 2.18 have been reversed so that the frequency of occurrence for a measurement is now vertically displayed on the left, whereas the measurement values are indexed along the bottom.

Histograms are some of the most commonly used graphical tools for illustrating the patterns of data measured within a process. They are graphical representations of frequency tables and provide a quick visual reference for the patterns of variation found within the measurements of specified process parameters.

Probability Plots and Control Charts

One of the drawbacks of a histogram is that the pattern of variation appears to change when selecting a smaller or larger measurement increment (that is, 0.5 in. or 0.005 in., rather than 0.05 in.). This can make interpretation of histograms somewhat difficult, particularly when trying to determine if the process measurements are normally distributed. For this reason, another type of graph known as the probability plot is used. Probability plots are graphical representations of cumulative percentages versus measured values on a modified logarithmic scale. The upper and lower 20% of the cumulative percentages are plotted on log scales, whereas the remainder of the data are on a normal scale. For normal distributions of data, this will result in a straight line. A chart of this type is illustrated in Figure 2.19. The data used were taken from the frequency table found in Figure 2.17.

In Figure 2.19, the vertical axis is labeled with measurement values. This allows the horizontal axis, which is a logarithmic scale, to indicate the range of cumulative percentages. The lower scale ranges from 0 to 100 and is typically used to plot the cumulative percentage data previously calculated on the frequency table. Once plotted, the lower scale provides an indication of the percentage of measured values that occur above a given

Measurement Value or Range					5					10					15					20					25	Frequency	Cumulative Frequency	Cumulative Percentage
5																												
10	X																									1	1	1.59%
15	X	X																								2	3	4.76%
20	X	X	X	X																						4	7	11.11%
25	X	X	X	X	X	X	X																			7	14	22.22%
30	X	X	X	X	X	X	X	X	X	X																10	24	38.10%
35	X	X	X	X	X	X	X	X	X	X	X															11	35	55.56%
40	X	X	X	X	X	X	X	X																		8	43	68.25%
45	X	X	X	X	X	X																				6	49	77.78%
50	X	X	X	X	X																					5	54	85.71%
55	X	X	X	X																						4	59	93.65%
60	X	X																								2	61	96.83%
65	X																									1	62	98.41%
70																												

Figure 2.17 Completed frequency table.

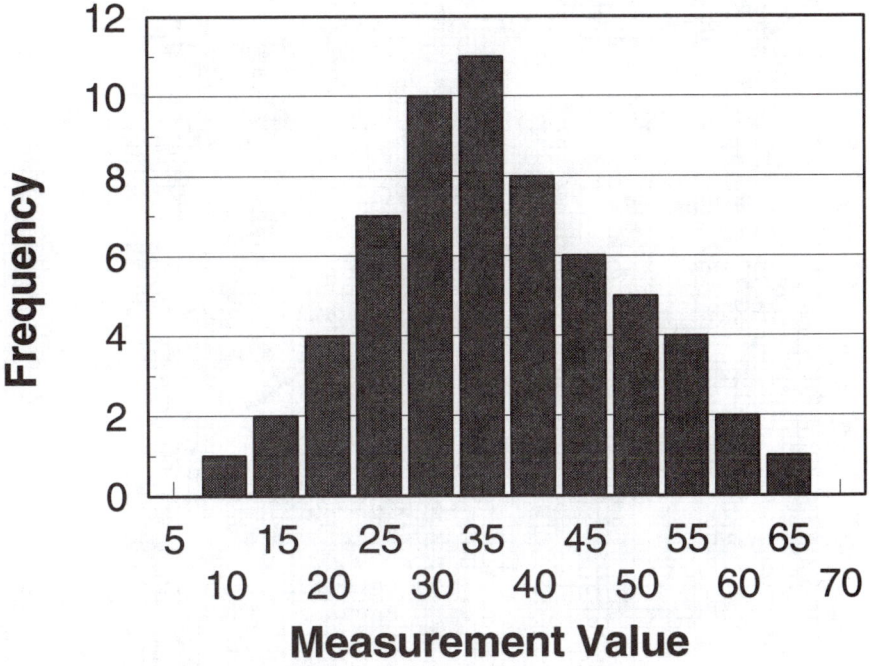

Figure 2.18 Sample histogram of compiled data.

measurement. The upper scale, which is opposite in direction, indicates the percentage of values that fall below a given measurement. After the cumulative percentages have been plotted, a straight, best-fit line is introduced. The closer the points are to the line, the more normally distributed the data. The dashed lines, as in all of the previous figures, are used to notate the upper and lower specification limits. In contrast to the histogram, interpretation of the probability plot is not affected by the selection of measurement value increments.

Thus far, only methods for constructing tables and graphs for measured data have been covered in this section. These data may also be expressed by numerical values that are derived from specific calculations. The primary numerical values are mean and standard deviation. The *mean* is typically referred to as the arithmetic average of a set of measurements. *Standard deviation* is a quantification of the raw data variability. There are two types of mean and standard deviation: population and sample.

The population mean and population standard deviation are frequently notated by μ and σ, respectively. These values are calculated from a set of measurements for an entire population. The sample mean and sample standard deviation are symbolized by $\overline{X}$ and S, respectively. The value of the sample standard deviation is sometimes referred to as σ_{n-1}. The sample mean and sample standard deviation, which are used most frequently, are an estimate of the population mean and population standard deviation. The values of the population mean and the population standard deviation are rarely calculated because an

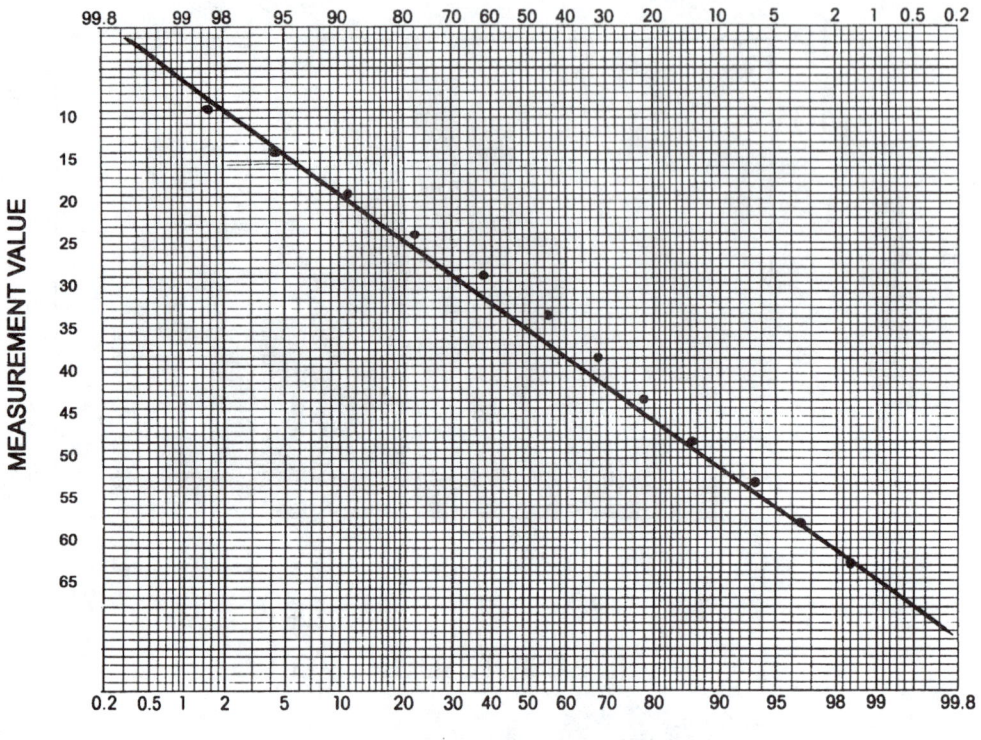

Figure 2.19 Probability plot of cumulative percentage data.

entire population of measurements is not needed to make predictions about the distribution of a process. In addition, it is generally not feasible to obtain such a vast amount of measurements. The equations used to calculate the sample mean and sample standard deviation are as follows:

$$\bar{X} = (\Sigma X_i/n) = (X_1 + X_2 + X_3 + \ldots\ldots + X_n)/n \tag{2.33}$$

$$S = [\Sigma(X_i - \bar{X})^2/(n-1)]^{-1/2} \tag{2.34}$$

Where

$\bar{X}$ = sample mean,
X_i = individual measurement values,
S = standard deviation,
n = number of measurement values, and
Σ = summation symbol.

The mean and standard deviation are useful when making predictions about a population of measurements, should the distribution appear to be normal. Figure 2.20 illustrates this relationship. Here, measured values falling within a population mean $\pm$ 1σ correspond

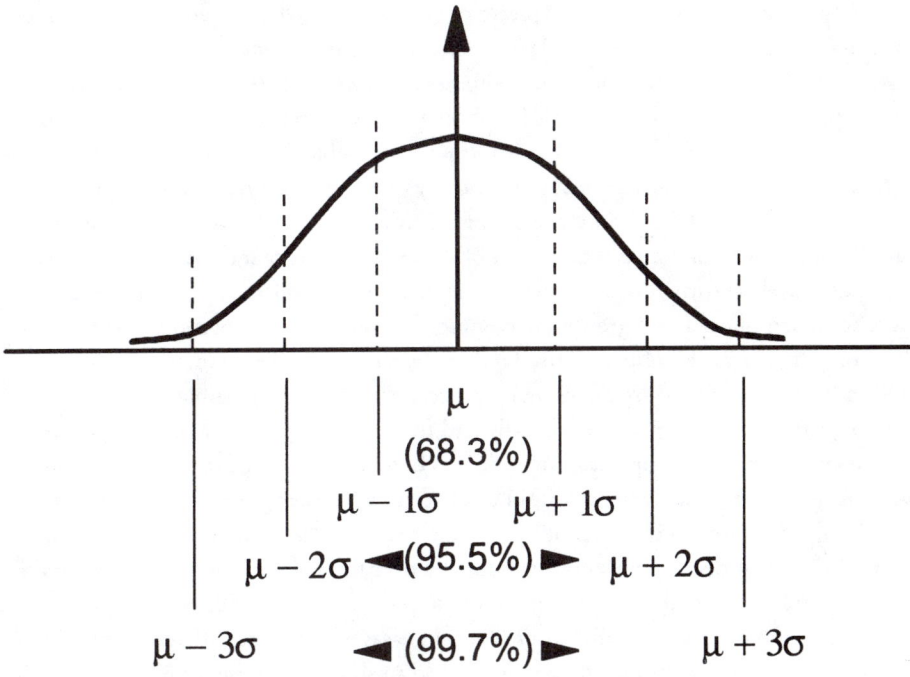

Figure 2.20 Standard deviation and mean for normal distributions.

to 68.3% of the areas under the curve. Similarly, values between a population mean $\pm$ 2σ and a population mean $\pm$ 3σ pertain to an area under the normal distribution curve of 95.5% and 99.7%, respectively. As the number of standard deviations increases beyond 3σ, the area under the curve changes only slightly. The curve asymptotically approaches the x-axis, but never gets there.

A good method for checking the calculated sample mean is to compare it to the plotted measurements on the probability plot. The point on the probability plot corresponding to the 50% mark should be nearly equal to the calculated mean. It is important that the distribution be normal to make accurate predictions using the mean and standard deviation. As previously discussed, if the probability plot is constructed on a log scale and it forms a straight line, then it may be assumed that a normal distribution exists. It is also important that a representative number of measurements be taken to construct a probability plot and calculate the mean and standard deviation. In general, the more observations made, the more accurate the predictions of the process. Therefore, at least 30 to 50 measurements should be taken to establish a reasonable representation of the process being evaluated. Without compiling a "critical mass" of data, incorrect assumptions or decisions on the capabilities of a process may be made.

Control charts are the next progression in summarizing basic information about a process. These charts differ from the data tracking methods previously discussed. In the previously discussed data tracking methods, no specific knowledge of the process was required; rather, only the measurement data were needed. Control chart theory takes into account parameters such as different pieces of equipment used, different operators, different batches of raw materials, and different periods of time. A benefit of control charts is that, rather than taking "snapshots" of a process or views which summarize data over extended periods, individual characteristics of the process are tracked allowing decisions to be made about the process as close to "real time" as possible. The sequence of events is illustrated in control charts, whereas, in previously discussed methods, this parameter could not be tracked. In addition, the use of control charts can create a more efficient sampling plan by minimizing the amount of data required to characterize the process.

To effectively construct control charts, it is necessary to collect data in smaller and more specific groups. These groups are commonly referred to as subgroups and may include only five to six sampling points. To create a meaningful representation of the process, these points must isolate as many process parameters as possible. For example, subgroups should be collected periodically and tracked in relation to a specific piece of equipment, an individual operator, or variances between batches of raw material. It is useful to select the measurements within a subgroup as close together as possible to provide data on the instantaneous variability within the process. By taking additional subgroups from the process, snapshots are compiled over time to construct a number of different control charts, each of which provides unique information. Subgroup data should not be taken from a stack of finished products in cases where these differences cannot be isolated because the results may be erratic or yield misleading information.

Individual measured values are not plotted on control charts. For the subgroup mean and range control chart, or the $\bar{X}$-R control chart, the mean and range for each subgroup are particularly important. The mean is plotted by taking the arithmetic average of the measured data of a subgroup. For five points, this would be the sum of the data values divided by five. The range is the difference between the highest and lowest values of a subgroup. Figure 2.21 contains a sample $\bar{X}$-R control chart.

The control chart does not indicate the specification limits for the products produced. This is typical of all control charts. Instead, upper and lower control limits are calculated. Control limits indicate whether process variation is distributed normally and stable over time (that is, predictable). If all points plotted on a control chart fall within the control limits and form a random pattern, the process is defined as being within statistical control. This is also an indication that the process variation is derived from common causes. In contrast, if the plot does not consistently fall within the control limits or displays unnatural patterns, the process is considered unpredictable and out of statistical control. The instability is typically a result of assignable causes.

Whether or not a process is classified as being within statistical control has no bearing on its ability to produce products within specification limits. However, if a process is not within statistical control, the long-term stability and predictability of the process to produce products within the defined specifications is compromised. Therefore, calculated control limits found on a control chart are boundaries placed on a process that establish stability, control, and reproducibility.

There are various types of control charts, each of which provide useful information. The primary categories of control charts include those that track variables data and attributes data. Examples of variables data include dimensions (length, width, height), weight,

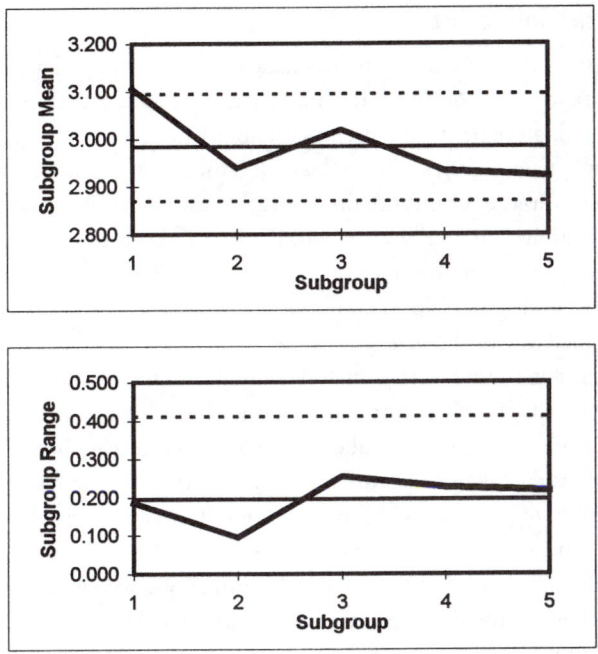

Figure 2.21 $\overline{X}$-R control chart.

hardness, and flexibility. Attributes data pertain to color, cracks, dings, and broken parts. The variability chart of primary use (previously illustrated as the $\overline{X}$-R control chart) is covered in more detail in the following section. Examples of attribute control charts, which are discussed in the section titled Quality Characteristics in this chapter, include C-charts, which track the number of defects; U-charts, which track the number of defects per part; NP-charts, which track the number of defective parts produced; and P-charts, which track the percentage of defective parts produced.

PROCESS CAPABILITY

A process capability study is a method for determining the ability of a process to produce products that meet specifications. The objective of the study is to find out how well a process can perform under the best of conditions. The tables and charts covered thus far in this section (that is, the frequency table, histogram, probability plot, and $\overline{X}$-R control chart) are the primary tools for conducting a capability study.

To create a set of conditions in which a process under evaluation can perform at its best, potential problems should be forecasted and eliminated. Compiling a list of critical operational parameters for a process or associated equipment is a good start. All assignable causes for variation should also be tracked down and minimized. If any modifications to a process must be performed during the process capability study, they should be recorded. This lends understanding to the occurrences of process anomalies or variations. The following subsections are the primary methods and sequences by which process capability studies are performed.

Variables Data Collection

Two distinct steps for collecting variables data are the planning phase and the collection phase. Before data are compiled, it is important that adequate time be spent preparing for and forecasting the needs of a process capability evaluation. The first step is to itemize the quality characteristics of a product. Because it is not practical or efficient to track these numerous pieces of information, selecting the items that deal with the most important specifications is typically sufficient. Here, the idea is to initiate a study that is not so onerous as to be counterproductive.

The selection of a subgroup size is an important parameter. As the tracking of variables data is conducted with the passage of time, data collection of each subgroup need not be large. Typically, subgroup sizes include only five to six measurements. This may vary depending on factors such as cost and time and the ease with which a measurement may be taken. It is important to remember that when data for variables are collected, subgroup measurements are taken consecutively. Measurements within a subgroup are an instantaneous snapshot of process variability. Variation between subgroups allows for the consideration of time.

Selecting the sampling interval between subgroups allows for viewing a process during a reasonable amount of time. In some cases, subgroups may be once per day. For processes with high production rates, subgroups may be once per hour. In either case, 20 to 30 subgroups should be compiled and taken with some degree of randomness to ensure the detection of process variation. This will yield in excess of 100 data points and provide a useful set of information that may then be plotted and interpreted.

Data collection is a fairly simple and straightforward process. However, when numerous measurements are required, the task can become tedious. This may increase the likelihood for data errors to occur during their transition from the measuring device to the recording logs. Errors in data entry are variables that are not process variations, and their occurrence can result in faulty conclusions in a process capability study. A practice commonly used to help streamline this task is to abbreviate numbers. This is particularly useful when the data values are carried to four or five significant digits. For example, if the specification values of interest are 2.500 0 in. (63.5 mm) and 2.590 0 in. (65.786 mm) and the measurement device's resolution is 0.000 5 in. (0.0127) mm, recorded data points could be the following:

2.500 5	2.550 0	2.510 5	2.525 5
2.530 0	2.534 5	2.500 0	2.505 5
2.575 0	2.570 0	2.545 0	2.510 0

Because only the last three digits of each value exhibit variation, the entire value does have to be recorded after each measurement. The prefix for all measurements, in the case above, is 2.5. The portion of the recorded values that remains fixed is the zero value. The resolution of the measurement device is the unit of measure. Recorded values are those portions of the measured values that are actually written down or stored. Whether abbreviating data point values or not, it is important that care be taken when transcribing data to eliminate or minimize errors.

Another useful practice when collecting data is to note the date and time of each subgroup measurement and a method for identifying each part measured. These practices allow for traceability and remeasurement if data are lost or appear questionable. Common methods for identifying parts include the tracking of part serial numbers (typically recorded on the part itself) or the assigning of subgroup numbers. When assigning a subgroup number, a numbering scheme should be set up and used to label parts via tags or indelible markers.

$\bar{X}$-R Charts

The first step in constructing an $\bar{X}$-R chart is to take the data collected and perform a sequence of calculations to determine the following key pieces of information:

$\bar{X}$ = the subgroup mean,
$\bar{\bar{X}}$ = the mean of all subgroups,
R = the subgroup range,
$\bar{R}$ = the range of all subgroups,
UCL_X = the upper control limit for subgroup means,
UCL_R = the upper control limit for subgroup ranges,
LCL_X = the lower control limit for subgroup means, and
LCL_R = the lower control limit for subgroup ranges.

The subgroup mean, or average, is determined by adding all the values in the subgroup and dividing by the number of values in the subgroup. The equation representing this operation is given below.

$$\bar{X} = \Sigma X_i / n \tag{2.35}$$

Where

X = subgroup mean,
Σ = summation,
X_i = individual measurement in subgroup, and
n = number of measurements in subgroup.

The subgroup range is the difference between the highest measured value in a subgroup and the lowest measured value, as the following equation indicates:

$$R = \text{Maximum value} - \text{Minimum value} \tag{2.36}$$

The mean of all subgroups, or the total average, is an average of the subgroup averages. This is found by adding all of the subgroup averages together and dividing by the number of subgroups. The equation for calculating the mean of all subgroups, which is similar to the equation for the subgroup mean, is as follows:

$$\bar{\bar{X}} = \Sigma \bar{X}_i / m \tag{2.37}$$

Where

$\overline{\overline{X}}$ = mean of all subgroup means,

Σ = summation,

$\overline{X}_i$ = individual subgroup mean, and

m = number of subgroups.

The range of all subgroups, or the average range, is an average of previously determined ranges and is given by the following equation:

$$\overline{R} = \Sigma R_i / m \qquad (2.38)$$

Where

$\overline{R}$ = average range of all subgroup ranges,

Σ = summation,

R_i = individual subgroup range, and

m = number of subgroups.

The only numerical manipulations performed in the previous set of equations involve taking the averages of sets of measured values and sets of calculated values. An example is provided at the end of this section to further illustrate the construction of an $\overline{X}$-R chart. However, before this can be done, upper and lower control limit calculations should be addressed.

Upper and lower control limits are calculated to show the extent to which subgroup averages and ranges vary if only common causes of variation are present. To calculate the upper and lower control limits for subgroup averages and ranges, the following set of equations may be used. The values for A_2, D_3, D_4, and d_2, given in Table 2.3, are based on the number of measurements in a subgroup. The derivation of these constants is arrived at through various statistical analyses and is beyond the scope of this text.

$$UCL_X = \overline{\overline{X}} + A_2(\overline{R}) \qquad (2.39)$$

$$LCL_X = \overline{\overline{X}} - A_2(\overline{R}) \qquad (2.40)$$

Table 2.3 Factors for determining control limits.

n	A_2	D_3	D_4	d_2
2	1.88	0	3.268	1.128
3	1.023	0	2.574	1.693
4	0.729	0	2.282	2.059
5	0.577	0	2.114	2.326
6	0.483	0	2.004	2.534
7	0.419	0.076	1.924	2.704
8	0.373	0.136	1.864	2.847
9	0.337	0.184	1.816	2.97
10	0.308	0.223	1.777	3.078

Table 2.4   Sample data.

Subgroup	Subgroup measurements				
1	3.036	3.124	3.007	3.191	3.166
2	2.927	2.882	2.928	2.976	2.974
3	2.986	3.030	3.151	3.023	2.897
4	2.943	2.795	3.021	2.940	2.959
5	2.777	2.928	2.993	2.920	2.991

$$UCL_R = D_4(\bar{R}) \tag{2.41}$$

$$UCL_R = D_3(\bar{R}) \tag{2.42}$$

In addition, the sample standard deviation may be estimated using the following equation:

$$S = \bar{R}/d_2 \tag{2.43}$$

Contructing an $\bar{X}$-R Chart

Using the data in Table 2.4, calculate the subgroup mean and range for each subgroup, the mean of all subgroups, the range of all subgroups, and the applicable upper and lower control limits. On completion of the calculations, plot the values on an $\bar{X}$-R chart. A sample calculation for each piece of information needed to plot an $\bar{X}$-R chart is demonstrated below:

$$\bar{X} = \Sigma X_i/n$$

For subgroup 1,

$$\bar{X} = (3.036 + 3.124 + 3.007 + 3.191 + 3.166)/5$$
$$\bar{X} = 3.105$$

$$R = \text{maximum value} - \text{minimum value}$$

For subgroup 1,

$$R = 3.191 - 3.007$$
$$R = 0.184$$

$$\bar{\bar{X}} = \Sigma \bar{X}_i/m$$
$$\bar{\bar{X}} = (3.105 + 2.938 + 3.018 + 2.932 + 2.922)/5$$
$$\bar{\bar{X}} = 2.983$$

$$\bar{R} = \Sigma R_i/m$$
$$\bar{R} = (0.184 + 0.094 + 0.254 + 0.226 + 0.216)/5$$
$$\bar{R} = 0.195$$

$$UCL_X = \bar{\bar{X}} + A_2(\bar{R})$$
$$UCL_X = 2.983 + (0.577)(0.195)$$
$$UCL_X = 3.096$$

$$\text{LCL}_X = \overline{\overline{X}} - A_2(\overline{R})$$
$$\text{LCL}_X = 2.983 - (0.577)(0.195)$$
$$\text{LCL}_X = 2.870$$

$$\text{UCL}_R = D_4(\overline{R})$$
$$\text{UCL}_R = (2.114)(0.195)$$
$$\text{UCL}_R = 0.412$$

$$\text{LCL}_R = D_3(\overline{R})$$
$$\text{LCL}_R = (0)(0.195)$$
$$\text{LCL}_R = 0$$

Table 2.5 is a compilation of the calculated values as determined from the measured data found in Table 2.4. With the parameters calculated, the $\overline{X}$-R chart may now be constructed. The horizontal axis of the $\overline{X}$-R chart is labeled with the subgroup numbers and the subgroup mean and range values are plotted vertically (the subgroup mean and range values are plotted separately). It is important to select a centerline for each value that is equidistant from the maximum and minimum values. This will allow each plot to be evenly distributed in a vertical fashion. It will also allow the upper and lower control limits to be labeled more easily. The increment on the vertical axis should also be selected to allow all values to appear on the plot. This exercise was originally conducted manually. Today, however, there are many software applications available that can greatly simplify this task. Even simple spreadsheet packages can streamline this process. In these applications, parameters are automatically calculated and plotted after the original measured data are entered. Plots for the aforementioned calculated $\overline{X}$-R chart parameters are shown in Figure 2.21.

Types of Patterns

After an $\overline{X}$-R chart has been constructed, the pattern must be interpreted. The desired type of pattern to find is commonly referred to as the *natural pattern*. This indicates that the process is in statistical control and displays variability mostly from common causes. For natural patterns, the plotted points should fall near the center of the graph with some of the outliers approaching the control limits. Few points, if any, should fall outside the

Table 2.5   Sample $\overline{X}$-R control chart parameters.[a]

Subgroup	$\overline{X}$ (subgroup average)	R (subgroup range)
1	3.105	0.184
2	2.938	0.094
3	3.018	0.254
4	2.932	0.226
5	2.922	0.216

[a] $\overline{\overline{X}} = 2.983$; $\overline{R} = 0.195$; $\text{UCL}_X = 3.095$; $\text{LCL}_X = 2.870$; $\text{UCL}_R = 0.412$; and $\text{LCL}_R = 0.000$.

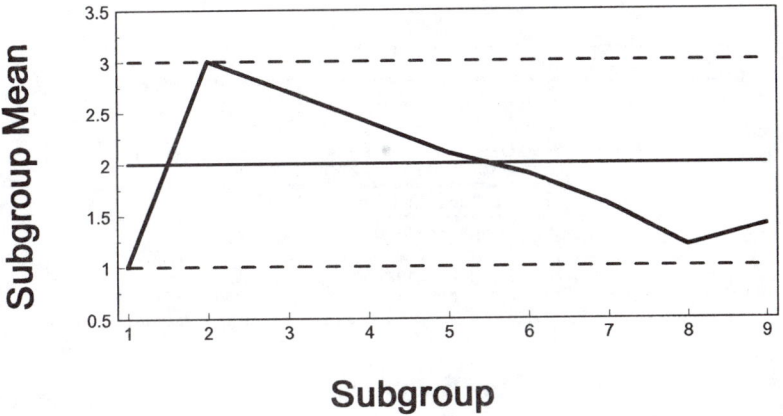

Figure 2.22 $\bar{X}$-*R* **chart trend.**

control limits. In addition, the plot should have an almost equal number of values on each side of the centerline. Although the subgroup mean and range portions of the chart should be evaluated separately, the procedure for interpreting patterns is the same.

Unnatural patterns, which may be caused by a host of assignable causes, indicate that a process is out of statistical control. It is important to note that even if a plot appears to demonstrate a natural pattern as described above, the following traits can indicate instabilities in a process: trends, runs, outliers, mixtures, and sudden changes in level.

Trends are patterns that migrate toward either the upper or lower control limits. The gradual wear of equipment can cause this behavior. This is typical of machining operations because of slow breakdown of the various bits used. Figure 2.22 is an example of an $\bar{X}$-*R* chart that displays a trend.

A *run* appears when several consecutive plotted values fall on a specific side of the centerline that depicts the total average across all subgroups or the average range. Probability calculations can determine how likely it is that a point will fall within a specified area of the chart. Figure 2.23 depicts the probability at which it is highly unlikely to find a specific number of consecutive points for typical subgroup sizes. It is this behavior that defines a run.

Outliers are points on a graph that fall beyond one of the control limits. These points can result from problems such as errors in measurement, errors in calculations, errors in plotting, or even damaged parts. By checking these possibilities for an outlier, it is highly likely that the cause will be found. However, if the causes are not easily found, the outlier may be the result of an anomaly in the process. This anomaly will be an assignable cause to the variability. When this is determined to be the case, the entire subgroup where the problem has been found must be eliminated from the plot because it is not representative of the capability of a process. Figure 2.24 illustrates an outlier.

In contrast to outliers, *mixtures* can be somewhat easily identified. These are typically

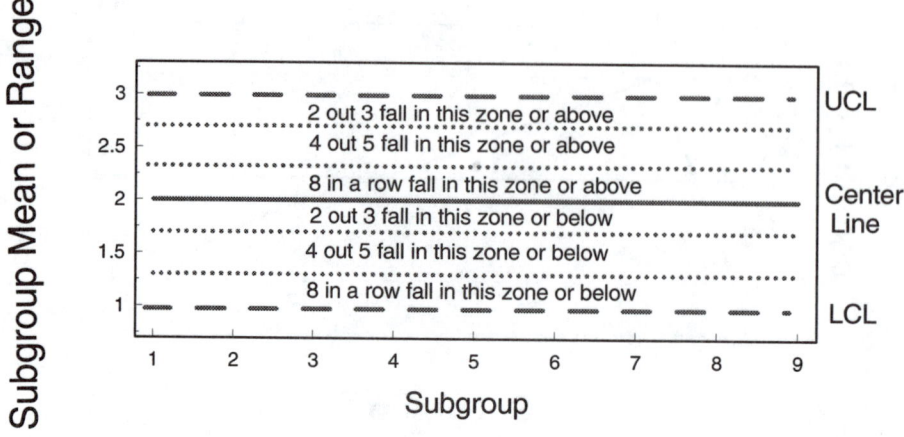

Figure 2.23 $\bar{X}$-R **chart runs.**

found where the process under study has not been completely isolated. A study conducted of a process that is manufacturing parts by several machines or a machine with several stations is an example of a mixture. The stability is determined by whether the calculated data fall within control limits. A mixture is one that displays the behavior depicted in Figure 2.25.

The last type of pattern discussed is one that indicates a *sudden change in level.* This type of plot indicates that a change in the process has occurred. Typical causes of this plot include changes in feedstock, machine/tool failures, and operators. As Figure 2.26 shows, it is not too difficult to identify a sudden change in level.

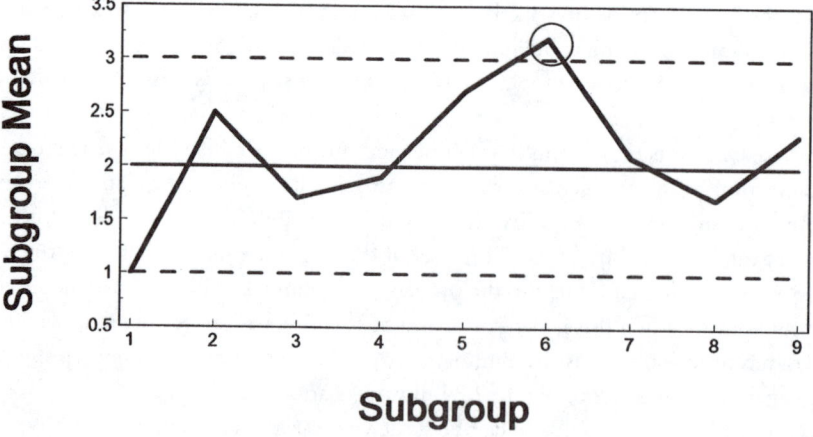

Figure 2.24 $\bar{X}$-R **chart outlier.**

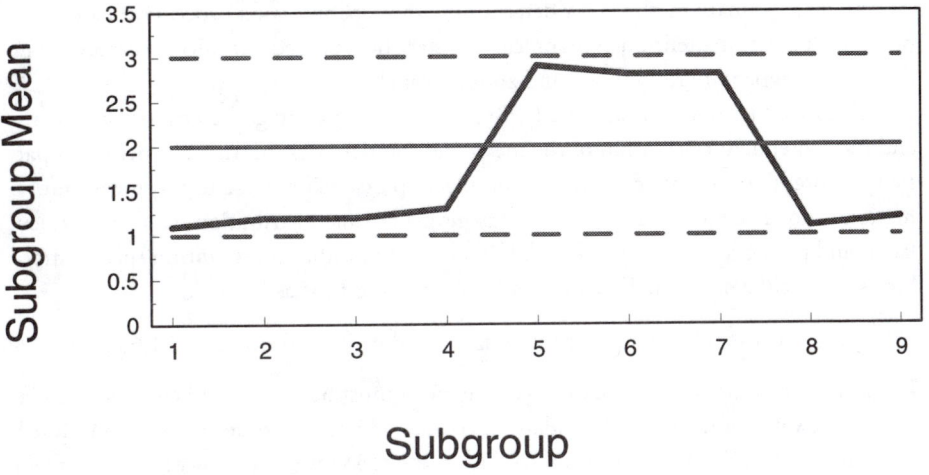

Figure 2.25 $\bar{X}$-R chart mixture.

Capability Determination Methods

Determination of the capability of a process is a method by which a process is measured for its potential for producing parts within specification limits. Control charts provide a means for evaluating this potential, with histograms and probability plots giving indicators of the shape and overall distribution of the manufactured products. It is important to remember that one of the first criteria for a process to be deemed capable is that the distribution of products be normal and fall within specification limits. There are several factors that should be investigated for the capability of a process to be properly quantified.

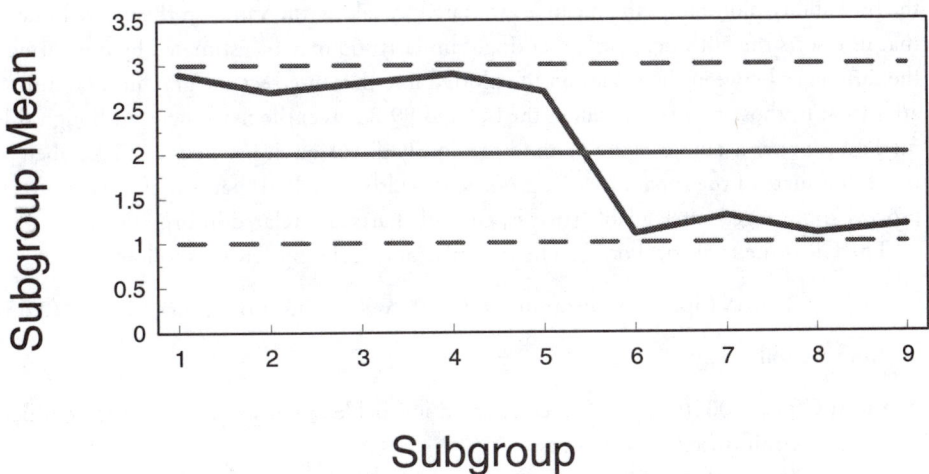

Figure 2.26 $\bar{X}$-R chart sudden changes in level.

The three primary methods for determining the capability of a process, when combined with the aforementioned graphical and statistical procedures, are the capability index, C_p, the capability ratio (CR), and another capability index, $C_p k$.

The capability index, C_p, is a method that uses information gathered from histograms and the subsequent calculation of the mean and standard deviation. This method is particularly useful when only data on the completed parts are known and in-process information is not available. If a histogram illustrates a normal distribution, C_p may be calculated, and predictions on the ability of the process to produce parts within specification limits may be determined. The formula for calculating C_p is as follows:

$$C_P = \text{(Upper specification limit − Lower specification limit)}/6S \qquad (2.44)$$

It is important to remember that the sample mean must be calculated before determination of the value of the sample standard deviation. The standard deviation is multiplied by 6 essentially to incorporate all values within the $\pm 3S$ range, which corresponds to all values within the 99.73 percentile. It is important to note that C_p is the ratio of the specification tolerances of a process divided by the statistical probability that 99.73% of the parts produced will fall within that range. Calculated index values are, therefore, typically interpreted as follows:

- At $C_p \leq 1.00$, the process is considered incapable of consistently producing products within specification limits;
- At $1.00 < C_p < 1.33$, the process is capable, but has the potential to become unstable; and
- At $C_p \geq 1.33$, the process is considered capable of consistently producing products within specification limits.

A more simplified method for determining the capability index that eliminates the need to perform calculations of the sample mean and sample standard deviation involves the probability plot. Here, the mean is estimated by taking the value on the plotted line that intersects the 50th percentile marking. Similarly, $6S$ may be estimated by measuring the difference between the values on the plotted line that intersect the graphical boundaries (that is, those points that fall at the 0.2 and 99.8 percentile markings). Although this method is not as accurate as the method previously discussed, it provides a quick, albeit rough, estimate of the capability of a process. It is also a method that is useful when in-process knowledge is unavailable (that is, control charts and related information).

The CR is the reciprocal of C_p. The formula for calculating CR is as follows:

$$CR = 6S/\text{(upper specification limit − lower specification limit)} \qquad (2.45)$$

Note the following:

- At $CR \geq 1.00$, the process is considered incapable of consistently producing products within specification limits;
- At $0.75 < CR < 1.00$, the process is capable, but has the potential to become unstable; and

- At CR ≤ 0.75, the process is considered capable of consistently producing products within specification limits.

In situations where only the histogram is used for conducting a process capability study, the normality of the distribution is of primary importance. Without this characteristic, predictions for consistency will be skewed. A simultaneous comparison with a probability plot of the data can help eliminate misinterpretations. A linear probability plot is an excellent indicator for normality and, with the upper and lower specification limits displayed, provides good information as to the overall capability of a process. As illustrated in Figure 2.27, the following patterns may occur by examining only histograms:

- Plot A—this pattern indicates a normal distribution and is centered within specification limits; it is an ideal situation.
- Plot B—although this pattern appears normal, distribution approaches or exceeds specification limits. Here, a probability plot would provide information on the percentage of parts falling outside specification limits.
- Plot C—this pattern is normal, but distribution is off-centered with specification limits. Process is probably capable: however, it may require adjustment to center the distribution.
- Plot D—here, two patterns are present, each of which may be normal. This is a classic indication that two processes have been included in the study. Any informa-

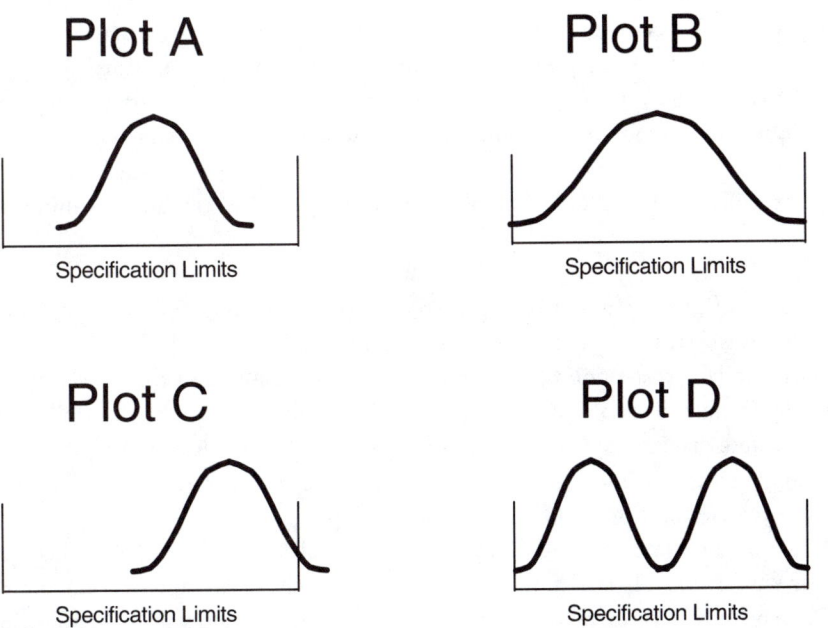

Figure 2.27 Capability ratio values and product specification limits.

tion derived from these data will be faulty until the two groups of data have been separated and evaluated independently.

An ideal way to determine process capability is with detailed process knowledge. This is provided by control charts and calculation of the capability index, C_pk. Whereas C_p and CR provide information on the spread of the distribution only, C_pk indicates this and the position of the spread within the specification limits. The capability index, C_pk, is determined by the following equation:

$$C_pk = \text{the lesser of } (USL - \bar{\bar{X}})/3S \text{ or } (\bar{\bar{X}} - LSL)/3S \qquad (2.46)$$

Where

USL = upper specification limit,
LSL = lower specification limit,
$\bar{\bar{X}}$ = total sample mean, and
S = sample standard deviation.

The value of C_pk indicates process capability in the same fashion as C_p. In general, if the value is greater than 1.33, the process is capable. However, in addition to this information, the following C_pk values determine whether the products produced fall within specification limits:

- A negative C_pk value means the total sample mean falls outside specification limits;
- A C_pk value equal to zero means the sample mean is equal to one of the specification limits;
- A C_pk value between zero and one means the sample mean falls between the specification limits, but a portion of the distribution is outside the limits;
- A C_pk value equal to one means the sample mean is centered between the specificating limits and the distribution ends on one of the limits; and
- A C_pk value greater then one means that the sample mean is centered within the specification limits and all points of the distribution are contained within the limits.

Therefore, C_pk provides a more complete picture of the capability of a process than C_p and CR. However, to ensure that an analysis is as complete as possible, the comparison of all pertinent data is paramount. Histograms, probability plots, and control charts should be used in conjunction to provide the best available information on the capability of a process before making predictions. It is important to remember that the topics discussed regarding process capability are not useful if the distribution is not normal. The best test available to determine normality is the probability plot.

Constructing $\bar{X}$-R charts and calculating the capability indexes and ratios determines a process's ability to be in statistical control and capable. If these two conditions are met, SPC can be fully implemented and used. (These events are discussed in the following section.) If the conditions are not met, it is probable that the process was not properly isolated or that specific parameters of the process have not been eliminated to achieve the desired consistency.

Processes that are in statistical control but deemed incapable are generally the result of high variability. High variability yields widely spread control limits on an $\bar{X}$-R chart, thus indicating that the process is in statistical control, but not capable.

Another condition is one in which a process is thought to be capable, but not in statistical control. Situations such as this are usually the result of a process that exhibits low variability (often the result of unnatural $\bar{X}$-R chart patterns). This is a condition that is difficult to remedy.

The final scenario is one in which the process is neither in statistical control nor capable. This is typically the result of a loosely managed process. Major problems are easily found and, on correction, yield improvements rather quickly. Variability is typically reduced by these actions and statistical control achieved. In some situations, however, solutions are not as easily found and making the most of the process becomes an ongoing, intensive process itself.

After completing an initial $\bar{X}$-R chart of a process, it is not uncommon to make changes to the process. Two common approaches are to modify the manufacturing process or alter the specification limits. If these two methods fail, product sorting may be initiated, although this is labor intensive and, therefore, an inefficient process. Each method has its costs and benefits, both of which should be weighed carefully. In either case, the process itself will need to be monitored using an ongoing capability study to determine whether the modifications were beneficial.

QUALITY CHARACTERISTICS

After the process's ability to achieve statistical process control and capability has been investigated, further methods using product attributes can be initiated. Up to this point, only control charts that exhibit variability have been discussed. Attribute control charts are different in that data are collected in subgroups of varying sizes and intervals and are based on product defects or nonconformities to specifications. Defects and nonconformities are defined as those characteristics of a product that fall outside specification requirements or result in a condition that prevents the product from being used in the manner for which it was designed. Examples of defects and nonconformities include color variations, scratches, dents, chips, and broken parts. Attribute control charts help determine the ability of a process to produce defect-free products, even when the process has been shown to yield products within statistical control and is capable of yielding products that may be used in the fashion for which they were designed.

Patterns such as those discussed in relation to variability charts are exhibited in attribute control charts as well. Moreover, the causes of these patterns are essentially the same. Determining process capability, however, is a process more ideally suited for analysis using variability charts because there are no upper and lower specification limits defined for quality characteristics. Instead, evaluating process capability using attribute charts is accomplished by defining an acceptable yield for a process. Yield is defined as the percentage of parts produced that passes inspection criteria. For example, an 80%

yield means 20% of the products produced failed inspection criteria. Attribute charts are a tool for performing evaluations of this type. The four types of attribute control charts that will be discussed are C-charts, which track the number of defects; U-charts, which track the number of defects per part; NP-charts, which track the number of defective parts produced; and P-charts, which track the percentage of defective parts produced.

All attribute charts are plotted in the same manner. However, the data are collected differently depending on the type of chart selected. In contrast to variability control charts where the subgroups are somewhat small (that is, approximately five), attribute subgroups are large. For C- and U-charts, the subgroup size may be based on a unit of products such as 100 bolts, or 50 personal computers. For P- and NP-charts, the subgroup size should be large enough so that at least five or more defective units are encountered. In general, the larger the subgroup, the more data that are assembled, which often makes chart interpretation an easier process. As with all endeavors, cost and time must be weighed against the benefits of large sampling events. A typical attribute control chart is a plot that labels the x-axis with the subgroup number against the following values of the subgroup on the y-axis: the number of defects in a subgroup, C; the number of defective parts, NP; the fraction of defective parts in a subgroup, P; or the number of defects per part, U.

C-Charts

A C-chart tracks the number of nonconformities found in a given subgroup. Because the number of nonconformities can vary based on the size of the subgroup selected, it is important to keep the subgroup size fixed. The most effective method is to track only one type of defect per plot. Plotting numerous types of defects on one chart tends to create a situation that is difficult to interpret. The subgroup size should be selected so that at least four to six nonconformities are detected. The date and time are also useful information, as they provide historical data on the subgroup intervals. In contrast to $\bar{X}$-R charts, the calculations for a C-chart are relatively simple. The equation for determining the average number of nonconformities per subgroup and the upper and lower control limits is as follows:

$$\bar{C} = \Sigma C_i / m \tag{2.47}$$

Where

$\bar{C}$ = the average number of nonconformities per subgroup;
C_i = the number of nonconformities found in a subgroup; and
m = the number of subgroups.

$$\text{UCL}_C = \bar{C} + 3(\bar{C})^{1/2} \tag{2.48}$$
$$\text{LCL}_C = \bar{C} - 3(\bar{C})^{1/2} \tag{2.49}$$

Plotting data in a C-chart is similar to plotting data in an $\bar{X}$-R chart. A sample C-chart is shown in Figure 2.28.

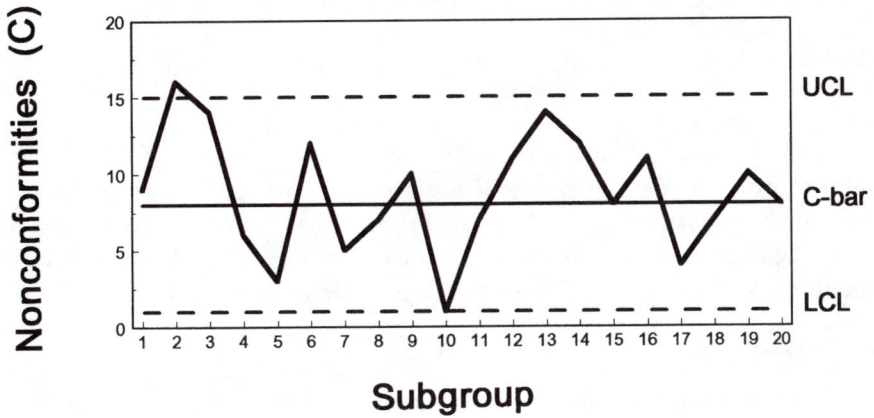

Figure 2.28 *C*-chart illustration.

U-Charts

A *U*-chart is used for tracking the average number of defects per part. The most useful plots are those that incorporate data of a single type of defect rather than numerous types. In general, the more wide ranging the information included on an individual plot, the harder interpretation becomes. Selection of a subgroup size is similar to that for collecting data for a *C*-chart. Typically, large subgroups are used so that a sufficient number of defects can be found. The equation used to create *U*-charts and plot individual points of the graph is as follows:

$$U = C/n \tag{2.50}$$

Where

U = the number of defects per part,
C = the number of defects in a subgroup, and
n = the number of parts in a subgroup.

The centerline of the plot and the upper and lower control limits are given by the following equation:

$$\bar{U} = \Sigma C_i / n_i \tag{2.51}$$

$$\text{UCL}_U = \bar{U} + 3[(\bar{U})^{1/2}/(\bar{n})^{1/2}] \tag{2.52}$$

$$\text{UCL}U = \bar{U} - 3[(\bar{U})^{1/2}/(\bar{n})^{1/2}] \tag{2.53}$$

Where

$\bar{n} = \Sigma n_i / m$,
$\bar{U}$ = the average number of defects per part,
C_i = the number of defects in a subgroup,
n = the number of parts in a subgroup,

$\bar{n}$ = the average number of parts in a subgroup, and

m = the number of subgroups.

The average number of parts in a subgroup is used in the control limit equations only if the variation in subgroup size is held fairly constant. If wide variations in the subgroup size are incurred (that is, greater than $\pm 25\%$ of the average subgroup size), then the number of parts in a subgroup must be substituted for the average number of parts in a subgroup. The plotting format of U-charts differs only slightly from C-charts in that the y-axis is labeled as the average number of nonconformities per subgroup rather than as nonconformities per subgroup. The methodology, however, is the same.

NP-Charts

An NP-chart plots the number of nonconforming products. This type of chart provides useful information when the specific number of defective units is of interest. In contrast to C- and U-charts, NP-charts may track more than one type of defect. For example, if a specific lot of parts is inspected to see how many of them meet specifications (that is, can actually be sold to customers), an NP-chart will help provide information. Subgroup sizes for NP-charts must be large enough to uncover a minimum of four to six defective parts and be held constant between subgroups. The only calculations required for NP-charts are the determination of the average number of nonconforming parts per subgroup (the centerline of the plot) and the upper and lower control limits as illustrated below:

$$\bar{N}\bar{P} = \Sigma NP/m \tag{2.54}$$

$$\text{UCL}_{NP} = \bar{N}\bar{P} + 3\{\bar{N}\bar{P}[1-(\bar{N}\bar{P}/n)]\}^{1/2} \tag{2.55}$$

$$\text{LCL}_{NP} = \bar{N}\bar{P} - 3\{\bar{N}\bar{P}[1-(\bar{N}\bar{P}/n)]\}^{1/2} \tag{2.56}$$

Where

NP = the number of defective parts, and

$\bar{N}\bar{P}$ = the average number of defective parts per subgroup.

If the lower control limit is calculated as a negative value, it must be set equal to zero because it is impossible to have fewer than zero defective parts. Figure 2.29 depicts a sample NP-chart.

P-Charts

The last type of attribute control chart discussed is the P-chart. A P-chart is similar to an NP-chart; however, rather than plot the actual number of defective parts found in a subgroup, the percentage is plotted in an NP-chart. This allows subgroup sizes to vary. However, the subgroup size must always be selected that will result in at least four to six defective parts to obtain useful information. The percentage of defective parts in a subgroup is determined by the following equation:

$$P = NP/n \tag{2.57}$$

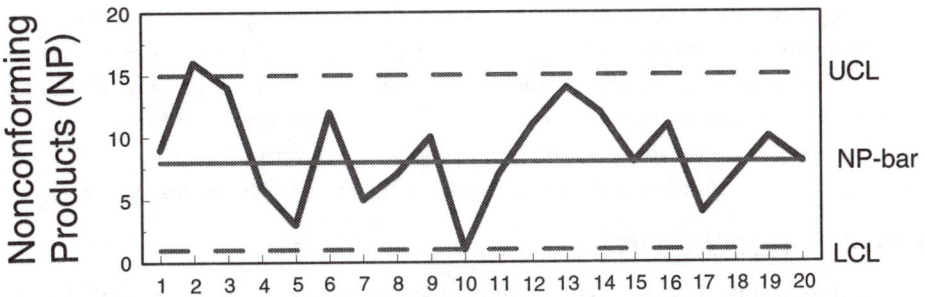

Subgroup

Figure 2.29 *NP*-chart illustration

The average percentage of nonconformities per subgroup is determined by the following equation:

$$\bar{P} = \Sigma NP_i / \Sigma \, n_i \tag{2.58}$$

Where

P_i = the fraction of defective parts in a subgroup (multiply by 100 to yield the percentage), and

$\bar{P}$ = the average fraction of defective parts per subgroup (multiply by 100 to yield the percentage).

Upper and lower control limits are determined by the following formulas:

$$UCL_p = \bar{P} + 3[\bar{P}(1 - \bar{P})/\bar{n}]^{1/2} \tag{2.59}$$

$$UCL_p = \bar{P} - 3[\bar{P}(1 - \bar{P})/\bar{n}]^{1/2} \tag{2.60}$$

As with *U*-charts, the average number of parts in a subgroup is used in the control limit equations only if the variation in subgroup size is held fairly constant. If wide variations in the subgroup size are incurred (greater than ±25% of the average subgroup size), then the number of parts in a subgroup must be substituted for the average number of parts in a subgroup. In addition, if the lower control limit is found to be a negative number, it must be set at zero as is the case with *NP*-charts. Moreover, the only difference in the plotting approaches of *NP*- and *P*-charts is that *P*-charts track data on the *y*-axis as the percentage of nonconformities per subgroup.

Data Acquisition Systems

The previous sections of this chapter have dealt with the identification of the mass flow characteristics of a process and the statistical control of its outcome. To achieve desired results, reliable data are required. Mass balance exercises can be performed to a certain degree based on process knowledge; however, for the majority of cases, in-process measurements are necessary to quantify governing parameters. These in-process measurements will frequently gauge whether the process is in statistical control or not. Thus, the need for data acquisition becomes apparent. This section discusses the most common methods and process variables measured in a process. The information provides an overview of data acquisition and an additional step toward efficient process control.

INSTRUMENTATION

The most common parameters governing processes involve physical attributes. These attributes include dimensional measurements—a frequently used tool in SPC. In many cases, however, this approach only provides information after a manufacturing operation is conducted rather than controlling the environment in which the process is conducted. Therefore, instrumentation topics covered herein elaborate on in-process physical quantities such as liquid flow measurements, liquid level control, conductivity, pH, oxidation-reduction potential (ORP), temperature, and pressure.

Liquid Flow Measurements

Liquid flow measurements are a basic requirement in a variety of industrial processes. Some applications are installed to provide only historical data, whereas others are pivotal to achieving critical process control. The latter of these types of applications are the most common and range from the precise metering of chemicals, to controlled reactions, to the physical limitation of flow as a result of the design constraints of process equipment. The majority of flow measurement tools measure liquid velocity and convert the data into volumetric flow rates. This velocity is the result of pressure changes that cause variations in the kinetic energy of the liquid of concern. Converting velocities to volumetric flows is a simple task and is expressed in the following equation:

$$Q = V \times A \tag{2.61}$$

Where

Q = volumetric flow rate of the liquid (for example gpm or L/min),
A = cross-sectional area of the pipe (ft^2, m^2), and
V = velocity of the liquid in the pipe (ft/sec, m/sec).

Although this is a somewhat crude simplification of the technique for measuring liquid flows in a pipe, most direct measurements are achieved using the average velocity by means of positive-displacement flow meters. Common flow meters of this type act as counters. As the liquid passes through the orifice in the meter, either a physical or electrical mechanism reacts to the movement of the liquid, providing instantaneous and/or total flow rate readings.

Most flow meters are designed to perform within specific operating parameters. This implies that a flow meter purchased for internal pipe diameters of 25.4 mm (1 in.) should not be installed on pipes of considerably greater or lesser diameters. The location of the meter should not be close to elbows, valves, and others obstructions to liquid flow. This applies to both upstream and downstream anomalies. As a general rule, because of frictional forces acting on the liquid, the location of a flow meter should be five pipe diameters downstream and two pipe diameters upstream from any fixtures that may disturb liquid flow. The term that takes this behavior into account is the *Reynold's number*. Its value provides information on whether flow is laminar or turbulent. The equation for determining the Reynold's number is as follows:

$$R = (3\ 160 \times Q \times \rho)/(\mu \times D) \tag{2.62}$$

Where

- R = Reynold's number;
- Q = volumetric flow rate, gpm (L/s);
- ρ = specific gravity of liquid;
- μ = viscosity of liquid, centipoise (cP) (Pa·s); and
- D = diameter of pipe, in. (mm).

At low velocities or high viscosity values, the flow exhibited is typically laminar and moves smoothly with a parabolic distribution in velocities across the pipe's diameter, with the highest velocity occurring at the center. The lower velocities occurring off the center line are caused by the frictional forces acting on the liquid from the pipe walls. Reynold's numbers corresponding to these conditions are typically less than 2 000.

The most common flow pattern exhibited, however, is turbulent flow. This condition occurs at Reynold's numbers greater than 3 000 and is seen with liquids traveling at high velocities or low viscosities. Turbulent flow has less of a velocity profile because of the random eddies and currents that are indicative of this type of behavior. Both turbulent and laminar flow patterns are illustrated in Figure 2.30. The variety of flow meters used in the wastewater treatment industry today prohibits a lengthy discussion of each within this chapter. Instead, a brief overview of the types of flow meters and their uses is presented in this section. In general, there are five types of flow measurement equipment:

Laminar Flow   Turbulent Flow

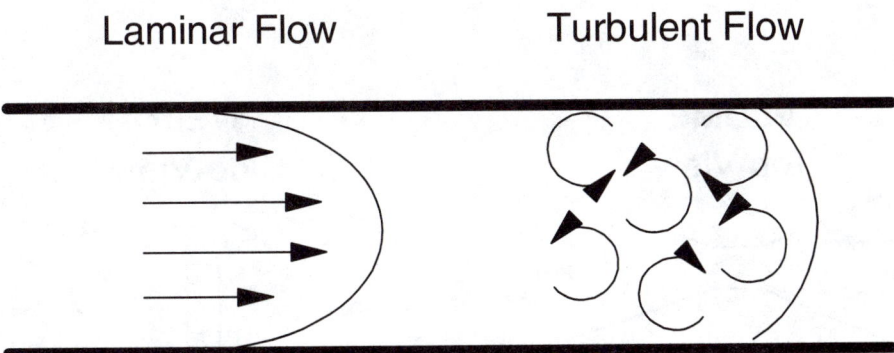

Figure 2.30 Laminar and turbulent flow patterns.

open-channel, positive-displacement, velocity, differential pressure, and mass devices. The first of these types of devices, open-channel, involves flumes and weirs.

Flumes measure liquid flow with little pressure drop. This helps to prevent the buildup of solids prior to the device in streams with large amounts of suspended solids. Common schematics indicate a converging upstream section, a vertically mounted flume, and a diverging downstream section. Weirs, in contrast to flumes, operate by means of a vertical wall in which the flow of liquid builds up prior to the device and, after sufficient accumulation, passes over the top of the plate. The pressure drop across weirs is considerably higher than flumes. The discharge from both flumes and weirs is a function of liquid level. Therefore, level measurements are used to determine flow rates. Figure 2.31 gives examples of the two types of open-channel flow devices.

Positive-displacement flow meters operate by breaking up the flow of liquid into discrete segments or volumes and allowing their passage. The number of segments corresponds to the amount of flow. These types of meters are ideally suited for metering specific quantities of a liquid to a process. Rotary vane meters are quite common and consist of a rotating impeller that may be thought of as an enclosed water wheel. As the wheel rotates, a specific volume of liquid passes through the device. These revolutions are counted and provide direct readings in the form of a volumetric flow rate. Other types of positive-displacement meters exist, such as the oval-gear meter and reciprocating piston meter. Although their configuration and function may be slightly different, the principles of their operation are similar.

There are several types of velocity flow meters, the primary of which are turbine, ultrasonic, and electromagnetic. These types of meters operate by measuring the volume of liquid passing through the meter and yielding a linear relationship. Turbine flow meters are generally the most common and consist of a multiple-bladed rotor. As the liquid passes through the blades, the rotor spins. The speed at which the rotor spins corresponds to the flow rate. Ultrasonic flow meters measure liquid flow by means of a Doppler signal or sound wave. Doppler flow meters consist of two transducers, with one acting as a transmitter and the other as a receiver. As the signal is sent out it is reflected by solids, bubbles, or other anomalies. Because these anomalies are moving downstream, the transmitted signal of known frequency is shifted. This frequency shift is proportional to the

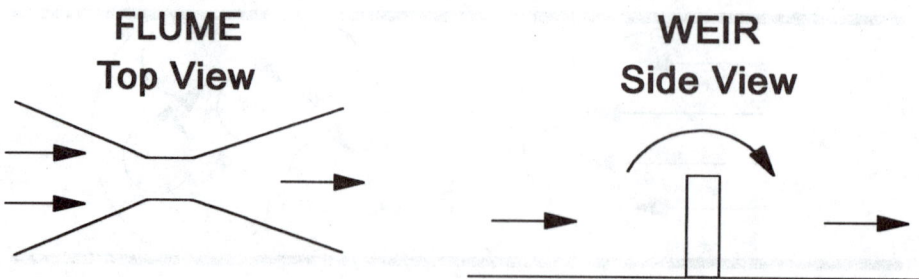

Figure 2.31 Open-channel flow devices.

liquid's velocity. Electromagnetic flow meters consist of a flow tube and a voltmeter. These devices are combined to measure the electrical conductivity of a liquid as it passes through the flow meter. This type of meter employs the Faraday principle, which states that a voltage will be induced when a conductor moves through a magnetic field. Therefore, as a conductive liquid passes through a magnetic field created by energized coils within the meter, a voltage is produced. The amount of voltage produced is directly proportional to the flow rate.

Differential-pressure flow meters are the most common type of flow meter in use today. These meters operate by measuring the pressure drop across the unit. The pressure drop is assumed to be proportional to the square of the flow rate and is produced by a fixture that causes a change in the kinetic energy of the liquid. Once this change has occurred, a differential pressure is detected and converted into a readout that corresponds to the actual flow rate. The mechanism for achieving a pressure drop or altering the kinetic energy of the liquid is well known and consists of orifices, flow tubes, Venturi tubes, nozzles, pitot tubes, and others.

The last flow meter introduced is the mass flow meter. Whereas the other flow meters discussed base their measurements on volumetric flow rates, mass flow meters are unique in that they measure mass, not volume. Mass flow meters are useful in applications in which chemical reactions must be accurately controlled. The most common device uses a principle referred to as the *Coriolis force*. A design that is frequently used consists of a U-shaped tube mounted in a housing connected to an electrical unit. The U-shaped tube is vibrated at its natural frequency by a magnetic device positioned at the bend in the tube. As liquid passes through the meter, it is forced to move vertically because of the movement of the "U-tube." The resistance to the liquid in vertical motion causes a twisting action to occur in the U-tube. It is this amount of twist that is directly proportional to the mass of liquid passing through the meter.

Liquid Level Measurements

Liquid level measurements are an important part of establishing process control and consist primarily of two types: point level measurement and continuous level measurement. Point level measurements are generally employed in situations where knowledge of a specific liquid height is desired. In most cases, this corresponds to a high-level alarm in which overflow is to be avoided, or a low-level alarm to keep the process operating within specified limits or parameters. Continuous level measurements are used in situations where knowledge of liquid levels is needed throughout a range, rather than merely at high and low levels. Accomplishing this type of measurement is more complicated than with point level and is traditionally integrated with other process controls that respond to various liquid level conditions.

The most common type of point level measurement is achieved through the use of a float switch. The most basic of all switches, the float switch responds to rising liquid levels by means of a material that floats on the surface of the liquid. When the float reaches a specified level, the mechanical action physically triggers a switch. A similar type of switch is accomplished through the use of a proximity sensor. This type of sensor is designed with a magnetic field switch. The magnetic portion of the switch is mounted on a float. When the liquid rises or falls to predesignated levels, the magnet actuates a hermetically sealed switch. These popular switches require little to no maintenance, provided the materials of construction are compatible with the environment in which they are employed. Figure 2.32 displays examples of the two types of float switches.

Simple Float Switch

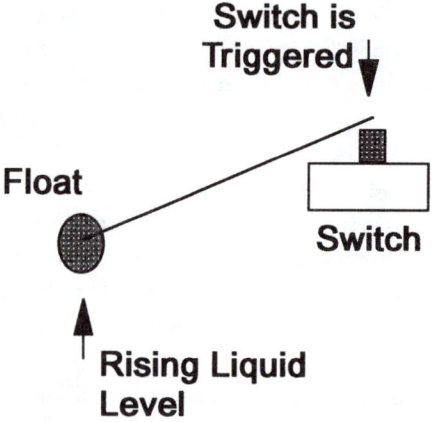

Magnetic Proximity Switch

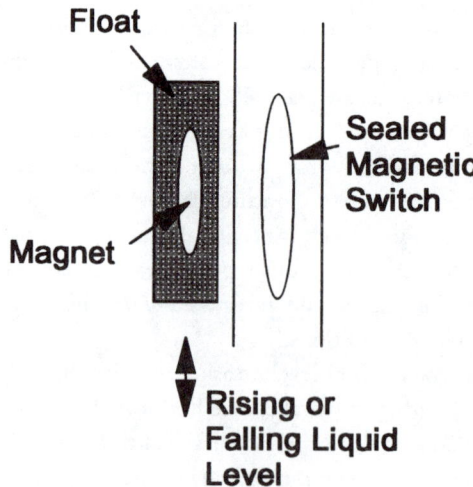

Figure 2.32   Float and magnetic proximity switches.

Continuous level measurements may be accomplished by a host of technologies. Although they provide data over a continuous range, they may also be used for point level control. The most common techniques for this type of data gathering incorporate the following types of sensors: capacitance, radio frequency, and ultrasonic.

Capacitance liquid level devices are set up to detect changes in the dielectric constant of the material surrounding the probe or electrode. A metal rod or probe that is inserted into a vessel or tank acts as the electrode, and the wall of the tank (if conductive) acts as the opposing plate of a traditional capacitor. When the tank liquid levels are low, air surrounds the electrode and a reference value is determined. As the liquid level rises, the dielectric constant of the medium between the two plates changes. The value of the dielectric constant depicts the ability of a material to store an electrical charge and will vary depending on the type of material being measured. A typical configuration for capacitance level measurements is shown in Figure 2.33. As the relationship of capacitance is examined, it becomes apparent that changes in probe size, distance from the electrode to the vessel wall, and dielectric constants of the fluids will cause variation in the capacitance values measured. It is these changes that allow continuous level monitoring to occur. However, if several parameters are not held fairly constant, the dielectric constant of the material measured will change, thereby yielding erratic or faulty measurements. Temperature swings, compositional variations, moisture content, and the deposition or buildup of material on the electrode are examples of factors that lead to process control problems caused primarily by changes or inaccurate detection of the dielectric constant of the medium involved. The potential for these variations should be considered when selecting a capacitance level monitor.

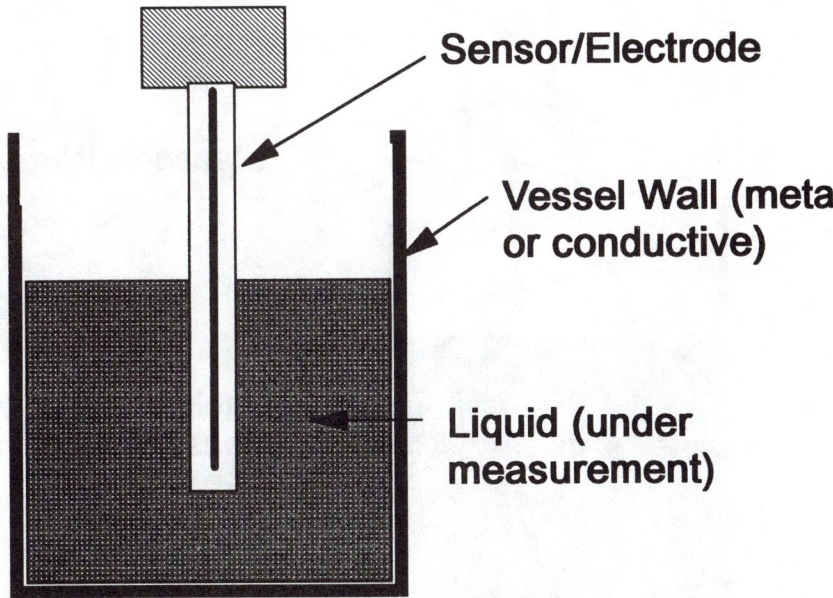

Figure 2.33 Capacitive liquid measurement.

Radio frequency measurements are conducted in a fashion that is not entirely different from that of capacitance. A sensor is emersed in a tank, a signal is transmitted, and the return signal is received. The primary difference is signal transmissions. However, similarities arise when taking into account the dielectric constant of the medium. Radio frequency signals are attenuated to varying degrees by the dielectric constant of the liquid measured. Therefore, all of the aforementioned situations in which inconsistencies may occur because of variations in dielectric constants also apply to radio frequency techniques.

Another method for detecting changes in liquid level is through the use of ultrasonic sensors, which come in two varieties. Ultrasonic sensors that come in contact with the material being measured are used primarily as point level control, while those that do not come in contact with the material being measured are used for continuous level monitoring. The science behind both varieties is essentially the same. An ultrasonic signal is generated from the probe, transmitted to the surface of the liquid to be monitored, and returned to the probe. As the distance from the probe to the liquid changes, so does the travel time for the signal. It is this change in time, albeit small, that provides an indication of the status of the liquid level. Ultrasonic probes are quite reliable. However, as with the selection of any measurement technique, a sound evaluation of the environment involved should always be conducted. For example, an ultrasonic sensor may provide erroneous results if the medium measured is topped by a frothy, foamy layer. A configuration for ultrasonic measurements is illustrated in Figure 2.34.

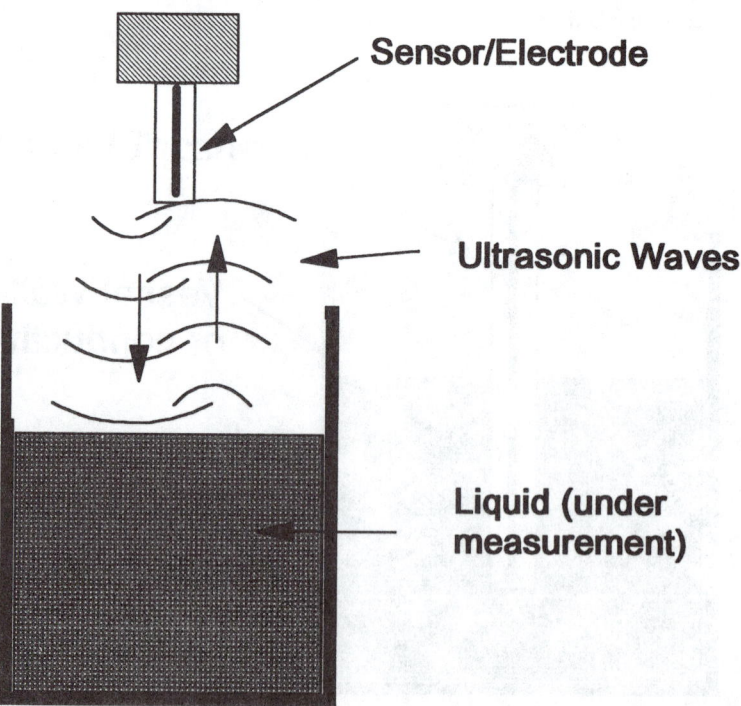

Figure 2.34 Ultrasonic liquid measurement.

pH Measurement

pH is a method of quantifying the degree of acidity of a solution. pH values range from 0 to 14: the lower the value, the more acidic the solution. *Acidity* is defined in several ways, although the simplest and most widely used definition is that acidity is the capacity of a solution to neutralize hydroxide ions (OH⁻) or, specifically, the concentration of hydrogen ions [H⁺] in a solution. pH is a mathematical expression denoting the negative logarithm of the hydrogen-ion concentration as given by the following equation:

$$pH = -\log[H^+]$$

The product of the hydrogen- and hydroxide-ion concentrations in moles per litre in any given solution must always be equal to 1×10^{-14}. Therefore, as the concentration of hydrogen ions in a solution increases, the concentration of hydroxide ions must decrease. For pH values less than 7, the predominant species is the hydrogen ion and the solution is considered acidic. At pH values greater than 7, the hydroxide ion is dominant and the solution is considered basic or alkaline. When the concentration of both species is equal, the pH is 7 and the solution is considered neutral. Table 2.6 illustrates the relationship of the hydrogen- and hydroxide-ion concentrations at varying pH levels.

Now that pH has been covered, a simple review of concentration values should be included as well. Concentration is simply the amount of a given substance in a specified volume. This can be expressed as mass per unit volume (that is, grams per cubic centimetre [pounds per cubic foot]) or as a moles per unit volume (that is, moles per litre). The latter of these two approaches is used in most chemical manipulations because it relates

Table 2.6   pH versus hydrogen- and hydroxide-ion concentration at 25°C.

pH	Hydrogen ions (H⁺), mol/L	Hydroxide ions (OH⁻), mol/L
0	1	(10^{-14}) 0.000 000 000 000 01
1	0.1 (10^{-1})	(10^{-13}) 0.000 000 000 000 1
2	0.01 (10^{-2})	(10^{-12}) 0.000 000 000 001
3	0.001 (10^{-3})	(10^{-11}) 0.000 000 000 01
4	0.000 1 (10^{-4})	(10^{-10}) 0.000 000 000 1
5	0.000 01 (10^{-5})	(10^{-9}) 0.000 000 001
6	0.000 001 (10^{-6})	(10^{-8}) 0.000 000 01
7	0.000 000 1 (10^{-7})	(10^{-7}) 0.000 000 1
8	0.000 000 01 (10^{-8})	(10^{-6}) 0.000 001
9	0.000 000 001 (10^{-9})	(10^{-5}) 0.000 01
10	0.000 000 000 1 (10^{-10})	(10^{-4}) 0.000 1
11	0.000 000 000 01 (10^{-11})	(10^{-3}) 0.001
12	0.000 000 000 001 (10^{-12})	(10^{-2}) 0.01
13	0.000 000 000 000 1 (10^{-13})	(10^{-1}) 0.1
14	0.000 000 000 000 01 (10^{-14})	1

individual molecules rather than substance densities. In the equation given for pH, the hydrogen-ion concentration is expressed in moles/litre.

One mole of a compound is defined as 6.023×10^{23} molecules of the substance. Avogadro's number of molecules of a given substance has a mass approximately equal to its molecular weight when expressed in grams. In the case of hydrogen chloride, the molecular weight is equal to $1.01 + 35.45$, or 36.46. Hence, the weight of a mole of hydrogen chloride is nearly 36.46 g. To determine the molecular weight of a molecule, atomic weights are added. In the previous example, this corresponds to 1.01 for the hydrogen atom and 35.45 for the chlorine atom. The concepts of mole and molarity (moles per litre) were covered in the section titled Chemical Equations and Stoichiometry earlier in this chapter. However, to ensure that these principles and relationships are understood, an example is provided in the following section.

pH AND HYDROGEN-ION CONCENTRATION. Determine the hydrogen-ion concentration and corresponding pH of a 0.050*M* solution of sodium hydroxide. The dissociation of sodium hydroxide in solution is given by the chemical equation

$$NaOH \vee Na^+ + OH^- \tag{2.63}$$

Therefore, 1 mole of sodium hydroxide yields 1 mole of Na^+ ions and 1 mole of hydroxide ions. Hence, the concentration of hydroxide ions in the aforementioned solution is 0.050 mol/L. Because the product of hydrogen ions and hydroxide ions must equal 1×10^{-14}, the following relationship may be established:

$$[H^+][OH^-] = 1 \times 10^{-14} \tag{2.64}$$

$$[H^+][0.050 \text{ mol/L}] = 1 \times 10^{-14}$$

$$[H^+] = 2 \times 10^{-13}$$

Now that the hydrogen-ion concentration is known, the pH may be calculated as follows:

$$pH = -\log[H^+] = -\log[2 \times 10^{-13}]$$

$$pH = 12.7$$

In the prior example an assumption was made that all the sodium hydroxide dissociated in water. This is true of sodium hydroxide because it is a strong base. However, not all molecules form strong acids or bases. This property is dependent on the ability of a compound to gain or lose electrons. An ion is a charged atom or molecule that has either gained or lost electrons. When ions are present in a solution, electrical energy may be conducted. It is this ability of a compound to dissociate into ions that determines the relative hydrogen- and hydroxide-ion concentrations in a solution and therefore the resulting pH.

The more easily a substance dissociates in water, the stronger are its acidic or basic tendencies. A strong base was illustrated in the previous example. However, a strong acid is given by the following equation:

$$HCl \rightleftharpoons H^+ + Cl^-$$ (2.65)

Compounds that dissociate little are considered weak acids or bases. This makes sense because the less hydrogen ions or hydroxide ions that are set loose in a solution, the less effect on pH they will have. An example of a weak acid is acetic acid. This compound will only dissociate 1 molecule in water for approximately every 100 present. The dissociation reaction is given by the following equation:

$$CH_3COOH \rightleftharpoons H^+ + CH_3COO^-$$ (2.66)

Water is another example of a weak acid/base and yields a neutral solution with pH 7 and $(1 \times 10^{-7})M$ hydrogen ions and $(1 \times 10^{-7})M$ hydroxide ions. It is the product of these two ion concentrations that yields the dissociation constant, K_w, for water (1×10^{-14}) that was used in the previous example.

Now that a foundation for the concept of pH has been constructed, methods and techniques for measuring these values in a solution will be discussed. The most basic system for measuring hydrogen-ion concentrations is through the use of a pH meter. A pH meter typically consists of three components—a measuring electrode, a reference electrode, and an input impedance meter. The measuring electrode is typically a glass bulb that is sensitive to hydrogen ions and provides an electrical output that varies with the changes to the ion concentration inside and outside of the bulb. The reference electrode does not vary with changes to hydrogen-ion concentration. As the voltage varies across the measuring pH electrode, the impedance meter picks up these small changes and translates them into a display reading. In essence, a pH meter is an electrical cell. The hydrogen-ion concentration differentials between the solution and the measuring electrode provide an electrical potential (that is, voltage) that is amplified to detect small changes.

As is the case in all chemical equilibria reactions, temperature changes can play a significant role. If temperature is not compensated for, measurement error will arise. To account for these changes, the Nernst equation is used to characterize the behavior of solutions with variations in temperature:

$$E = E^\circ + [(2.3RT_K)/nF]\log[a_i]$$ (2.67)

Where

E = electrical potential as a function of temperature, volts;

E° = standard electrode potential of the reference electrode, volts;

R = molar gas constant;

T_K = temperature, K;

n = charge of the ion;

F = Faraday constant; and

a_i = activity of the ion.

Fortunately, computations to account for temperature variation are performed within most pH meters. Few, if any, lack this capability.

The science of pH measurement is an established technique. An area of pH measurement in which problems typically arise is in the use of methods to control the pH of a process. Because there are a number of variables that exist from process to process, this book will not attempt to cover the range of possibilities and schematics that may be used to govern pH. However, several basic pieces of equipment are needed. These are a solution for measurement, a pH meter, a pH transmitter/controller, and concentrated feed solutions (acidic and/or caustic) for performing pH adjustments. A simplified diagram for a pH control system in a continuous neutralization process is illustrated in Figure 2.35.

Although the concepts of pH measurement are simple, there are a number of problems that may arise when designing pH control systems. As a result, careful thought should be put into designing these systems.

First, for a process that requires tight control on pH fluctuations, direct addition of a strong acid or base may cause wider swings in pH than those induced by the process itself. Therefore, additional stages on control may be required to premix dilute pH adjustment solutions. This, in turn, creates additional requirements for controlling pH of the feedstock solutions. To be more effective, flow rates must also be compatible. This re-

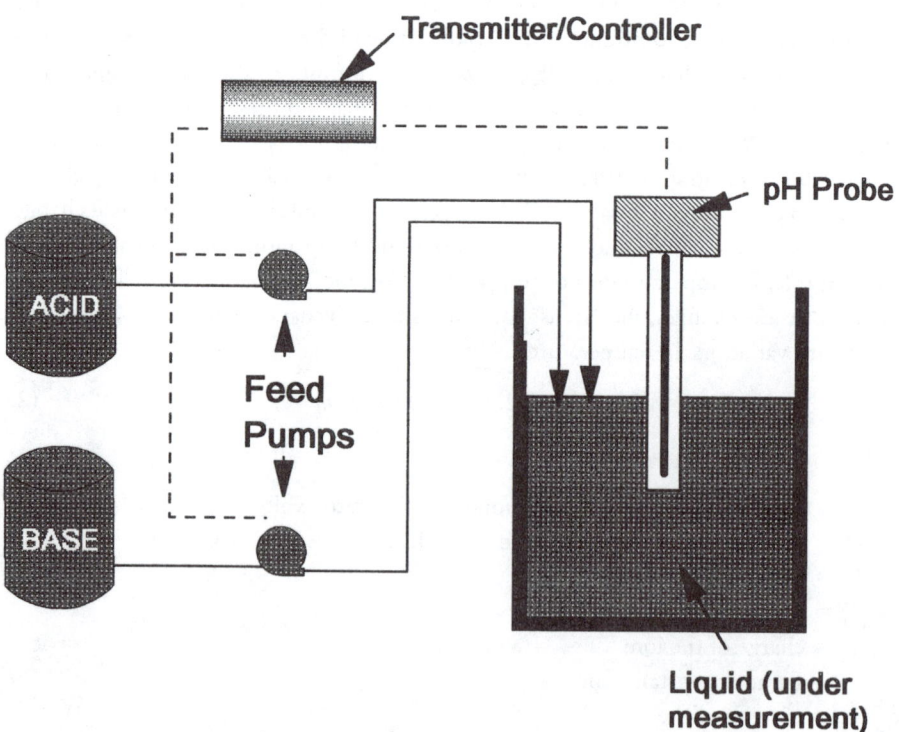

Figure 2.35 Simplified pH control system.

quires additional controls to monitor fluctuations in process flow rates. With all the aforementioned functions in place, each unit must be tied into a central analyzer to achieve the proper level of control.

Second, if pH tolerances are set too wide for a process, unnecessary additions of acid and/or caustic will occur. This yields a sinusoidal wave of pH that could lead to further complications in the overall manufacturing process downstream. Additionally, the scenario is not cost effective, as it may tend to result in more voluminous waste streams.

Third, some processes involve highly viscous materials or numerous feed solutions. In these situations, how is proper and efficient mixing achieved to control pH? Although this problem has nothing to do with the instrumentation for controlling pH, if mixing issues are not addressed adequately, the entire setup for monitoring and controlling pH may be compromised.

Finally, placement of the pH probes may also become a process variable. This may create problems from a measurement standpoint. Rather than measuring the pH of incoming or outgoing solutions in a process tank, the pH of the concentrated adjustment solutions skew the readings, thereby yielding no process control at all.

Conductivity Measurements

Conductivity measurements are an indication of a solution's ability to conduct an electrical current. Measuring conductivity is accomplished through the use of a probe with two electrodes. Because the current of a solution is performed by the transport of its ionic constituents, the higher the ion concentration, the higher the conductivity. The term used for a solution with low conductivity is *resistivity*, which is the reciprocal of conductivity. Alternating currents may be used to minimize ion migration or buildup on the electrode surfaces. Figure 2.36 illustrates this concept.

$$\text{Conductivity} = 1/\text{Resistivity} \tag{2.68}$$

The traditional units for measuring conductivity are the *mho* and the *Siemen* (S). A mho may be thought of as the reciprocal of an ohm (Ω), which is the designation for resistivity. One mho equals one Siemen. Many measurements are listed as Siemens per centimetre (S/cm) or mhos per centimetre (mho/cm). These units are in reference to parameters revolving around the definition the "probe constant." Depending on the distance between the two electrodes, the current signal will vary. Therefore, this parameter should be taken into account. Solutions with low conductivities (that is, high resistivities) typically require a sensor with a probe constant less than 1.0. Solutions with high conductivities require sensors with probe constants greater than 1.0.

Conductivity probes are useful in several applications. The most common conductivity probes are related to the measurement of specific chemical concentrations and dissolved solids. Probes used to measure specific chemical concentrations correlate the conductivity readings of solutions containing sulfuric acid, hydrochloric acid, or sodium hydroxide to chemical concentration values that may be used for dilution control. This is possible because many acids and bases produce solutions with increased levels of hydrogen ions and hydroxide ions that are highly mobile and yield detectable increases in solu-

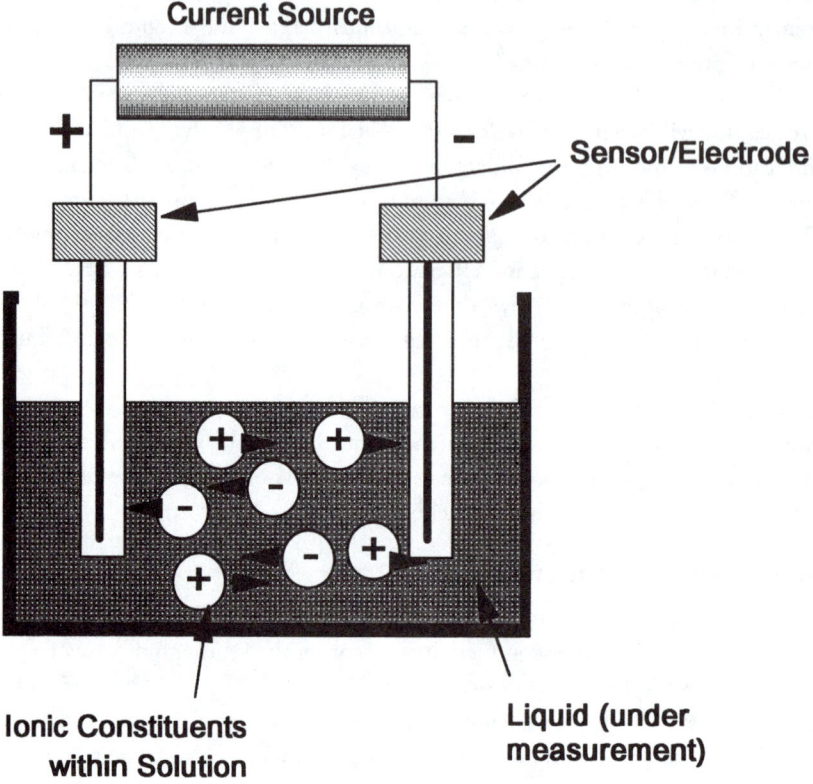

Figure 2.36 Conductivity measurement

tion conductivity. These readings indicate conductivity values that are typically much higher than those of the corresponding salts.

In many cases, specific ions that are monitored drastically affect the conductivity of a solution. In plating operations, an example may be illustrated by a countercurrent rinsing operation. Imagine a plating bath with a high level of metal ions in a solution (copper, nickel, and so on). The rinse tank following this process is contaminated with the drag-out of concentrated metal ions. By using conductivity probes to monitor the rinse tank metallic ion concentration, a solenoid valve may be triggered at a predefined point that signals that the rinsewater is becoming too contaminated to provide sufficient rinsing. The solenoid value initiates the flow of fresh water, thereby reducing the concentration of metal ions in the rinse tank. When the conductivity reading of the rinse tank drops back below a specified setpoint, the flow of water is shut off. This example is somewhat brief: as with the control of pH, there are many process control issues that must be addressed. However, the example serves to illustrate a useful application of conductivity meters in rinsewater control.

Outside the monitoring of acids, bases, and specific chemical species is the ability to perform conductivity measurements to evaluate the concentration of dissolved solids in a solution. The amount of dissolved salts and minerals in natural waters contributes to conductivity in a uniform manner. Therefore, conductivity readings serve as a tool in the measurement of dissolved material in a given solution. This is useful in systems in which the flow of water in a process is a closed-loop flow and where fresh water is only administered because of evaporative losses or high levels of dissolved solids. This may be commonplace in boilers, cooling towers, or even swimming pools. If the water used in a boiler is recirculated to minimize water usage and maximize conservation, the levels of solids will increase in the solution over time because of evaporative losses. This higher level of solids will cause scaling and precipitation to occur, reducing the efficiency of the process and generating additional maintenance problems. Through the use of conductivity measurements to track the increase in dissolved solids, the flow of fresh water may be administered to the process on only an as-needed basis.

The implementation of conductivity measurements to control a process can be quite useful. As in the case of monitoring pH, instrumentation systems must be designed carefully and must take into consideration criteria specific to the application. The most notable advantage of conductivity measurement applications is that the data obtained are typically accurate and repeatable. The process is applicable over wide ranges of temperature and pressure with the ability to determine the concentration of almost any ionizable solute. However, temperature effects must be compensated for because the behavior of a chemical may change with temperature. The primary drawback to conductivity measurements is that the process is not ion selective. Contributions to these values may result from a range of chemical species or factors. Knowledge of measurement processes and the ability to isolate solution parameters may help eliminate complications.

Oxidation-Reduction Potential

The measurement and control of ORP is a commonly used tool in many wastewater treatment applications. Some of the most typical chemical processes monitored include the reduction of hexavalent chromium, the oxidation of cyanide, and the bacteriological degradation of organic matter. The measurement of ORP is similar to that of pH. However, there are intrinsic differences that make ORP an excellent qualitative instrument.

Measurement of ORP is achieved through the use of a galvanic cell constructed of a reference electrode, typically calomel, and an indicating electrode of a highly noble metal, usually platinum or gold. The platinum or gold electrode functions as the anode and the calomel as the cathode. The anode is made of a highly noble metal because the potential for oxidation is less than that of any of the oxidizable components within the solution under evaluation. Therefore, the anode is the location of oxidation reactions and is typically unaffected.

Oxidation and reduction reactions are chemical reactions in which specific atoms are either losing electrons (oxidation) or gaining electrons (reduction). As electrons are given up by one chemical species, electron uptake is occurring by another. Each half-reaction exhibits an electrical potential (E^0) when compared to a reference. This reference is the reaction of hydrogen ions reduced to hydrogen gas:

$$H^+ + e^- \rightarrow 1/2H_2(g) \ E^0 = 0.00 \text{ V} \tag{2.69}$$

If the reduction potential is greater than that of the hydrogen reaction (that is, greater than zero), then the reaction will favor reduction. If the reduction potential is less than zero, the reaction will favor oxidation. Consider the following reduction potential for silver (Ag):

$$Ag^+ + e^- \rightarrow Ag(s) \; E^0 = 0.80 \text{ V} \tag{2.70}$$

Because this reduction reaction yields a positive E^0 value, the reaction will proceed as written. Silver ions will deposit on the cathode. The cathode represents a location in which electrons are given up or released into solution. The anode is the opposite and represents a location where electrons are consumed or removed from solution. Cathode is to reduction as anode is to oxidation. In what direction will the following reaction proceed when referenced to the hydrogen reaction: to the right for reduction or to the left for oxidation?

$$Fe^{2+} + 2e^- \rightarrow Fe(s) \; E^0 = -0.41 \text{ V} \tag{2.71}$$

The reduction potential for this is negative. Hence, the reaction will favor oxidation and proceed in a direction opposite of that which is written.

With these basic concepts in hand, ORP measurements can be discussed. Oxidation-reduction potential measurements are the net electrical potential of reactions occurring at the cathode and at the anode. The value of these measurements provides an indication of the extent a desired reaction has run to completion. In a wastewater treatment process in which hexavalent chromium is reduced to trivalent chromium, an ORP meter will exhibit a range of voltage readings depending on the state of conversion that has been achieved. At the beginning, readings of electrical potential will be higher in comparison to those experienced as the conversion process nears completion.

An ORP meter is essentially a voltmeter. Electrical potential is registered on a digital readout typically in millivolt (mV) values. A device of this type is included with other instrumentation to control chemical feed rates when a deviation from a low or high setpoint is detected. The high setpoint is commonly configured to control the addition of reagent (that is, the oxidizing or reducing agent), and the low setpoint will trigger an alarm that the system is falling out of specified parameters.

Although some systems may only be set up to respond to ORP values by turning chemical feed pumps on and off, others may employ proportional controllers. Proportional controllers create an output signal that is proportional to the deviation from a setpoint. The larger the distance a reading occurs from a setpoint, the stronger the output signal will be. The faster the ORP of a solution changes, the faster a chemical reagent is added to the treatment process. A controller signal is fed to a variable-speed control pump or similar device. The quantity of chemical delivered over time is determined by the strength of the output signal. The greater the output, the faster the chemical is metered into the process solution. This type of system can respond to changes in solution characteristics much more quickly than a traditional on–off configuration and helps prevent swings in process characteristics.

Temperature

Temperature scales, assorted unit conversions, and calculation methods were discussed in the section Chemical Engineering and the Material Balance, presented earlier in this chapter. This section takes these concepts a step further by identifying methods for measurement. The four primary techniques used for temperature measurement are thermocouples, resistance temperature detectors (RTDs), thermistors, and infrared thermometry.

Thermocouples are one of the most common methods for temperature measurement. The concept was discovered in 1821 by Thomas Seebeck, who noted that when two wires of dissimilar metals are connected at both ends, and one of the ends is heated, a continuous current is generated. If this circuit is broken, the open-circuit voltage becomes a function of the temperature applied at one of the junctions. The voltage is also dependent on the metals used. This principle is referred to as a *thermoelectric circuit* and is illustrated in Figure 2.37. Measuring the voltage across the junction is not as simple as it appears because there are other thermoelectric circuits formed when a voltage meter is attached to the process. Depending on the configuration, various thermocouple types are established. Software compensation is typically used to convert voltage readings from a thermocouple into actual temperature values. However, voltage–temperature correlation is not linear. To offset this effect, mathematical methods are used to increase the accuracy of the temperature measurement.

Resistance temperature detectors represent another useful method for measuring temperature. The basic concept of RTDs was discovered around the time of the thermocouple by Sir Humphrey Davy, who related the resistivity of various metals to temperature. The most common metal used in this type of device is platinum. Over the years, several configurations of RTDs were developed. Initially, RTDs consisted of a fragile, helically wound coil of platinum wire around a glass tube. Eventually, RTDs graduated to more high-tech versions that integrated modern manufacturing techniques consisting of ceramic substrates and various metal deposition processes. However, the principle for measurement has remained essentially unchanged: that is, the resistivity of a metal increases

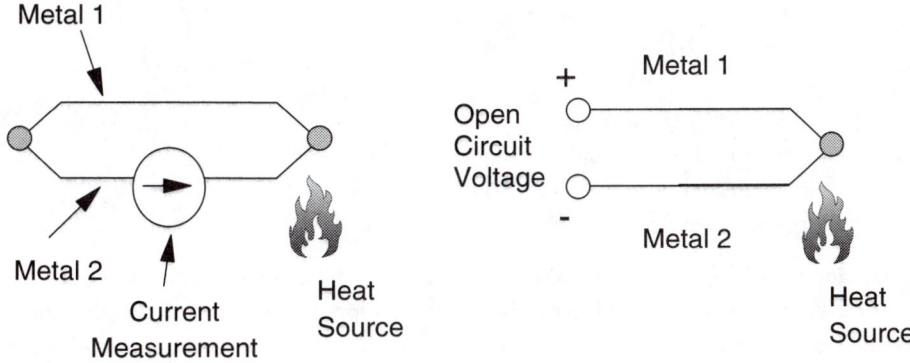

Figure 2.37   Thermocouple circuits.

as temperature increases. Measurement of this parameter yields a nonlinear curve over a wide temperature range. This nonlinearity is less pronounced than that exhibited by thermocouples, but it too must be offset by software to provide the desired level of accuracy.

Thermistors are also temperature-sensitive resistors, but their sensitivity is much greater than that of RTDs. Thermistors exhibit much wider swings in resistivity values over a given temperature range than either thermocouples or RTDs. Rather than being constructed of various metals, thermistors consist of semiconductor materials. In contrast to RTDs, this type of temperature-measurement device decreases in resistance as temperature increases. These characteristics allow for small temperature changes to be detected. One drawback of thermistors is that the nonlinearity relationship between resistance and temperature is much higher than that of the aforementioned techniques. However, with computer controllers and software compensation, accurate measurements can still be achieved.

The last type of temperature measurement is infrared thermometry. Infrared thermometry is different from the other temperature measurements discussed in that it is accomplished without contacting the medium under evaluation. This approach is particularly useful when a quick response is needed or the targets are moving, inaccessible, or exhibit other restrictions that are not amenable to more conventional measurement techniques. Infrared radiation was discovered by Sir Isaac Newton, who noted that the electromagnetic energy of light could be separated into distinct bands or colors when passed through a prism. These bands were later discovered to possess specific energies by Sir William Herschel. Expansion of these concepts over the centuries has led to the modern approaches in infrared thermometry.

Infrared energy measurement is based on what is referred to as *blackbody emittance*. *Emittance*, or *emissivity*, is the ratio of thermal radiation emitted from a graybody (non-blackbody) to that from a blackbody, at identical temperatures. Blackbodies do not reflect energy. This concept is perhaps more easily understood by using a simple relationship that references the law of the conservation of energy. For any given object, the incident radiation energy of a specified wavelength must equal the sum of the energy that passes through the object, is reflected, and is absorbed. This is best demonstrated through Figure 2.38 and the following equation:

$$E_I = e_t + e_r + e_a \qquad (2.72)$$

Where

E_I = incident energy (J) or (btu),
e_t = energy transmitted through the object (J) or (btu),
e_r = energy reflected from the object (J) or (btu), and
e_a = energy absorbed by the object (J) or (btu).

In Equation 2.72, *emissivity* is the term corresponding to the energy absorbed (e_a). In most cases, the object that is subject to incident radiation is opaque; therefore, the energy transmitted is minimal. Hence, the emissivity of an object may be set equal to the inci-

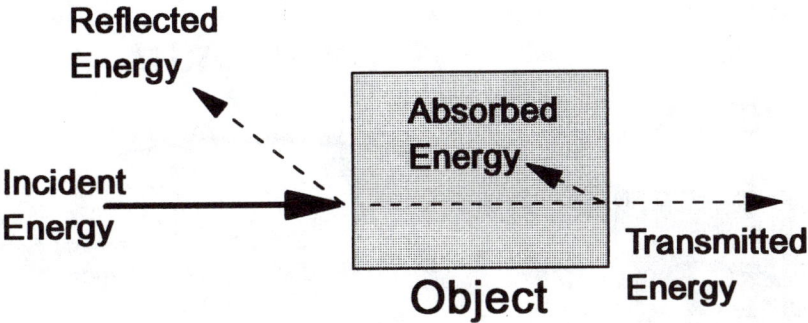

Figure 2.38   Radiant energy principles.

dent energy minus that which is reflected. Most devices of this type use a lens to focus
the infrared energy emitted from a target onto a detector. This energy is converted into
electrical impulses that are then translated into temperature units for display. A considerable amount of data manipulation via computer control is required to achieve these results. Although infrared thermometry is sometimes thought to be inaccurate because of
the lack of contact with the object under measurement, the technique, in general, is
highly reliable and is in widespread use throughout industry and research. Further elabo-
ration on the specific types of infrared measurement devices and their applications is beyond the scope of this text.

Pressure

Pressure is defined as the amount of force applied over a unit area. Regardless of the
magnitude of the force, deflections, distortions, or dimensional changes are produced on
the object subjected to that force. Pressure measurement devices are designed to sense
these changes and correlate the data into meaningful information such as pressure or
strain readings. *Strain* is the quantification of a deflection or distortion that is normally
recorded as the ratio of the change in length of a specimen, either in tension or compres-
sion, to the initial unstressed reference length. Pressure instrumentation equipment may
be used to maintain process pressure, trigger a process control parameter, or set off alarms
to indicate critical conditions. The units of pressure and a general overview of this topic
were presented earlier in this chapter in the section Chemical Engineering and the Material Balance.

The most important element in a pressure control system is the sensing device. The
sensing element, which acts in conjunction with an opposing spring, determines the overall sensitivity and range of pressure that may be measured. Once the pressure is measured,
it is this movement that transfers a signal to a switch that may be set up to provide specific process control. The most common types of pressure-sensing devices are the piston,
diaphragm, and bellow. Illustrations of each are provided in Figure 2.39.

Pistons are traditionally used for high pressures most commonly above 6.9×10^7 Pa
(10 000 psi). Their configuration is somewhat self-explanatory. The piston rod is set up

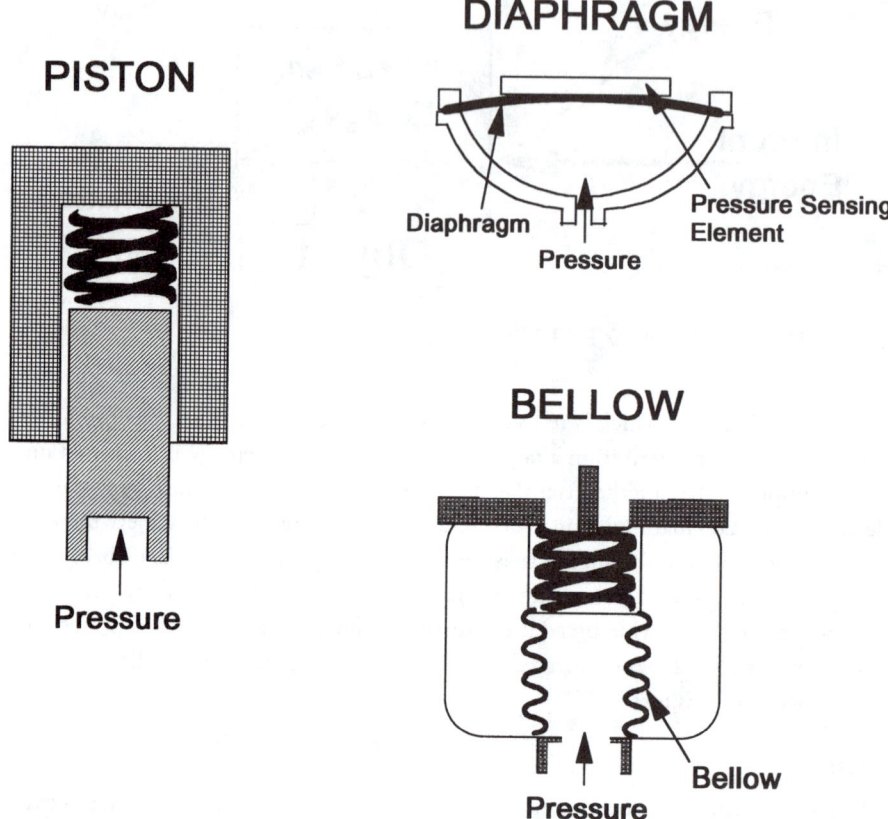

Figure 2.39 Pressure-sensing devices.

to act under pressure against an internally mounted spring. An O-ring seals the internal portion of the chamber from the exterior. The more restrictive the O-ring, the less sensitive the element is to pressure changes.

Diaphragms provide a smooth, flat surface on which pressure is exerted. The deflection of this surface displaces a plate that is in contact with the diaphragm surface, which, in turn, is mounted to a spring. Most uses of this device are restricted to pressures ranging from atmospheric to nearly 4.1×10^7 Pa (6 000 psi).

Bellows consist of an element that expands or contracts with changes in pressure. This is in contrast to deflections, as is the case with pistons and diaphragms. The most typical configuration employs a cylindrical compartment closed at one end that has several folds throughout the material. The folds allow for expansion and contraction. The sensitivity and range of the element may be tailored by a spring that is mounted against the bellow to measure these movements. These devices are best suited for environments that do not exceed 2.8×10^6 to 3.4×10^6 Pa (400 to 500 psi). Although the aforementioned sensors are the most traditional techniques for measuring pressure, developments in printed circuit technologies have created enhanced alternatives. *Pressure transducers* are integrated circuits that have pressure-sensitive diaphragms and strain gauges etched into the substrate of specified materials. In this application, strain gauges are bonded to the diaphragm and are able to measure the external influences acting on the device. These forces, as discussed earlier, are registered as tensile or compressive strain and correspond to the expansion or contraction of the pressure transducer materials. Unfortunately, the types of influences that can result in strain are not restricted to pressure. Strain can occur because of temperature changes that result in thermal expansion, other external forces, or internal effects experienced by the unit's structure, such as fatigue or material degradation because of hostile environments. Effective pressure transducers take these factors into account. Almost all devices of this type are coupled with compensation networks to reduce errors.

With the development of pressure transducers, reliable process control systems may be designed. As previously discussed in liquid level control applications, liquid levels act as a trigger for sensor output. However, some environments may exhibit conditions that are not amenable to this type of measurement. In these situations, pressure transducers mounted below the liquid surface can be used to measure pressure changes that result from the changing amount of head produced with changing liquid levels. Pressure transducers may also be of value monitoring paint-spraying operations. Process control in this situation is achieved by monitoring the pressure experienced at the spray gun nozzle. If the pressure passes a specified limit, it may indicate that the nozzle is clogged. Conversely, lower pressure limits may indicate the nozzle is worn out from spraying excessive amounts of paint.

Applications for pressure transducers are widespread. To ensure reliability, the keys for proper selection of this equipment should include the following characteristics: accuracy over temperature ranges of concern, durability within the environment of use, compatibility with external chemicals, and interchangability with replacement transducers.

COMPUTER CONTROL AND INTEGRATION

A complete data acquisition system employs the use of signals fed from a process via any of the aforementioned instrumentation devices and the power of a computer. Once the data are in a computer, information may be stored for future reference, displayed for immediate interpretation, or analyzed through the use of various software packages to yield spreadsheets, charts, plots, and other useful manipulations. Depending on the level of control desired, a computer can also be used to send signals back to the source or process to provide "remote control." An example of this might be the need to control the pH of a process solution. The pH probes mounted at a process tank send signals to a computer that correspond to instantaneous readings. If the reading falls outside a specified range, a chemical feed pump is activated administering acid or caustic depending on the situation. A computer-based data acquisition system is not required to accomplish this particular example; however, it illustrates how remote display and control of liquid level, pH, and other process parameters can be valuable. Each system must have the necessary hardware and software to perform its function.

There are many software packages available today that can accomplish a host of tasks and include a plethora of features. In most cases, some programming skills are required to tailor the software to fit the specific needs of individual process control problems. Knowledge of computer languages such as BASIC, FORTRAN, or C can prove to be beneficial, especially in the case of stand-alone systems.

The hardware systems used for data collection typically consist of two choices—stand-alone systems or what is referred to as "plug-in" types. The concept of these two systems is illustrated in Figure 2.40. Stand-alone or communications-based configurations are independent of the computer type selected for achieving process control. Communication between the stand-alone configuration and computer is accomplished through data transfer that occurs via a cable hooked into the computer's serial or parallel port. These systems provide additional benefits to those of plug-in systems because of their ability to maintain higher accuracy, the ease by which systems may be modified and expanded, and the wide range of sensor inputs/outputs. These systems are also traditionally installed at the process area, with data transferred via phone lines, using a modem.

COMMUNICATIONS BASED SYSTEMS

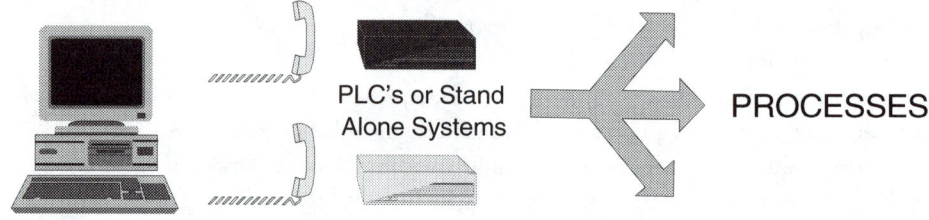

PLUG-IN SYSTEMS

Figure 2.40 Hardware configurations for data collection (PLC = programmable logic controller).

In communications-based systems, some equipment may be required to be in constant contact with a remote computer or host to operate properly, while others can operate independently for a specified period of time. The latter of these two options allows a process control system to acquire data and provide system control without the direct use of a remote computer. Data that are accumulated may be "dumped" to the computer at intermittent or specific time intervals for later interpretation and manipulations. This type of stand-alone configuration is flexible in its operational capabilities because of its independent sources for data and program storage. Increased flexibility, however, typically comes at increased expense.

An example of a stand-alone system integrates the use of programmable logic controllers. Programmable logic controllers incorporate numerous elements such as power supplies, memory, input/output capabilities, a central processing unit, and a communications capability. All of these features are traditionally housed within a convenient package that promotes their widespread use and ability to be located at or near a specific manufacturing process or operation.

Plug-in systems can be economical as long as a computer is available. However, with the cost-competitive nature of the personal computer today, plug-in systems may prove to be an inexpensive option, regardless of whether an existing computer is available or not. Plug-in systems, which are designed to be integrated with process control schemes, are easy to use. For example, data collection may commence after a computer card is simply plugged in. In most applications, software is provided or easily available so that little-to-no programming is required. A significant difference between plug-in and communications-based systems is that each channel for a plug-in system may be tailored to fit the type of signal input that is required. For example, on a plug-in system with eight channels, each channel may be set up to accept data as voltage, current, or some other type of signal. Stand-alone modules, which typically have multiple channels as well, have no flexibility in selecting the type of input desired. A stand-alone system with eight channels will typically be of the same type (that is, voltage, current, and so on) that is specified at the time of purchase. An additional benefit to plug-in systems is that the speed at which data may be obtained is much faster than with communications-based alternatives.

The two types of plug-in options commonly used are those that acquire data and those that communicate with other pieces of equipment. Data acquisition boards are configured with analog inputs/outputs, digital inputs/outputs, and an option for the number of channels needed for a specific application. Communication boards provide a highway for information transfer from remote devices such as analytical equipment, printers, and other computers. Proper attention should be given to the selection of a board for a specific application. Parameter definition is also an important step when designing a data acquisition system because of the numerous configurations that are possible. The requirements for data acquisition/collection, data analysis, process control, and/or alarms are different and should be specified before selection of plug-in hardware.

SUGGESTED READINGS

Borer, J. (1985) *Instrumentation and Control for the Process Industries*. Elsevier Applied Science Publications, New York, N.Y.

Feigenbaum, A.V. (1983) *Total Quality Control*. 3rd Ed., McGraw-Hill, Inc., New York, N.Y.

Himmelblau, D.M. (1974) *Basic Principles and Calculations in Chemical Engineering*. 3rd Ed., Prentice-Hall, Inc., Englewood Cliffs, N.J.

Koronacki, J., and Thompson, J.R. (1993) *Statistical Process Control for Quality Improvement*. Chapman & Hall, Inc., New York, N.Y.

Oakland, J.S. (1986) *Statistical Process Control*. John Wiley & Sons, Inc., New York, N.Y.

Omega Engineering, Inc. (1995) *The Omega Handbook Series*. Stamford, Conn.

3 Chemical Alternatives and Process Modifications

Industrial accidents that once occurred and caused significant environmental damage were precursors to the evolution of the current regulatory framework for controlling the release of hazardous chemicals and for managing their ultimate disposal. Slow or intermittent releases are also important because they are notorious for contaminating groundwater, which is a primary source of drinking water for a large percentage of the population. Therefore, not only is the prevention of accidents important, but so is the manner in which daily operations are conducted. This chapter provides examples of opportunities for better managing some widespread industrial manufacturing processes. The goal of people who manage these operations is to provide efficient and reliable control for those processes that must use hazardous chemicals while continuously searching for nonhazardous alternatives. Technology plays an important role in these efforts; good housekeeping and common sense initiatives also provide a strong foundation.

A proactive examination of methods to reduce wastes while increasing process efficiency and, hence, boosting profits is given for metal finishing, solvent-cleaning and -degreasing, and surface-coating operations. To provide a basic understanding of how statutes and regulations can affect these business operations, a brief review is provided. In addition, each industrial segment discussed within this chapter will provide an in-depth review of the most pertinent regulations governing the operations specific to that type of industry. Although those discussed are federal regulations, additional state and local extensions also exist. In many instances, the federal government, specifically the U.S. Environmental Protection Agency (U.S. EPA), grants state and local regulating bodies authority to administer federal programs, provided their regulations are at least as stringent as federal versions. Thus, there can be considerable overlap and, therefore, multiple or duplicate requirements on a given subject.

Regulations and agencies that play significant roles in the way daily operations are conducted and to some extent govern the equipment to be employed include the Resource Conservation and Recovery Act (RCRA), the Clean Air Act (CAA), the Clean Water Act (CWA), and the Occupational Safety and Health Administration (OSHA).

The Resource Conservation and Recovery Act, as it relates to mandatory pollution prevention programs, was discussed briefly at the beginning of Chapter 1. However, RCRA goes beyond these requirements. The Resource Conservation and Recovery Act deals with the portion of solid wastes that is considered hazardous by specific definitions.

This statute—and ensuing regulations—plays a major role in the daily operations of a number of different industries because it defines specific waste-handling procedures. Wastes are considered to be hazardous by one of two classification schemes. The first is by specific identification using various U.S. EPA–published lists within the Code of Federal Regulations (CFR). The second is by physical or chemical characteristics as defined by the categories of corrosivity, reactivity, ignitability, and toxicity. Appendix A contains a list of various U.S. EPA waste codes.

Permits are not required for most manufacturers as long as hazardous wastes are not stored on site for more than 90 days (180 days for small-quantity generators). A typical process for generators is to identify all wastes produced within a facility and then to classify these wastes, including identifying those that are hazardous with the proper U.S. EPA waste code. The next step for generators is to obtain a U.S. EPA identification number for use when shipping hazardous wastes off site using a U.S. Department of Transportation (U.S. DOT) approved uniform hazardous waste manifest. The shipping of hazardous wastes is also regulated under U.S. DOT regulations but is not discussed here.

In addition to the primary process elements of RCRA, there are numerous other requirements that come into play. Some of these include storage practices, inspection logs, training, contingency plans and emergency coordinators, waste minimization plans, and recordkeeping and reporting.

Additional discussion is not provided here because the original source should be consulted to ensure accuracy, as regulations are continually subject to revision. Additional information on these topics may be found in 40 CFR 260–268.

The CAA is one of the most regulatory-intensive of all the statutes discussed in this chapter because it contains provisions dealing with a wide range of operations such as boilers, painting, degreasing, and ventilation. Other regulations are based on new source construction and protect relatively clean air areas from significant deterioration while creating initiatives to prevent more polluted areas (nonattainment zones) from becoming any worse, with the intention of making improvements. While permits are required for some activities, registration and/or notification may be a requirement of various regulatory bodies on other occasions.

Specific chemical compounds are also scrutinized in the CAA. These include criteria pollutants, hazardous air pollutants (HAPs), ozone-depleting compounds (ODCs), or a class of chemicals that can include almost anything—volatile organic compounds (VOCs). Appendix B provides a list of those compounds classified as HAPs. Processes incorporating HAPs are also regulated by the National Emission Standards for Hazardous Air Pollutants (NESHAP). The National Emission Standards for Hazardous Air Pollutants regulate individual compounds and also specific manufacturing practices through the implementation of maximum achievable control technologies (MACTs). Depending on the case, a MACT can be achieved through the implementation of specific equipment, raw materials, and management practices.

The CWA and its associated regulations were devised to protect numerous types of water bodies in the U.S. by setting controls and limitations on the types of wastewaters discharged from a range of industrial and municipal activities. Discharges that go directly to the environment are permitted through the National Pollutant Discharge Elimination

System (NPDES), which is a federally run program, except in states that have received authorization to administer it. These types of discharges also include those for stormwater runoff. Runoff from some industrial activities may carry high amounts of contaminants that can build up on roofs, raw material storage and containment areas, parking lots, or other areas.

For discharges that are routed to publicly owned treatment works (POTWs) via sanitary sewers, no NPDES permit is required. Instead, the POTW must be permitted because it is the body that is discharging directly to the environment. Industrial activities generating wastewaters sent to a POTW must pretreat the wastewater to meet specific discharge limits that pertain to the type of processes employed. Metal finishing is a category that has discharge limits for total cyanide, various individual metals, total metals, total toxic organics, and, in some cases, total suspended solids. Although not permitted by NPDES, the discharge of wastewater to a POTW is a permitted activity conducted most often at local agencies. However, because each municipality may operate under different conditions, the potential for local governments to require more stringent limits than those set at the federal level is apparent.

The Occupational Safety and Health Administration also has its own set of regulations. Its regulations, however, relate to the protection of workers rather than the environment. Some of the most important OSHA regulations that pertain to virtually all industrial activities are the Hazard Communication Standards (HCS). The purpose of this set of standards is "to ensure that the hazards of all chemicals produced or imported are evaluated, and that information concerning their hazards is transmitted to employers and employee" (29 CFR 1910.1200[a][1]).

The chemical categories identified include carcinogens, corrosives, highly toxic materials, toxic materials, sensitizers, and chemicals with target organ effects. As with RCRA, numerous components are involved. When integrated, these components can help achieve the aforementioned goal. The key elements of HCS are the written program, chemical container labeling (different requirements than RCRA), hazard determination, chemical inventory, material safety data sheets, and employee information and training.

Documentation is a primary element of this regulation. Without it, none of the listed items could be effectively implemented, tracked, or presented during a regulatory compliance audit. In addition to HCS, several other regulatory areas exist, such as process safety management of highly hazardous chemicals, hazardous waste operation and emergency response, confined space entry, lockout/tagout, personal protective equipment, and electrical safety.

Metal Finishing

The Federal Water Pollution Control Act (FWPCA) of 1972 established the regulatory basis by which water pollution in the U.S. was initially addressed. This act established pretreatment programs for wastewater discharge and defined the standards and effluent characteristics of wastewater for a variety of industries. Pretreatment programs are designed to remove harmful contaminants that, if discharged to a sanitary sewer for subsequent treatment at a POTW, would cause significant upsets to the biological treatment processes and result in highly contaminated discharges to surface waters. It was not until the 1980s that specific standards for the metal finishing industry were implemented. Costs associated with compliance of the FWPCA and its associated standards are derived from the capital required to construct pretreatment facilities, operations and maintenance (O&M) of the pretreatment facilities, administration associated with the application and upkeep of discharge permits, and the potential for fines and lawsuits resulting from discharges falling outside the specific parameters defined in the permits.

As a result of regulations and an increasing awareness of the need for environmentally and economically sound practices, many metal finishing shops are implementing measures to contain, conserve, and recycle their process chemicals and waste products. Most of the process management techniques and technologies covered in this subsection are easily implemented and can result in immediate payback with proper coordination and backing from both management and shop floor personnel. Benefits derived from these activities can include reductions in contaminants at process areas, significant savings incurred through lower usage rates of treatment chemicals, reduced wastewater treatment sludges, lower labor costs, and lower utility costs. While these benefits are secondary effects, they should not be overlooked when conducting waste minimization projects. Moreover, they can help provide the drive to reduce wastes by demonstrating larger savings and faster payback periods. While most production processes are segregated and thought of individually, waste minimization requires that processes be defined from start to finish. To capture the benefits, the entire cycle—from materials purchased to waste disposal—must be considered.

The first two of the following metal finishing topics covered, drag-out reduction and rinsewater management, represent common-sense alternatives to directly reducing the generation of waste. Implementation costs are minimal with these approaches, and, as a result, success stories have been provided that may be used to encourage additional waste minimization projects. The remaining metal finishing sections, Solution Recovery and Recycling and Chemical Replacement, represent higher cost initiatives because of the larger capital expenditures or significant process changes required. These approaches have also proven to be successful, although their payback periods are somewhat longer.

DRAG-OUT REDUCTION

Withdrawal Rates and Drain Time

The concept of drag-out (Figure 3.1) is common to all metal finishing processes and is the primary cause for the contamination of rinsewater. If the amount of contaminants transferred from chemical process tanks to rinsewater tanks can be minimized, rinsewater may be used for longer periods of time, thereby reducing the volume required. Another beneficial effect is that the amount of chemical concentrates required to "top off" the process tanks that have lost valuable solution may be reduced as well.

The most common methods for reducing drag-out are achieved through withdrawal rate reduction, drain time increases, drain boards, improved racking configurations, and product modifications to reduce the retention of process chemicals.

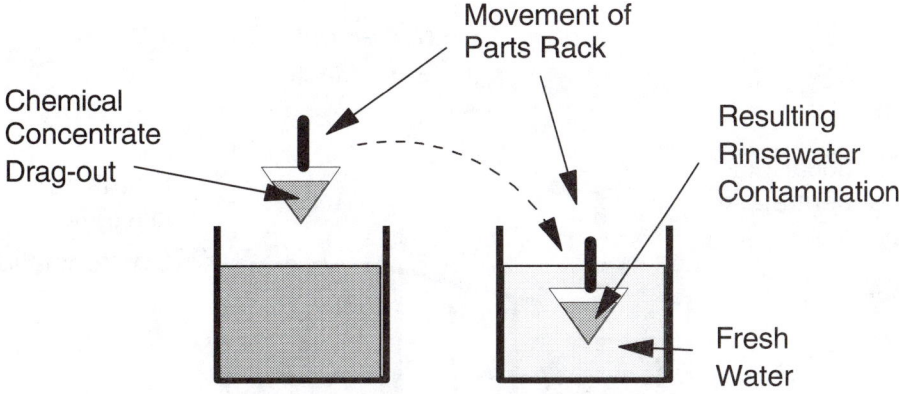

Figure 3.1   Conceptualization of process chemical drag-out.

Reducing the rate at which parts are withdrawn from process tanks can have a profound effect on the quantity of concentrates that are retained on part surfaces. This particular option is not viable if racked parts are handled manually because there is no measurable way to control the rate of withdrawal. However, in situations where mechanical hoists are employed to raise, lower, and transport the racked products, the speed of the motor may be adjusted to reduce the rate at which the parts are withdrawn from solutions. Increased drain time is a secondary effect. The primary mechanism for reducing contaminant drag-out is the viscosity of the process bath solution and its associated surface tension, which aids in the removal of the solution by pulling the contaminants from the parts in a "squeegee-like" fashion. This procedure of slowing the withdrawal rate of parts out of process baths coupled with increases in drain time (frequently as little as 10 to 20 seconds) over the process baths prior to transferring the racks over and into adjacent rinse tanks has been proven to reduce contaminant drag-out by as much as 50% (U.S. EPA, 1992).

Racking and Drain Boards

The use of drain boards also provides an excellent method for minimizing drag-out while conserving chemicals in process bath solutions. As shown in Figure 3.2, drain boards allow drag-out to be redirected to chemical concentrate tanks. If properly oriented and constructed of compatible material, this simple modification can increase drain time while preventing chemicals from reaching the floor. Housekeeping issues can also be significantly improved.

Additional methods for reducing drag-out include part racking techniques and the inclusion of special design features into the parts themselves, which allows for enhanced drainage for process solutions. When racking parts, pieces should be oriented to minimize the amount of contaminants that will be trapped in cavities. If possible, cavities should be positioned facing downward. In situations involving circuit boards or barrels, draining efficiency may be increased by having the parts canted at a small angle toward one corner or edge. This adjustment provides a point at which the solutions may aggre-

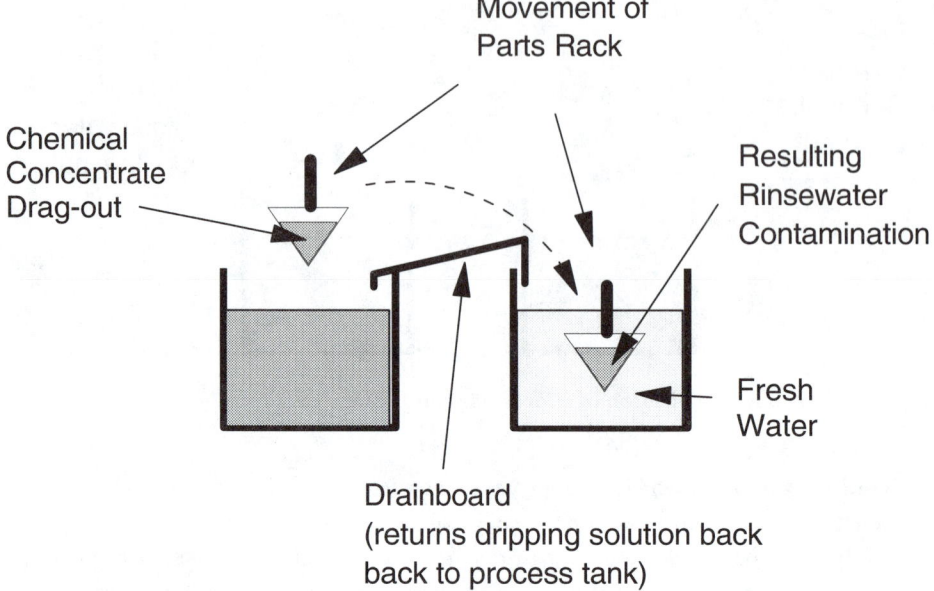

Figure 3.2   Drain board configuration.

gate quickly and drip off. Figure 3.3 provides an illustration of how part orientation during racking can assist in drag-out reduction.

Because most products are not conceived with the manufacturing process in mind, designing drag-out reduction features is the most difficult option to implement. Designing a product to fit consumer demands while optimizing the production methods used is a concept discussed in the section Life-Cycle Assessment, in Chapter 5. Consideration of every step in a product's life span, which includes design methodologies, manufacturing processes, O&M, and, ultimately, disposal, is a key factor in waste minimization. This

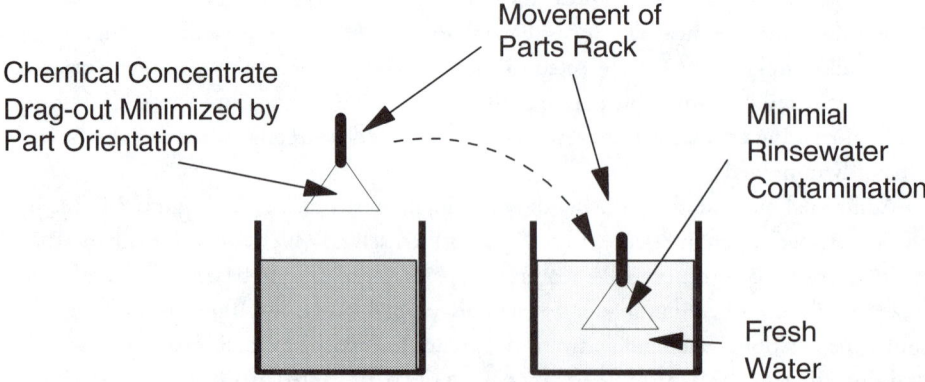

Figure 3.3   Drag-out reduction and racking.

path requires considerably more up-front thinking and can result in higher design costs, but downstream benefits frequently demonstrate sufficient payback. Consumers are beginning to demand these features be considered as well.

RINSEWATER MANAGEMENT

There are numerous methods for using and conserving rinsewater. While some methods require additional plant space and capital expenditures, others require minimal effort. In either case, the benefits of implementation are typically well worth it. These methods achieve their goals by cost avoidance. Therefore, it is important to track process information to quantify the benefits. For example, if one of the following rinsewater management techniques is selected for implementation, water usage should be tracked beforehand. Costs associated with the contaminated water should also be identified. Such costs include items such as potable water purchase, pretreatment costs (that is, demineralization, deionization, or reverse osmosis), and subsequent wastewater treatment, which includes treatment chemical usage costs, utility costs, labor, facility upkeep, and others. While these may be indirect costs, they all should be considered, particularly in cases where plant operators are required to provide cost justifications to management to receive adequate funding. The following case study further demonstrates this viewpoint.

Countercurrent Rinsing Case Study

At Bell Helicopter Textron, Inc., a countercurrent rinsing station was proposed for a chemical processing line in which only single-stage rinsing was employed. Theoretical reductions in water use were estimated to be 97%. To take into account "real-life" operational inconsistencies and fluctuations, however, a conservative estimate of approximately 50% reduction was presented to management. Despite engineering calculations and conservative estimates, initial reactions from management included the thought that because water is inexpensive, money should not be spent. Indeed, at a cost of fractions of a cent to purchase city water, this belief would appear to be true. However, with minimal additional effort, a financial package was prepared yielding a strong marketing strategy. In the end, capital was approved and the installation of the rinsing project proved financially advantageous. The itemized costs and savings used to provide justification for the project are shown in Table 3.1. Note that items such as water purchase, water preconditioning (deionization, reverse osmosis), and wastewater treatment (chemicals and labor) were included in the justification. The result of this project was a payback period of approximately 3 weeks and an annual savings exceeding $200 000. As a result of the success of the pilot project, additional funding was provided for other rinse stations with no financial justification.

Countercurrent Rinsing

Countercurrent rinsing is one of the most effective methods for reducing water usage in metal finishing processes. The procedure consists of a series of rinse tanks in which the flow of rinsewater is opposite in direction to the flow of parts, hence the term *countercurrent*. Fresh water is fed to the final rinse tank and overflows into each preceding tank until the water reaches the rinse tank immediately following the process tank. The con-

Table 3.1   Costs and savings of the countercurrent rinsing case study.

Item description	Unit savings, gal[a]/wk	Unit cost, $/gal[a]	Total cost/ (savings)
Purchase and installation of additional rinse tanks, pumps, and filters			$13 151
Rinsewater purchase, preconditioning, and wastewater treatment	61 440	0.07	($4 300) per week
Payback period		3.06 weeks	
Annual savings (48 weeks/year)		$206 400	

[a]gal × 3.785 = L.

centration of contaminants in the final rinse is maintained at a specified/acceptable level—typically 0.1% of that in the prior process tank. Fresh water is added to maintain final rinse quality. Figures 3.4 and 3.5 provide a comparison of a traditional single-tank rinsing methodology with a two-tank countercurrent configuration. Calculations illustrating the benefits of countercurrent rinsing are discussed in the Material Balances section in Chapter 2.

Note the amount of rinsewater required for each configuration. Rinsewater usage is reduced almost 97% by installing an additional tank and implementing countercurrent rinsing techniques. Conventional two-tank configurations where each individual tank is injected with fresh incoming water may also be employed to achieve a nearly 94% reduc-

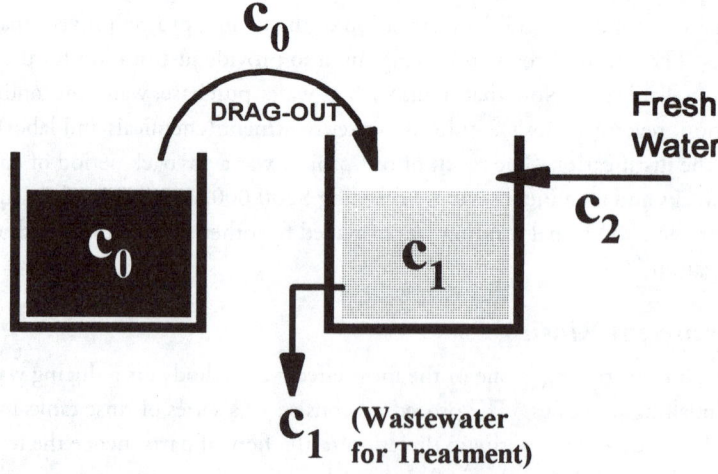

Figure 3.4   Traditional single-tank rinsing configuration.

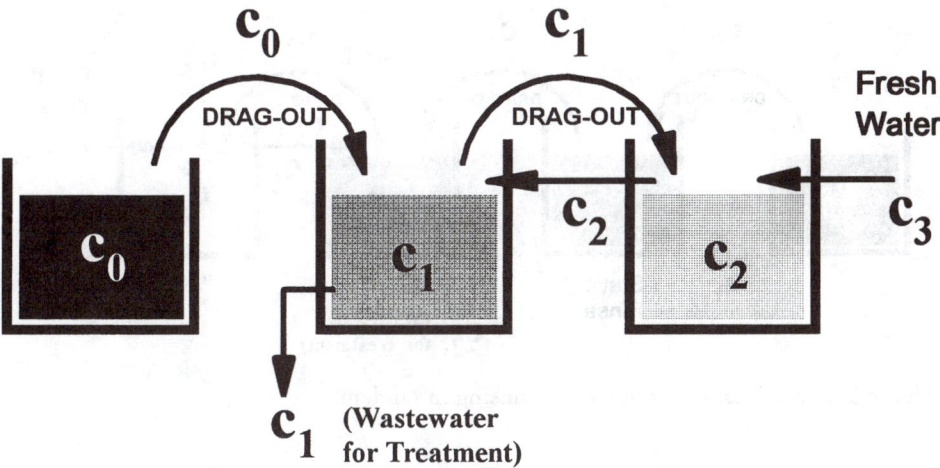

Figure 3.5   Countercurrent rinsing two-tank configuration.

tion in water usage. However, a primary difference between the two-tank options is that only the final rinse in the countercurrent configuration must be monitored or controlled. This is a result of the automatic overflow of rinsewater from one rinse tank to the next, which is accomplished by ensuring that there is adequate head from the final rinse to the initial rinse. It is more difficult to ensure consistent control of the conventional two-tank layout because each rinse tank must monitored. In addition, lack of proper control can lead to increased water usage.

Reviewing the concepts of mass balance provides the understanding that regardless of the number of rinse tanks and their configuration, the mass of chemical drag-out remains unchanged. Only the concentration of contaminants in the wastewater is altered. In general, the less rinsewater used, the higher the concentration of contaminants that need to be treated. In most cases, this is preferable because low contaminant concentrations are typically more difficult and less efficient to treat than waste streams with higher levels. However, this is not always the case because process-tank dumps frequently posses levels that are too high.

Static Rinsing Versus Spray Rinsing

It may occasionally be desirable to direct overflow from the initial rinse tank of a countercurrent rinsing configuration into the process tank to regenerate the concentrated solution. Frequently, however, this rinse bath may not be of sufficient concentration to make the practice feasible. If this is the case, static rinsing may be a viable option. Static rinsing is frequently employed in the first rinse tank adjacent to a chemical process bath. No water flows through the rinse tank to minimize contaminant levels. The idea is to allow the rinsing solution to become concentrated with the process chemicals caused by the steady influx of drag-out. At specific intervals, this solution may be used to replenish the preceding process tank, recycled to recapture valuable constituents, or reused after purifi-

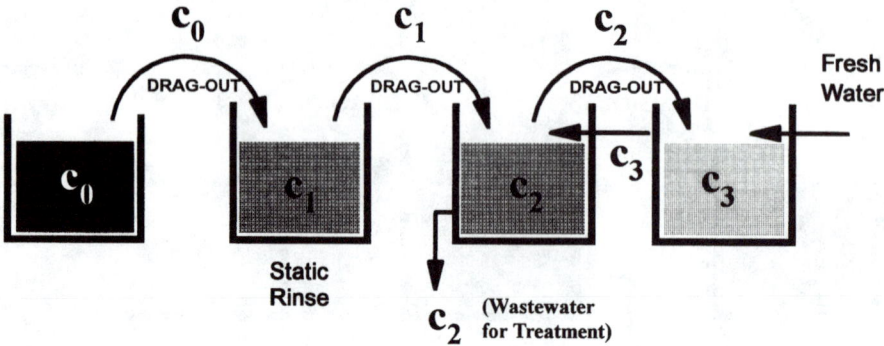

Figure 3.6   Static and countercurrent rinsing in tandem.

cation as clean water in a final rinse tank. Static rinsing configurations can remove up to 90% of drag-out contaminants before final rinsing. The point at which water is typically added to the static tank to maintain the rinsing efficiency is when the tank reaches approximately 10 to 20% of the concentration of the process bath solution. To achieve good results or a higher quality solution, deionized water in static rinse stations should be used. Because continuous "makeup" water is not used to maintain a low level of contaminants, static rinses generate little wastewater. Figure 3.6 illustrates a modified countercurrent rinsing station employing a static rinse tank.

Another option that may be used in a configuration similar to static rinsing is spray rinsing. Spray rinsing is an efficient process that can greatly conserve water usage, although it is not amenable to every process line because it is sensitive to part configurations. As water impinges on the surface of the part, it flows downward increasing in contaminant concentration along its path. As additional water is used, runoff becomes more dilute. In essence, this process is similar to a multiple-station countercurrent rinsing configuration. However, no steps in contaminant concentrations occur because the moving of parts from tank to tank is eliminated.

The pressure of water that is sprayed plays an important role in a spray rinsing configuration. As the atomized water strikes the part surface, the energy creates considerable turbulence, yielding rapid diffusion of the adhering process solution. As previously mentioned, part configurations greatly determine the ability of spray rinsing to be effective. This rinsing process is particularly effective with work that contains mostly flat surfaces and little to no cavities. If parts contain numerous holes, cavities, or shielded areas from the sprayed water, it is conceivable that water usage will be higher than with other available methods.

The number and positioning of spray heads is also an important parameter. The most common setup employs horizontally mounted fan jets, some of which are operated at specific intervals and duration. As with most rinsing systems, deionized water is preferred. Its usage will minimize the blockage of spray nozzles and allow for a more efficient operation of water recycling technologies.

Flow Restrictors and Conductivity Cells

Regardless of the rinsing systems employed, uncontrolled use of water negates the benefits of their design. As with countercurrent rinsing, if the flow of water is the same as that for a conventional single-tank setup, no improvement will be realized. Therefore, it is imperative that the flow of water be administered on an as-needed basis to maintain a final rinse concentration of 0.1% of the concentration of the immediately preceding process bath.

Two methods for controlling the use of water include flow restrictors and instrumentation such as conductivity meters. Conductivity meters are effective at measuring the level of contaminants in a rinsing solution, thereby triggering the opening of a fresh-water valve.

A flow restrictor is a device that can be installed at a rinse tank's water inlet. Its purpose is to control the flow of water at a predetermined rate. For this function, flow restrictors are reliable devices. However, because flow restrictors are typically set at specified flow rates, changes to part flows may not be taken into account. Conversely, when part flows increase, final rinses will not be as effective. As part flows decrease, unnecessary amounts of water will be used. Therefore, it is a prudent practice to cut off the flow of water if parts are not being processed.

Methods that employ switches, foot peddles, or timers to administer fresh water to rinse tanks are similar in use to flow restrictors. However, these methods are considered less reliable in maintaining final rinse characteristics because they trigger the discharge of predetermined volumes of water into the rinsing system based on the passage of a full rack of parts. Because these methods are based on the theoretical flow of parts, which, by experience, is known never to be constant, problems can arise, the worst of which is a rinse with high contaminant levels. High contaminant levels can lead to corrosion or other degradations in quality that may increase scrap and/or rework. Therefore, careful application of these methods is required.

Conductivity meters, on the other hand, can take into account variations in part flow and the drag-out experienced by differences in individual parts. These probes provide an indication of the level of contamination in a rinse tank by measuring conductivity: the higher the conductivity reading, the higher the level of contamination. With a sensor of this type interfaced with a transmitter set to trigger a solenoid valve at specified conductivity levels, water will be introduced into a rinse tank on an as-needed basis only. As the conductivity reading is reduced to acceptable levels, the same process control system stops the flow of fresh water. Conductivity meters are one of the most effective means for minimizing the use of rinsewater. However, conductivity meters also require maintenance. If not properly cleaned and calibrated at regular intervals, equipment can become useless. Figure 3.7 provides a diagram of a rinsing system that uses conductivity meters.

Rinse Bath Agitation and Contact Time

Solution agitation and contact time are two factors that come into play regardless of whether products are immersed in process baths or rinse tanks. Solution agitation and contact time are interdependent. The better the agitation, the less contact time required.

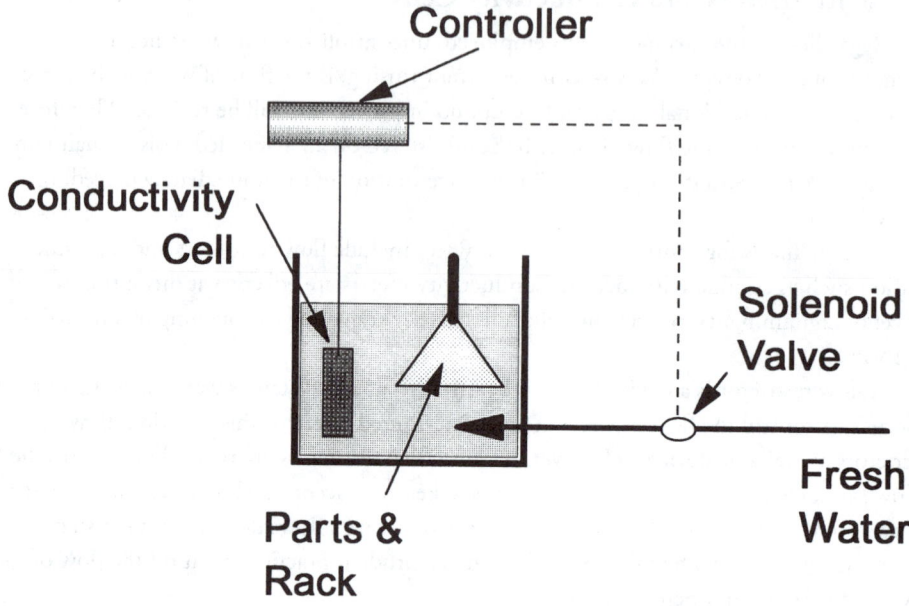

Figure 3.7   Conductivity cells.

Although the impact of these factors is apparent, it is not easily quantified. The longer a rack of parts remains in a rinsing bath, the more efficient the removal of contaminants will be. In addition, the degree of cleansing is only as effective as the quality of the rinse-water. Production rates can also have an effect. As production rates increase, contact time typically decreases. Therefore, operators must determine through trial and error or experimentation the length of time a rack of parts must remain in a rinsing solution to provide the highest degree of rinsing quality while not interfering with production requirements.

As previously mentioned, contact time may be reduced with improved agitation. Several of the common methods used are more effective than others. Air sparging, for example, is not considered an effective method for mixing regardless of whether the tank is used for rinsing or for blending process chemicals. In addition, air sparging creates tank losses by generating fine mists, which are the result of bubbles breaking on the solution surface. These airborne particles can also cause increased corrosion rates for surrounding equipment and facility fixtures, and pose significant health hazards because of the increased concentration of respirable particles. As a general rule, air agitation is not a preferred method for mixing.

Mixing achieved by solution recirculation or the use of a type of propeller system is generally the most effective technique. Recirculation of tank solutions is similar to the mixing process of a propeller, although the current and flow of the solution is slower and less turbulent. Therefore, propeller configurations yield the most efficient rinsing while minimizing contact time. In any event, the optimization of a rinsing system necessitates the balancing of production rate requirements, capital costs, and O&M costs.

An additional technique commonly employed in printed circuit board plating shops to provide an effective method for rinsing involves moving the work rather than the solution. This technique uses a set of rocker arms to gently move the rack of parts in a horizontal or vertical motion while it is immersed in the rinsing solution. In some cases, motion in a particular direction is preferred because of the part configurations. For example, when trying to rinse the plated holes of a printed circuit board, it is more effective to move the boards side to side rather than move the solution randomly in all directions.

SOLUTION RECOVERY AND RECYCLING

Until recent years, plating shops consisted of two distinct processing areas—the plating or chemical processing lines and the wastewater treatment facility. However, because of tightening regulatory requirements in virtually all media (solid waste, water, and air), the generation and disposal of waste products is no longer inexpensive and must be reduced. This is the essence of waste minimization, a requirement of RCRA. As discussed in Chapter 1, waste minimization programs are mandated for generators of hazardous wastes. In addition to RCRA, CWA has placed specific discharge limits on the metal finishing industry. A violation of these limits, which are typically overseen by local municipalities that are granted authorization to manage the industrial wastewater pretreatment programs, can result in fines or closure for a plating facility. In short, without a permit to discharge treated wastewater, these facilities cannot operate.

During the last decade, the metal finishing industry has made dedicated efforts to minimize waste and has increasingly looked to recovery and recycling technologies. Many companies began implementing these technologies as a cost of doing business to ensure that regulatory requirements could be more easily met. However, since this time it has become apparent that recovery and recycling are not costly. Rather, these approaches typically create more consistent and controlled processes while reducing costs. The incentives for developing new processes will become apparent as the following technologies are discussed and compared to examples of conventional treatment methods.

The technologies covered in this section are divided into three categories. The first category addresses the capturing of used or contaminated rinsewater and its subsequent purification for reuse. This approach generates concentrates that must be treated before disposal or reclaimed and used for regenerating process baths, or that must recapture the metals for subsequent sale outside of the plating shop. Examples of these technologies include ion exchange and electrolytic recovery techniques.

The second technology also purifies rinsewater for reuse, but simultaneously concentrates the drag-out constituents for reuse in the process baths. Examples of this technology include membrane and evaporative processes.

The last category addresses process bath management rather than rinsewater management. By maintaining good controls over the concentrates, tank dumps may be greatly reduced or eliminated. Although the volume of waste concentrate solution is typically much lower than that of spent rinsewaters, the potential for savings, increased process control, and efficiency is always present.

Ion Exchange

Ion exchange is a process in which dilute aqueous solutions containing metal salts can be purified through the use of porous organic beds. The beds consist of resin beads that may be cationic or anionic. Contaminated water passing through these beds is purified by the adsorption of the metal ions onto the resin surfaces. Once the bed becomes saturated, the metal ions can be recovered as a concentrate through chemical stripping using a concentrated acid or base, as appropriate.

Although used for many years as a method for purifying feed water, the ion exchange process has found its way into waste treatment. The ion exchange process as it relates to treating wastes includes the recirculation of rinsewater, recovery of heavy metals from rinsewater, regeneration of process bath solutions, and use as a final polishing step for treated wastewater.

This process, in contrast to those subsequently discussed, is ideally suited for removing low concentrations of metal salts from dilute rinsewaters under high-flow-rate conditions. Many techniques are designed for situations involving low flow rates in which high concentrations of recoverable materials is of primary interest. Most ion exchange systems consist of a prefilter (typically a gravel or carbon-gravel filter), two cation exchangers, two anionic exchangers, and various instrumentations. Figure 3.8 illustrates a typical ion exchange configuration for treating wastewater solutions.

Ion exchange uses a synthetic resin that contains active functional groups that can be substituted for ions with a greater affinity for the resin. The material, which swells in water and is insoluble, is a copolymer of resins such as styrene, acrylate, acrylamide, and reactions products with divinylbenzene. Divinylbenzene is the agent that produces the zeolitic structure that makes molecular filtration and the exchange of ions possible. Selection of a particular resin may be based on the metal or constituent desired for removal. Consider a wastewater stream containing 99% calcium ions and only 1% copper ions. If the solution is routed through a nonselective cation ion exchange column, the medium may become saturated quickly. However, if only the copper ions are required for removal, an ion "selective" resin can be employed that allows the calcium to pass. This greatly extends the time period between resin regeneration.

The first rinse after a process tank can contain nearly 90% of the contaminants resulting from drag-out. Routing the solution through ion exchange beds can effectively return clean rinsewater to the system. In most cases, the level of cleanliness is more than sufficient with traditional resins and contains low levels of pollutants. Note that water consumption from processes implementing ion exchange is mostly caused by evaporative or regenerative losses.

Ion exchange processes consist of two stages. The first stage occurs when units are in service and spent rinsewater is passed through the resin beds. Ions in the rinsewater that have a greater affinity for the resin are exchanged for similarly charged ions attached within the beds, typically hydrogen or alkali/alkaline earth metals. Because they are positively charged, cationic resins exchange metal ions. Anionic resins remove negatively charged species such as sulfates and nitrates. The second stage represents a regenerative phase.

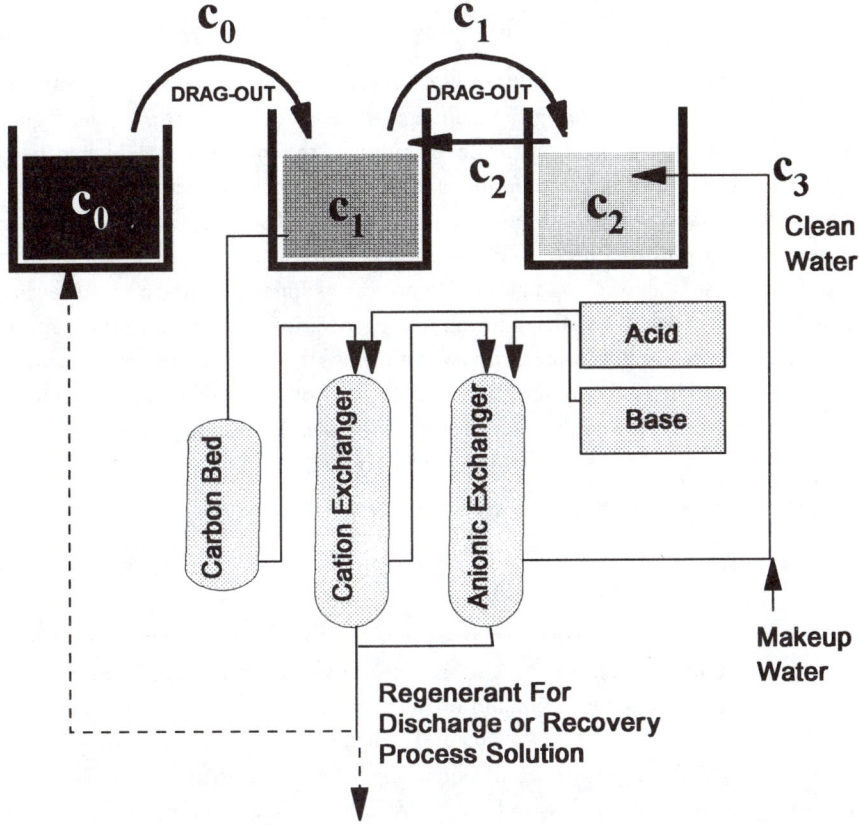

Figure 3.8 Ion exchange rinsewater recovery.

There are two primary types of cationic exchange resins: strong acid and weak acid. Strong-acid resins use sulfonic acid groups by which cations in the waste stream, typically metals from a process bath, are exchanged with the easily removed hydrogen atoms. The resulting effluent is a clean stream of water ready for reuse. As the site for the exchange of hydrogen for metal cations is depleted, the bed is considered saturated and must be regenerated. Regeneration with sulfuric or hydrochloric acid will reverse this reaction and yield an acidic waste stream of highly concentrated metal cations that may also be suitable for recycle or reuse. However, because sulfonic acid groups easily dissociate their hydrogen atoms, regeneration is somewhat more difficult to accomplish than when using weak-acid resins. The following reactions demonstrate the reversible process by which cationic exchange is accomplished. The reaction proceeds to the right as the cations from the waste stream are removed from solution. The reaction moves to the left during regeneration. The resin is depicted by R, which is the normal designation in chemistry for organic chains, and H denotes the easily exchangeable proton from the sulfonic functional group of the resin. The process is demonstrated for a waste stream containing copper (Cu) ions typical of many plating and printed circuit board manufacturing baths:

$$Cu^{2+} + 2RH \leftrightarrow R_2Cu + 2H^+ \tag{3.1}$$

Depending on the cation to be removed from solutions, strong-acid cation resins have a range of affinities for performing the exchange process. For the most part, the higher the oxidation state (charge) and the larger the radius of the metal ion, the higher the potential is for exchange. Table 3.2 demonstrates these tendencies (Freeman, 1989).

The performance of weak-acid resins is similar to that of strong-acid resins, although their functional groups are different. For weak-acid resins, carboxylic acid groups are employed. In this case, the functional groups do not easily donate hydrogen to the exchange process. This is somewhat beneficial during the regeneration stage because the resins reprotonate more efficiently. However, this benefit is offset by the fact that weak-acid resins do not function well at low pH levels. This is in contrast to the wide range of pH that may be effectively flowed through strong-acid resins. The affinity of weak-acid resins for various cations is as follows (Hartinger, 1994, and Kauczor, 1969):

$$H^+ > Cu^{2+} > Pb^{2+} > Fe^{2+} > Zn^{2+} > Ni^{2+} > Cd^{2+} >$$

$$Ca^{2+} > Mg^{2+} > NH_4^+ > K^+ > Na^+ \tag{3.2}$$

Anionic resins exhibit a similar exchange process to their cationic counterparts, although there is a difference in the functional groups used to aid in the exchange of anions. Strong-based anionic resins contain quaternary ammonium groups that easily exchange hydroxide ions (OH^-) over a wide pH range, removing acid and aiding in the dissociation of salts (NaCl) as shown in Equations 3.3 and 3.4. Regeneration is accomplished using a strong base such as sodium hydroxide (NaOH). As previously mentioned, with strong-acid cationic resins the regeneration process is not accomplished as easily as with weak-resin alternatives because of the ease with which hydroxide ions are given up in the process:

$$ROH + H_2SO_4 \leftrightarrow R_2SO_4 + H_2O \tag{3.3}$$

$$ROH + NaCl \leftrightarrow RCl + NaOH \tag{3.4}$$

The relative affinity for strong anionic-based resins for the aforementioned constituents are as follows (Hartinger, 1994, and Oehme, 1971):

Table 3.2   Strong-acid cationic exchange affinities.

Cationic oxidation state	Strong-acid cationic exchange preference
Monovalent	Ag>Cu>K>NH$_4$>Na>H
Divalent	Pb>Hg>Ca>Ni>Cd>Cu>Zn>Fe>Mg>Mn
Trivalent	Fe>Al

$$NO^{3-} > CrO_4^{2-} > PO_3^{3-} > oxalate > NO_2^- > Cl^- > formate >$$

$$citrate > tartrate > phenolate > F^- > acetate > HCO_3^- > HSiO_3^- > CN^- >$$

$$H_2BO_3^- > OH^- \qquad (3.5)$$

Weak-based anionic resins use tertiary ammonium groups that readily associate with hydroxide ions. This characteristic yields a resin bed that is easily regenerated; however, the characteristic limits the range of usefulness from mid-to-high pH solutions and cannot split salts as readily. The relative affinities for weak anionic-based resins are as follows (Hartinger, 1994, and Oehme, 1971):

$$OH^- > [Fe(CN)_6]^{4-(3-)} > [Cu(CN)_4]^{3-} > [Ni(CN)_4]^{2-} > anionic\ surfactants >$$

$$CrO_4^{2-} > SO_4^{2-} > HPO_4^{2-} > NO_3^- > NO_2^- > SCN^- > Cl^- > formate >$$

$$complex\ anions > citrate > tartrate > oxalate > F^- \qquad (3.6)$$

As resin beds become saturated with contaminants, they must be regenerated. A common part of this process is to backwash the system to remove particulate matter and resuspend the bed for efficiency. As the regenerate solution—typically sulfuric acid (H_2SO_4) for cationic columns or sodium hydroxide for anionic columns—is passed through the beds, the concentrated metal ions and other species are removed and replaced with the original ions. The resulting regenerate solution of interest will now contain a high metals content at low pH. The options are to batch-treat the concentrated solution (using precipitation methods, in most cases) or to employ electrolytic recovery techniques, which are covered in the following subsection. After the beds have been regenerated, they are rinsed to remove any remaining regenerate solution and placed into service again. This step can either be accomplished on site for the recovery of metals that are destined for reuse or sent off site to a licensed facility to streamline operations. Table 3.3 lists various types of ion exchangers and some of their most useful applications (Hartinger, 1986 and 1994).

Table 3.3   Ion-exchange applications.

Type of ion-exchange unit	Ion-exchange functional groups	Potential applications
Strong-acid resin, cationic exchange	$-SO_3H$	Rinsewater recirculation; purification of acidic process solutions
Weak-acid resin, cationic exchange	$-COOH$	Recovery of nonferrous metals
Strong-based resin, anionic exchange	$-N(CH_3OH$ or $-N[(CH_3)(C_2H_4OH)]OH$	Rinsewater recirculation; recovery of precious metals; separation of free acids
Weak-based resin, anionic exchange	Tertiary amino groups	Rinsewater recirculation; recovery of precious metals

The benefits of ion exchange include minimization of rinsewater usage and recovery of process solutions, coupled with the use of little space and operator maintenance. By conducting a thorough investigation of a wastewater-generating process, performing waste stream segregation where feasible, and selecting the best suited resin for a given application, ion exchange can be an extremely effective and proactive measure to increase recycled rinsewater quality, minimize the quantity of wastewater generated, and reduce the potential for permit excursions.

The primary drawbacks of ion exchange systems are the loss in throughput capacity and/or the failure to produce water of a specified quality. Capacity losses can occur because of the loss of resin within the system, thereby yielding fewer sights for ion exchange. Capacity losses can also occur because of inefficient processes in the regeneration stage in which captured species are not adequately removed. Mechanical damage of the resin caused by high pressure and temperature variation are also a concern. Pressure can squeeze the beads together, producing larger overall particles and creating a situation in which water passage is difficult. Lower output and increased regeneration times are the typical results. In addition, temperature swings in which repeated freezing and thawing occur can crack the resin beads, while high temperatures can commence regeneration because of the reversible nature of the chemical reactions taking place.

Failure to produce water at a specified level of quality can occur because of a number of issues. Organic compounds and excessive concentrations of iron, bacteria, and certain salts that readily precipitate are a few of the most common constituents that can foul ion exchange resin beds. These issues should be considered when determining the application and location for ion exchange systems. For widely mixed wastewater streams, fouling will be more prevalent. For closed-loop applications on rinsing stations or individual process baths, wastewater parameters can be more tightly scrutinized (less variation will result in fewer O&M problems). Minimizing the amount of compounds that are known to cause fouling and ensuring that regeneration procedures are conducted properly can provide a smooth operation with low O&M costs.

Electrolytic Recovery

Electrolytic recovery techniques are important elements in many recycling programs. The primary types of electrolytic recovery processes include cathodic, anodic, and electrodialysis. These technologies do not require the addition of chemical reagents to perform recovery. Instead, they achieve results through the transfer of electrons. Electrolytic recovery processes are electrochemical reactions that consist of oxidation and reduction events driven by the application of electrical potential. The cathode of a cell is the location where reduction occurs, yielding metal in its elemental form. This is commonly referred to as *cathodic electrolytic recovery* or *electrowinning*. Gases liberated during this process are dependent on the chemical solutions involved, and include nitrogen or oxygen from the anode and hydrogen from the cathode. Anodic electrolytic recovery can also be performed. This approach is particularly desirable for the regeneration of spent oxidants such as cyanide solutions, chromium electroplating, or anodizing solutions.

Cathodic electrolytic recovery is one of the most commonly used electrolytic techniques for the removal of metals ions from concentrated solutions resulting from ion exchange regenerates, static rinses, and spent process baths, as demonstrated in Figure 3.9.

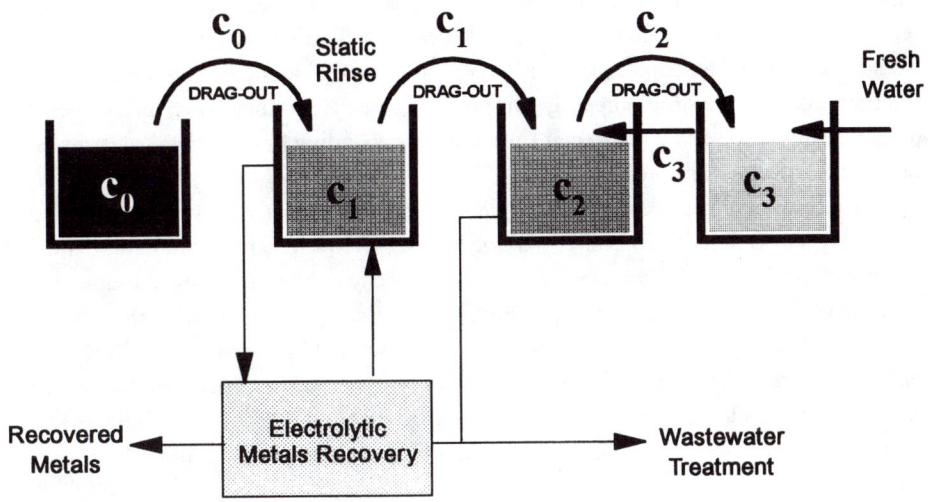

Figure 3.9   Electrolytic recovery.

This technique removes plateable, concentrated metal ions from solutions by plating the ions onto an electrode. Some of the most common metals recovered for monetary reasons include gold, silver (Ag), and copper, although cadmium, zinc (Zn), and nickel are also recovered. After sufficient quantities of metals are accumulated on the electrode, metals may be removed and sold or recycled into the various finishing processes from which they originated. Consider the following standard reduction potentials for metals dissolved in an acidic sulfate solution:

$$Cu^{2+} + 2e^- \leftrightarrow Cu \; E^0 = +0.34 \text{ V} \tag{3.7}$$

$$Ag^+ + e^- \leftrightarrow Ag \; E^0 = +0.80 \text{ V} \tag{3.8}$$

For a set of reactions to occur spontaneously, the net electromotive force (E^0) must be positive. In a galvanic cell (that is, one in which voltage is generated rather than applied), these reactions will occur spontaneously in the direction written, provided hydrogen gas is bubbled across the anode given by the following reaction:

$$H_2(g) \leftrightarrow 2H^+ + 2e^- \; E^0 = +0.00 \text{ V} \tag{3.9}$$

The overall cell potential would be +0.34 V for the plating of copper and +0.80 V for silver. If, however, these metals were replaced by a solution derived by immersing a strip of zinc in a sulfuric acid solution, the reaction and reduction potential would be

$$Zn^{2+} + 2e^- \leftrightarrow Zn \; E^0 = -0.76 \text{ V} \tag{3.10}$$

In this case, the opposite would occur. The zinc reaction would occur to the left (dissolve), and hydrogen gas would evolve to yield a cell potential of +0.76 V.

In electrolytic cells, an external current source is used to drive the reactions in a direction that would not occur spontaneously. This allows high-purity metals to be plated on various conductive substrates that are functioning as the cathode. In cathodic electrolytic recovery, the reactions involving copper and silver in Equations 3.7 and 3.8, respectively, would occur in the direction as written to deposit elemental metal on the cathode. Other reactions taking place simultaneously can produce gaseous products, such as nitrogen or oxygen at the anode and hydrogen at the cathode, depending on the chemical composition of the solutions involved.

The rerouting of concentrated metal ion solutions into beneficial uses can significantly reduce the generation of wastewater treatment sludge. However, it should be noted that because electrolytic recovery techniques are not effective at low metal concentrations, effluent from the process is generally not suitable for discharge without prior treatment.

Anodic electrolytic recovery techniques are also useful. The overall process is essentially no different than that for cathodic recovery except that the desired processes are occurring at the anode. Common examples of anodic electrolytic recovery techniques include the regeneration of spent oxidants containing cyanide and chromium plating and etch solutions.

Another type of electrolytic process is electrodialysis. Electrodialysis is also an electrochemical approach to waste minimization. This process, in contrast to the aforementioned electrolytic techniques, handles streams with relatively low ionic concentrations and serves as a recovery tool to concentrate dilute streams for reuse. Methods for concentrating a solution are achieved by applying a direct current across a series of alternating anodic and cathodic exchange membranes. This act removes various dissolved metal species and other ionic constituents. The process, as illustrated in Figure 3.10, is not only electrolytic in nature, but represents a membrane technology as well.

As a wastewater solution is fed through the device, anions will pass through the anionic membranes and an identical action will occur for the cations. A direct current is applied across the membrane and provides the mechanism to drive the electrochemical process. Ions migrating toward a membrane that is selective for ions of opposite charge are thus redirected for concentration in the appropriate channel.

An example of effective electrodialysis is the regeneration of hexavalent chromium plating baths. Routing a solution of this type through semipermeable membranes removes unwanted cations such as iron and produces a relatively pure solution of chromate ions in the anodic compartment. The benefits of this process are that it consumes little energy and recovers only ionic materials. Therefore, unwanted impurities may be removed, leaving a regenerated solution that may either be returned to a chemical process bath or piped to additional recovery processes. One drawback of the process, however, is that membranes are used. By nature, membranes require careful process control and, if subjected to contaminant residues, high maintenance.

Membrane Processes

Membrane separation technologies can be divided into the following categories: microfiltration, ultrafiltration, nanofiltration or hyperfiltration, and reverse osmosis. Each process is designed to provide a physical mechanism of action to separate contaminants from an aqueous solution. The degree of effectiveness at removing dissolved and suspended solids is dependent on the size of the opening (pore) within each type of mem-

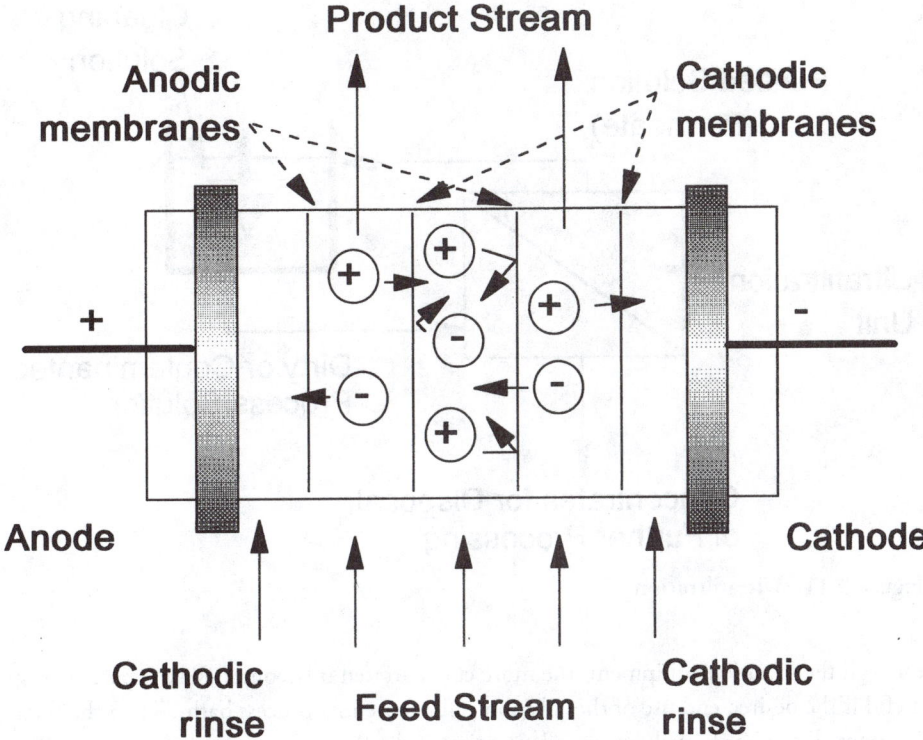

Figure 3.10   Electrodialysis.

brane and the pressure applied across its surface. Liquid that is able to pass through the membrane is referred to as *permeate*. In many situations, membrane filtration is used to remove contaminants and provide a relatively pure water stream. However, it may also serve to concentrate a dilute waste steam. For example, the more times a contaminated solution is passed through the membrane (that is, recirculated), the more concentrated it becomes. A benefit of concentrated solutions is that they may be reused in process baths.

Mircofiltration and ultrafiltration differ only in their ability to separate particles of different sizes. Both operate under relatively low pressures and are primarily used to separate organic constituents, complex metals, and finely dispersed solids from water. The molecular weights of compounds typically removed using ultrafiltration range from 500 to 1 000 000 (Freeman, 1989). Microfiltration is not as effective, but provides some overlap in this area as well. A typical configuration for a membrane filtration device is illustrated in Figure 3.11. Other configurations employing a plate and frame layout exist, but their mechanism of action is no different.

The membrane is mounted parallel to the flow of the contaminated feed stream. The water containing constituents of a size smaller than the pore openings of the membrane pass through the material and flow out as permeate. The portion of the feed stream that is held back is highly concentrated with contaminants and flows directly out of the device. As previously mentioned, the more times this concentrated stream is rerouted

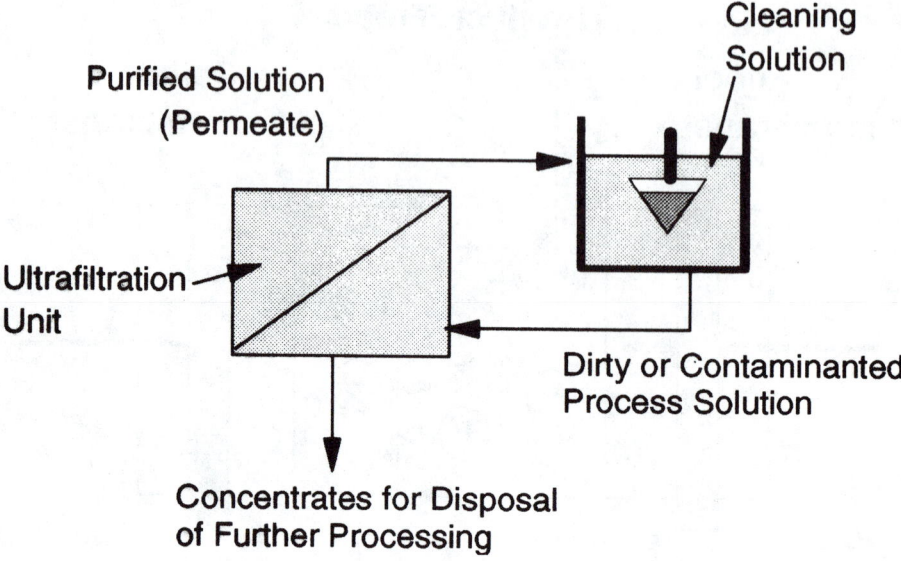

Figure 3.11 Ultrafiltration.

through the filtration equipment, the more concentrated it becomes. This is particularly useful if the desired end use of the solution is for reuse in a process bath. When the filter becomes clogged or fouled—as it will over time—the flow of water may be reversed. This action flushes a large portion of the contaminants from the membrane surface and extends its useful life. An excellent use for ultrafiltration is to regenerate degreasing baths by removing oil and grease buildup and to remove metal complexes, which is sometimes difficult when attempted in traditional wastewater treatment processes.

Nanofiltration or hyperfiltration represents an intermediate technology between ultrafiltration and reverse osmosis. This level of membrane separation technology is able to remove a portion of the dissolved solids in solution in addition to its organic and suspended constituents. The contaminants ordinarily removed from exhibit molecular weights in the range of 100 to 500 (Freeman, 1989). The design and operation of the system is no different than that previously covered.

The membrane materials most commonly used consist of cellulose acetate, polyamides, polyimides, and thin-film composites. The selection of the type of material is contingent on the conditions of the filtration operation. These conditions include pH, temperature, and the presence of aggressive constituents like chlorine and bacteria. In chemically aggressive environments, inorganic membrane materials, such as ceramics, may prove to be more beneficial because of their inherent ability to tolerate harsh conditions and periodic cleaning by concentrated acid or base solutions. Membrane configurations can consist of tubular, spiral-wound, hollow-fiber, and plate and frame modules.

Reverse osmosis is used to remove inorganic ions from an aqueous solution. Osmosis is a process that employs a semipermeable membrane to separate two solutions with different dissolved solids content and allows only water to flow through the membrane in a direction that will produce equal concentrations of salts on both sides. Reverse osmosis, in contrast, is run in the opposite direction driven by high-pressure gradients and produces water of a high degree of purity. An illustration of the difference between the two processes is provided in Figure 3.12.

Most reverse osmosis systems are used to provide process rinse or chemical bath make-up water where low salt content is required to ensure high quality control standards. Reverse osmosis is also used in the electroplating industry to recover valuable metals from rinse tanks that have become contaminated with plating solutions because of the drag-out associated with the processing of products and rack movement. A typical configuration for the use of reverse osmosis is shown in Figure 3.13.

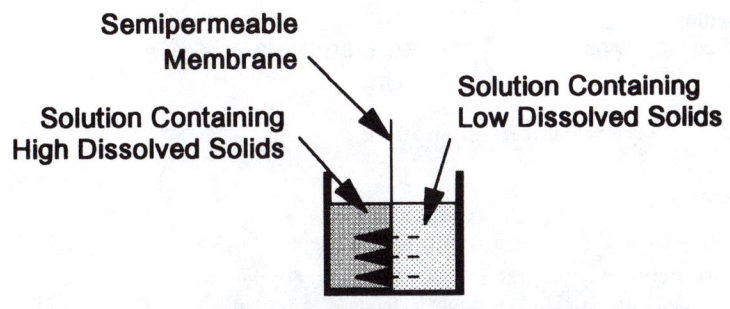

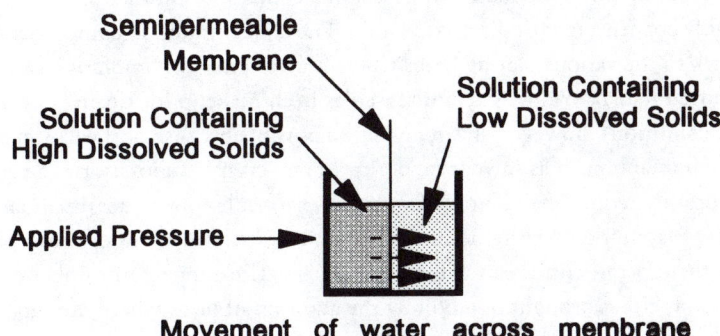

Figure 3.12   Osmosis and reverse osmosis.

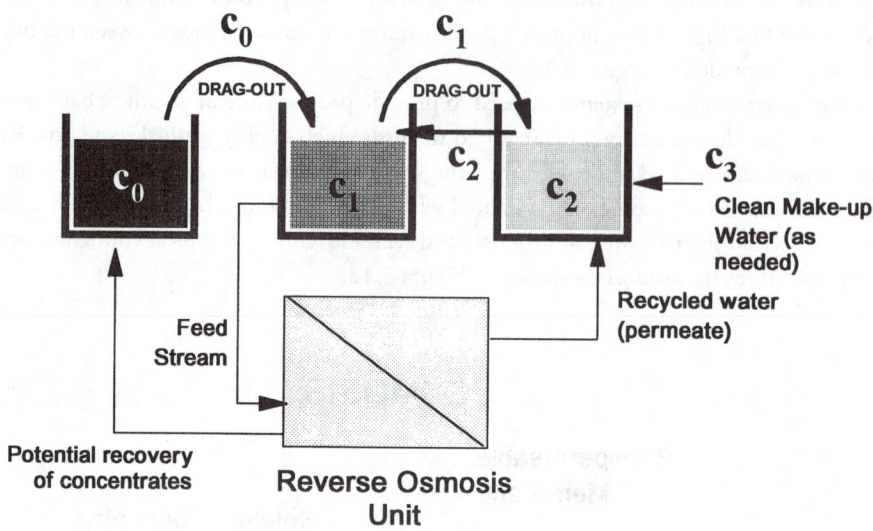

Figure 3.13 Reverse osmosis applications.

One drawback of reusing concentrates in chemical process baths is that although desired constituents may be present such as nickel, copper, and other metals, impurities that were present in the reverse osmosis feed stream will remain. The buildup of these impurities with the passage of time is a factor that should be considered before routing any regenerated solution to a production process bath.

Evaporation

Evaporation is a physical separation process that has been used successfully for years to aid in the recovery of plating bath chemicals. The principle of evaporation is simple. As water is evaporated from contaminated rinsewater, the solution that remains becomes more highly concentrated in dissolved solids. The concentrates, in some cases, may be recycled back to the various plating baths where they originated. Similarly, water that has been removed may be condensed and reused as fresh makeup for rinsing operations. Under some conditions, however, this may not be possible because of the existence of organic contaminants such as solvents and oils. In any event, evaporation is a process that can significantly reduce the volume of wastewater for subsequent treatment and disposal.

Primary evaporation techniques include one that operates at atmospheric pressure and the other that operates under a vacuum. For units performing at atmospheric pressure, the solution is either brought to a boil by the addition of heat, where the vapor pressure of water becomes equal to atmospheric pressure, or the solution is held at a temperature less than boiling, where evaporation is still achieved but the rate of vaporization is slowed. The latter process achieves evaporation with water while at a vapor pressure less than that of atmospheric pressure, thus vaporizing the water molecules at a slower rate. The vacuum process of evaporation is similar, except that boiling points are lowered because the heat of vaporization for water under vacuum or low pressures is less than that for water at atmospheric pressure. Although less energy consumption is required to boil water under a vacuum than under atmospheric pressure, evaporation units operating at atmospheric pressure typically have lower capital costs.

A common application for both types of evaporation techniques is the use of chromium electroplating process bath regeneration. A typical layout of this application is illustrated in Figure 3.14. The need for an additional purification system on the initial rinse tank should be noted. This is strategically located to minimize the amount of impurities in the metal drag-out that, if not removed via a method such as cation exchange, would have to be rerouted back to the process bath in an ever-increasing concentration. Without this purification step in the chrome plating process, a significant degradation in product quality would occur because of evaporation.

It is important to remember that evaporation techniques do not remove undesirable impurities from the chemical concentrates left behind. Therefore, as with reverse osmosis, the drawback of recycling the concentrates from this process must be evaluated carefully before implementation. In addition, because the evaporation process can consume relatively large amounts of energy, it is a common practice to use this method in conjunction with other rinsewater reduction technologies.

Bath Composition

This section is presented as a reminder of the importance of process control. The more consistent a process is, the higher the quality of the products produced and the

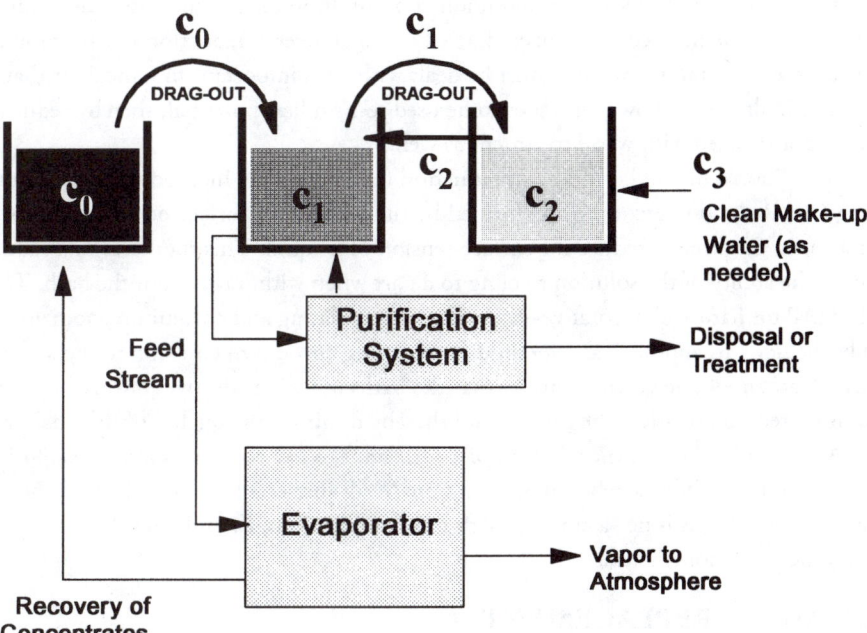

Figure 3.14 Evaporation for chrome plating bath recovery.

longer the life of the process baths involved. The section Statistical Process Control (SPC), in Chapter 2, introduced the concepts of measuring and predicting process variability and quality characteristics. By understanding process chemistries and determining optimum operating conditions, tolerances may be established that, when followed, can be successful. Parameters that are amenable for such analysis and can significantly affect the quality and consistency of the process baths associated with the metal finishing industry include pH, temperature, bath concentrations, and chemical additives. Establishing ranges that define optimum conditions is a good start, but not enough. Regular monitoring is essential to ensure that proper control is achieved. If anomalies are detected, corrective action must be initiated to ensure long-term effectiveness. Tracking these parameters using SPC techniques also allows determinations to be made on the capability of the process to consistently achieve results within specification limits.

For many plating solutions, pH and temperature must be maintained within a relatively narrow window. By installing instrumentation and control systems mentioned in the Data Acquisition Systems section in Chapter 2, acid or base may be added to plating baths automatically, producing higher quality solutions and improved part quality. Temperature control may also be achieved through thermostatically controlled heating elements. The only setback to this approach is that many shop operators may not believe that funding for such automation is warranted. In cases where proper O&M procedures are employed, implementation of such technologies provides a quick return on investment.

Controlling plating bath concentrations is also important. Proper maintenance at levels that can minimize drag-out without sacrificing plating quality is ideal. Although it may take some work to locate optimum ion concentration ranges, the effort can result in reductions in chemical costs, reduced drag-out, and an overall reduction in the amount of wastewater contaminants that must be dealt with. It is important to remember that reductions in drag-out allow rinsewater to be used to its fullest potential, thereby reducing the costs associated with waste treatment as well.

An additional method of drag-out reduction (other than the methods discussed in the Drag-Out Reduction section) is performed by the addition of surface-active agents, or surfactants. Surfactants reduce the surface tension of a liquid. This translates to a reduction in the ability of the solution to cling to a part when withdrawn from the bath. The NESHAP for hard and decorative chromium electroplating and chromium anodizing tanks include this approach as a method for reducing chromium emission to the atmosphere. Care should be taken to ensure that the baths in which additives of this nature are administered do not affect the process and thereby diminish the quality of the final product. An example of this is the addition of a surfactant to a chromium anodizing solution that is used before adhesive bonding operations. Will surfactant residues degrade the bond? Although this is not a concern with most applications, careful consideration should be given nonetheless.

CHEMICAL REPLACEMENT

Process control and good management practices are two important elements in the minimization of waste. Another approach, however, is to find chemical substitutes for those constituents that are the primary drivers in waste generation. Two of the most common and highly regulated substances in the metal finishing industry include chromium and cyanide related compounds. Because of the toxicity of chromium and cyanide related compounds, most wastewater pretreatment permits contain specific effluent limits for these substances, the values of which vary between regulated communities. If not adequately controlled, chromium and cyanide discharge products can cause disruptions in the biological treatment processes conducted at POTWs. Hence, the replacement of these compounds should be given priority with the introduction of process controls to minimize these wastes as a second alternative.

The search for chemical alternatives to hazardous constituents is not limited to chromium and cyanide, although considerable success has been achieved in this area. Research is currently being conducted by governmental bodies, universities, and private businesses to provide a path of continuous improvement in this area.

Chromium

As stated earlier, one of the most highly regulated metal finishing compounds is chromium. The most predominant of these is hexavalent chromium, which is considered a carcinogen. Hence, it is not surprising that air emissions are controlled, wastewater pretreatment discharges are permitted, and solid wastes must be classified as hazardous based on their presence.

Chromium air emissions must be controlled because of their classification as HAPs. An area of CAA that regulates chromium is the NESHAP for hard and decorative chromium electroplating and chromium anodizing tanks that stipulates emission limitations and process control techniques. The mechanism that creates airborne particles is the generation of gases within a solution. When bubbles burst, mist or suspended water particles can disperse the chromium constituents. Therefore, environmental contamination and worker health are considerable concerns.

Issues concerning wastewater treatment are also important. The CWA administered by many local governments places strict discharge limits on this compound via a permitting process. When discharged to an industrial wastewater pretreatment facility, high levels of hexavalent chromium must be reduced to trivalent chromium, precipitated, settled into a sludge, and dewatered before final disposal. Because of the presence of chromium, this sludge in most cases may also fall under the category of hazardous wastes defined in RCRA, which also places requirements on chromium-containing wastes.

For these reasons, eliminating chromium from metal finishing operations is desired. An alternative to the use of chromium that has successfully been implemented in metal finishing operations has been the replacement of chromic acid anodize with phosphoric acid anodize. However, this approach is not amenable to all metal finishing operations because it is dependent on the subsequent processes employed and the end use of the product. Although phosphoric acid anodize has been found to be acceptable as a preparation for structural adhesive bonding, it is generally inadequate for prepaint corrosion resistance. Where feasible, however, major benefits may be realized by the simplification of the wastewater treatment process and the significant reduction in constituents that are regulated. Phosphoric acid concentrates and associated rinsewater contain little heavy metal contamination. Therefore, a pH adjustment may be all that is required for treatment. Cost savings are typically incurred through reduced chemical usage and labor during the wastewater treatment processes, reduced volume and disposal, and the reduction in potential for permit excursions.

Switching from hexavalent chromium to trivalent chromium should be investigated in situations where the elimination of the element is not possible. A process area where this may be feasible is decorative chrome plating. Benefits of switching from hexavalent chromium to trivalent chromium include a reduction in the generation of hydrogen gas and a wastewater treatment process that can eliminate the chemical reduction step for chromium. Other process improvements may also be realized because bath concentrations are typically lower for trivalent chromium, which results in lower drag-out rates and, hence, a lower volume of spent rinsewater and sludge.

Cyanide

Cyanide is a concern from both a worker health and safety standpoint and an environmental standpoint. Primary environmental issues are regulated by CWA and RCRA in much the same manner as they are for chromium. This compound is used in a range of plating baths such as cadmium, copper, and other metals. Because of the complexibility of the cyanide anion, a high tolerance for variation in bath constituents and contaminants exists.

Recent developments in the area of noncyanide alternatives to traditional copper plating have been implemented in numerous shops. An example of such an application is in the printed circuit board industry, in which alkaline pyrophosphate baths are used. Another option is to use fluoboric acid as an ingredient. However, although the resulting plating process is efficient, it is also more hazardous and difficult to treat.

Zinc baths are also amenable to cyanide substitution using sodium hydroxide as the primary constituent. This modification can result in equivalent product quality while reducing costs. Other options that also provide an excellent finished product include the use of acid chloride baths that exhibit higher cathode efficiencies while using lower voltages.

Many commercially available process bath modifications consist of proprietary solutions. Some have been fully developed and provide cost-effective alternatives. Other process bath modifications have not been sufficiently debugged and may be useful in specific applications only. Nonetheless, investigations are warranted because regulatory controls are likely to become more stringent.

Metal Stripping

Metal stripping is typically accomplished by the immersion of parts into various chemically active solutions or by electrolytic techniques. The removal of metals involves a process that is similar to that of plating (that is, the deposition of metals). Although it might be assumed that strong acids would serve this task well, acids can also etch or damage valuable substrates. For this reason, a large number of effective chemical solutions used for metal stripping involve alkaline solutions.

One of the most desirable characteristics of a metal stripping process is the ability to dissolve coatings in a timely fashion while avoiding damage to the underlying substrate. It is also important to maximize the amount of time the dissolved metals may be held in a solution. This frequently involves the use of complexing agents, although their benefits should be weighed carefully when considering the effects that may be incurred during wastewater treatment processes.

The most popular metal stripping baths contain oxidizing agents. This approach is a chemical process in which the most predominant solutions integrate nitrates, nitric acid, chromic acid, and peroxides. Acids are generally able to maintain high metal concentrations without the use of various additives. Alkaline-based stripping solutions, however, must use complexing agents to keep metals dissolved. The most common complexing agents include sodium cyanide, ammonium salts, and various amine-based compounds such as ethylenediamine, diethylenetriamine, and others. A significant problem with alkaline strippers is that the best performers contain cyanide, which is not a desirable constituent. In the case of copper or nickel removal, however, ammonium ions or amines may be employed to provide a complexing environment. Chlorite anions and persulfates may serve this purpose as well. Table 3.4 demonstrates a range of noncyanide options for use in metal-stripping baths (Dammer, 1980; Hammel, 1982; and Hartinger, 1994).

Complexing agents are useful for keeping metals in a dissolved state when alkaline conditions are necessary. However, this usefulness becomes a problem once the waste solution is transferred to a wastewater treatment facility for subsequent removal of the metals before discharge. Complexing agents have caused many permit violations. Therefore, these solutions should be segregated for individual treatment. This will allow other operations to continue with few or no interruptions.

Table 3.4   Noncyanide alternatives for metal stripping.

Chemical-stripping bath constituents	Metals coatings removed
Amine (cyanide), nitroaromatics	Cd, Cu, Ni, Zn
Ammonium nitrate	Cd, Zn
Caustic soda	Zn
Caustic soda, nitroaromatics	Pd-Sn, Sn, Zn
Dilute hydrochloric acid	Cr, Zn
Dilute sulfuric acid	Zn
Chromic-sulfuric acid or nitric acid	Cu
Sodium chlorite, ammonia	Cu
Ammonium persulfate	Cu, brass

Solvent Cleaning And Degreasing

Cleaning and degreasing operations are an important aspect of many industrial manufacturing processes. Cleaning and degreasing is a process frequently conducted as a surface preparation for operations such as plating, painting, and bonding. The main objective of the process is to remove surface contaminants such as particulates, organic or inorganic films, and oxidations layers. Examples of particulates include dust, dirt, and metallic or plastic machining residues. Films can consist of oil, grease, hydraulic fluids, paint, adhesives, and inks. Oxidation layers can result from the exposure of metallic substrates to ambient conditions (as is the case of ferric alloys and the appearance of rust) or anodizing processes frequently conducted for corrosion protection for aluminum and other metal alloys.

There are as many substrates to be cleaned as there are contaminants to be removed. Metals, plastics, glass, and ceramics exhibit different physical properties and respond to cleaning in different ways. Therefore, it is highly unlikely that one cleaning technology, compound, or solvent will be useful in all applications. In addition, what is considered a "clean" state for one type of process may be considered dirty for another. This can be depicted by considering the difference in a clean surface for paint adhesion versus that required for structural bonding in a high-performance aircraft. Inadequate surface preparation in the aircraft, for example, may result in a failure during flight.

A solvent, by traditional chemistry definition, is the liquid portion of a solution that involves a liquid and a gas or solid. The gas or solid portion is referred to as the *solute*. Similarly, in terms of cleaning and industrial applications, a solvent is the main liquid constituent, and the contaminants removed from a surface are the solute. Solvents are commonly thought of as nonaqueous compounds and include organic liquids such as methyl ethyl ketone, methanol, and acetone, or chlorinated solutions such as methylene chloride, perchloroethylene, and trichloroethylene. Cleaners, in contrast, are often aqueous or alkaline in nature.

The mechanisms of action involved in solvent cleaning and degreasing include chemical and mechanical methods. The chemical activity of a solvent is that which dissolves, chemically changes, or removes a contaminant. Mechanical aspects involve interactions between a solvent and the surface of a part and include physical processes such as spraying, agitating, air sparging, and ultrasonic methods.

Some of the main solvent cleaning approaches include surface wiping, cold cleaning, and vapor degreasing. Surface wiping, the process of which is self-explanatory, is commonly conducted before painting, cleaning small tools and equipment, and making electrical connections. Cold solvent cleaning is conducted in tanks held at room temperature. The process involves the immersion and agitation of parts in a bath of solvent. As time passes, the solvent achieves a high concentration of contaminants. When this occurs, the solvent loses its ability to effectively clean and must be discarded as spent material. This operation is common and can be performed on small parts in wash bins or on a much larger scale, depending on the size of the cleaning tank.

Vapor degreasing is also performed in tanks, but at elevated temperatures. The temperature corresponds to the solvent's boiling point to create vapor. The main difference between vapor degreasing and cold solvent cleaning is that parts are suspended above the solution with vapor degreasing rather than immersed in it. The vapors perform a cleaning action by condensing on the surface of a part, dissolving the contaminants, and subsequently draining off. As the part approaches the temperature of the solvent, vapor no longer condenses on its surface and the cleaning process is complete. Cleaning solutions for this process also become contaminated. However, because the solvent vapors perform the cleaning action, contaminants are left behind, thus extending the life of the solvent vapors. Solvent emissions are controlled to a certain extent by cooling jackets surrounding the top of the degreaser. Vapors reaching this zone are condensed and returned to the tank. This process is used before a number of operations such as plating, anodizing, etching, adhesive bonding, machining, and painting.

In the past, chlorinated solvents were the cleaning materials of choice (particularly for the removal of organics) for a number of industries in the aerospace, automotive, machining, metal finishing, and electronics fields. The most common chlorinated compounds used in the past are listed in Table 3.5.

One of the desirable characteristics of these materials is that they are nonflammable. This is beneficial from a worker safety standpoint. However, the determination that these compounds cause destruction of stratospheric ozone has resulted in most of them being highly regulated in the CAA Amendments of 1990. As a result, most of these compounds are no longer allowed to be manufactured or used in industrial processes.

A brief history of the regulations developed for these compounds is provided. Figure 3.15 illustrates the uses and consumption amounts of chlorofluorocarbons (CFCs) in the U.S. in the late 1980s.

In assessing the risk of worldwide ozone depletion and the need for regulatory action, U.S. EPA took into consideration the actions of other nations while establishing solutions within the U.S.

In 1981, the United Nations Environment Program began negotiations to reduce the threat to the ozone layer. In response, the Vienna Convention for the Protection of the

Table 3.5   Chlorinated cleaning compounds and their traditional applications.

Chlorinated compound	Applications
Chlorofluorocarbon-113	Extremely versatile vapor degreasing solvent
Trichloroethylene	Efficiently removes varnishes, paint films, resins, and buffing compounds
Perchloroethylene	Good cleaner for high-melt waxes, wet parts (caused by limited hydrolysis), and light-gauge metals
111,Trichloroethane	Ideally suited for printed circuit boards and other electronic components
Methylene chloride	Applicable when an aggressive solvent is required or with substrates that are temperature sensitive

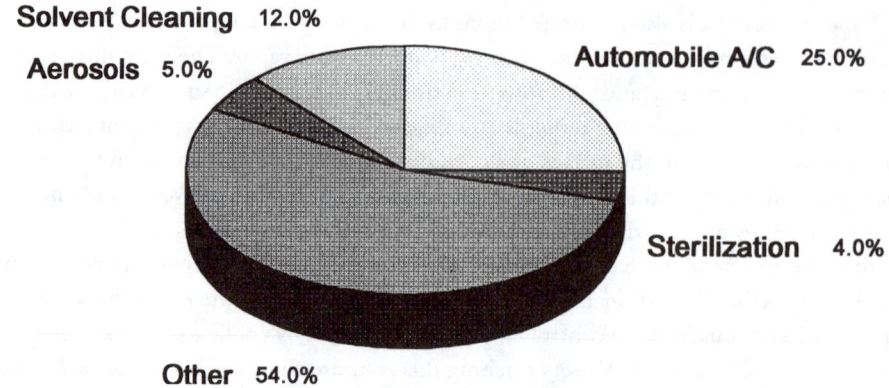

Figure 3.15   Use of chlorofluorocarbons in the U.S. (A/C = air conditioning).

Ozone Layer occurred in March 1985 and provided a framework for international cooperation in research, monitoring, and information exchange. These two efforts were the
first steps toward addressing the global ozone depletion problem.

However, in 1987, U.S. EPA agreed with the findings of the 1986 World Meteoro-
logical Organization that no significant change had occurred in global estimates of total
column ozone (that is, the amount of ozone from the earth's surface through the stratosphere in any given place). Although the existence of the "Antarctic hole" was accepted,
questions of whether the depletion was caused by manufactured chlorine or by natural
causes such as solar cycles or volcanic activity were raised.

Following an 18-month review involving more than 100 of the world's leading atmospheric scientists, the executive summary of the Ozone Trends Panel report was released
on March 15, 1988. In short, the final analysis provided "undisputed observational evidence" of increased atmospheric levels of CFCs, carbon dioxide, methane, nitrogen oxides, and halons because of human activities. The review showed an ozone depletion of
1.7 to 3.0% between 1969 and 1986 at latitudes between 30 deg and 64 deg in the
northern hemisphere. In addition, the report indicated an apparent ozone decrease of 5%
or more since 1979 at all latitudes beyond 60 deg south throughout the year.

By January 1989, the Montreal Protocol on Substances that Deplete the Ozone Layer,
which was established in September 1987, had participation from at least 68 nations.
The Montreal protocol requires nations who join to restrict their production and consumption of CFC-11, -12, -113, -114, and -115, and Halons 1211, 1301, and 2402 in
bulk form. The Montreal protocol does not place limits on the controlled substances;
rather, it places the chemicals in groups and limits the total ozone-depleting potential of
each group. As a result, the mix of controlled substances a nation produces and consumes
may change, as long as the total ozone-depleting potential of the mix does not exceed
specified limits. More than 75 nations representing approximately 90% of the world's
production capacity have signed the Montreal protocol (U.S. EPA, 1991).

The latest legislation to address ozone depletion is the CAA Amendments of 1990.
The 1990 amendments specifically address the significance of protecting stratospheric
ozone by establishing a "Stratospheric Ozone Protection Program." Title VI of the CAA
Amendments of 1990 includes requirements for ozone-depleting compounds that are

somewhat more stringent than those of the Montreal protocol. Title VI attempts to address four major areas of ozone depletion:

- Acceleration of the phase-out of CFCs and methyl chloroform;
- Control and ultimate elimination of the production and use of hydrochlorofluorocarbons (HCFCs), which are a family of chemicals that are serving as transition chemicals to replace some of the uses of CFCs;
- Elimination of ozone-destroying compound emissions; and
- Implementation of effective trade sanctions.

The Title VI phase-out requirements for chemicals of concern are shown in Table 3.6. The associated regulations and emission reduction programs are summarized in Table 3.7.

Also found under the CAA Amendments of 1990 is the requirement to develop NESHAPs, which are compound and industry specific. Details relating to this effort were discussed at the beginning of this chapter. Of particular interest is the NESHAP for halogenated solvent degreasing. This NESHAP is probably the most significant regulation driving the search for new and less hazardous solvent cleaning and degreasing alternatives. The purpose of this and other NESHAPs is to provide " . . . the maximum degree of reduction in emissions of the HAP . . . . that the Administrator (EPA), taking into consideration the cost of achieving such emission reductions, and any nonair quality health and environmental impacts and energy requirements, determines is achievable for new or existing sources in the category or subcategory to which such emission standards applies . . ." (CAA Amendments of 1990 Section 112[d][2]).

For halogenated solvent cleaning, equipment must meet specific design criteria and be operated in a fashion designated by the U.S. EPA to minimize emissions during the entry, cleaning, and exit of parts from the equipment. There are also a number of requirements for recordkeeping, training of employees, and reporting. These regulations apply to batch vapor, in-line vapor, in-line cold, and batch cold solvent cleaning machines that use solvents containing the following compounds, or any combination of these compounds in a total concentration greater than 5% by weight, as cleaning and/or drying agents:

- Methylene chloride (Chemical Abstract Number [CAS] No. 79-09-2),
- Perchloroethylene (CAS No. 127-18-4),
- Trichloroethylene (CAS No. 79-01-6),
- 1,1,1-Trichloroethane (CAS No. 71-55-6),
- Carbon tetrachloride (CAS No. 56-23-5), and
- Chloroform (CAS No. 67-66-3).

Another NESHAP that also controls a portion of the solvent cleaning business is that for aerospace manufacturing and rework facilities. The applicable section of this regulation is 40 CFR 63.744, Standards: Cleaning Operations. Requirements for the following procedures are identified in these standards: housekeeping measures, hand-wipe cleaning, spray gun cleaning, and flush cleaning.

Table 3.6   Title VI — chemicals of concern.

	Class I	Class II
	Chlorofluorocarbons	Hydrochlorofluorocarbons
	−11	−21
Group I	−12	−22
	−113	−31
	−114	−121
	−115	−122
		−123
	Halons	−124
		−131
		−132
Group II	−1211	−133
	−1301	−141
	−2402	−142
		−221
		−222
	Chlorofluorocarbons	−223
		−224
	−13 −211	−225
	−111 −212	−226
Group III	−112 −213	−231
	−214	−232
	−215	−234
	−216	−235
	−217	−241
		−242
		−243
Group IV	Carbon	−244
	tetrachloride	−251
		−252
		−253
		−261
		−262
		−271
Group V	Methyl chloroform	

In addition to recordkeeping and reporting, specifications are defined for cleaning equipment and raw materials. Under the category of hand-wipe cleaning, there are several options for selecting a solvent. The most prominent option is that the solvent have a composite vapor pressure of 45 mm Hg or less at 20°C or meet one of the composition requirements found in Table 3.8.

Table 3.7   Title VI—phase-out requirements.

Chemical compound	Phase-out requirement
Chlorofluorocarbons	Eliminate use by the year 2000
Halons	Eliminate use by the year 2000
Carbon tetrachloride	Eliminate use by the year 2000
Methyl chloroform	Eliminate use by the year 2002
Hydrochlorofluorocarbons	Production freeze by the year 2015; eliminate use by the year 2030

If nonhazardous or less hazardous alternatives can be developed and implemented, the ease with which businesses can perform cleaning and degreasing operations may be significantly improved. The following sections provide an overview of some new and existing approaches. Although options exist, there is no "one-size-fits-all" solution. Some of the factors to consider when evaluating a new material or selecting new equipment include

- Environmental regulations (federal, state, and local);
- Health and safety regulations (toxicity, odor, flammability, and so on);

Table 3.8   Composition requirement for approved cleaning solvents— NESHAP for aerospace manufacturer and rework facilities.[a]

Cleaning solvent type	Composition requirements
Aqueous	Cleaning solvents in which water is the primary ingredient (equal to or greater than 80% of solvent solution as applied must be water). Detergents, surfactants, and bioenzyme mixtures and nutrients may be combined with the water along with a variety of additives such as organic solvents (for example, high-boiling-point alcohols), builders, saponifiers, inhibitors, emulsifiers, pH buffers, and antifoaming agents. Aqueous solutions must have a flash point greater than 93°C (as reported by the manufacturer) and must be miscible with water.
Hydrocarbon-based	Cleaners that are composed of a mixture of photochemically reactive hydrocarbons and oxygenated hydrocarbons and have a maximum vapor pressure of 7 mm Hg at 20°C. These cleaners do not contain HAPs or ODCs.

[a]40 CFR 63.744 (b)(1).

- Physical and chemical properties (suitability/compatibility for a specific operation);
- Waste generation (types, treatment and disposal considerations); and
- Energy consumption and operating costs.

Before beginning a discussion of the alternatives, it should be pointed out that U.S. EPA intends to provide technical assistance to industries in addition to regulating them. To provide technical assistance, U.S. EPA has compiled the following resources:

- The Solvent Alternatives Guide,
- The Solvent Alternative Utilization Handbook,
- National Center for Manufacturing Sciences Solvent Database,
- Alternative Technologies Transfer Information Clearinghouse, and
- Pollution Prevention Information Exchange System.

The Solvent Alternatives Guide is a software program that requests some basic information about the processes and parts involved and then compares the data to information contained within the database. The result is a set of recommendations for chemical and process alternatives. The program is user friendly and available through U.S. EPA by calling one of their electronic bulletin boards at (919) 541-5742 and downloading the file (SAGE.ZIP) or accessing the information via the U.S. EPA's internet home page.

The Solvent Alternative Utilization Handbook and the National Center for Manufacturing Sciences Solvent Database are also electronic databases and contain detailed regulatory information, chemical and physical properties of various solvents, process applications and parameters, and recycling and recovery options. The Solvent Alternative Utilization Handbook, compiled in conjunction with the U.S. Department of Energy through the Idaho National Engineering Laboratory, may be obtained by calling (208) 526-7834. Copies of the National Center for Manufacturing Sciences Solvent Database may also be obtained by calling (313) 995-4910.

The remaining two resources, Alternative Technologies Transfer Information Clearinghouse and the Pollution Prevention Information Exchange System, are electronic bulletin board systems that have a range of databases and information that go beyond the scope of solvent cleaning and degreasing. The two bulletin boards may be accessed by calling (301) 670-3808 and (703) 506-1025, respectively. Most of these resources are also available on the internet.

AVAILABLE TECHNOLOGIES

Cleaning is an important process in the metal finishing industry. Performed incorrectly, it can nullify many chemical processes conducted downstream. A properly prepared substrate is key to successful and high-quality electroplating, painting, and other surface treatment technologies. The degree of cleanliness, however, varies for each application. For example, anodizing may not require that a surface be prepared as meticulously as that for electroplating or electronic subassembly processes.

The composition of the substrate and the compatibility between the cleaning media and the material to be cleaned are important considerations. Physical and chemical characteristics of the substrate, surface contaminants/residues, and the cleaning solutions should be evaluated alongside the desired effects, including those that may result outside of the parameters. Melting points, porosity, conductivity, specific heat, and hardness are some of the issues that should be considered. Organic and inorganic properties of contaminants such as paints, oxides, polishing compounds, fingerprints, and oils should also be evaluated. Some cleaning solutions may dissolve a contaminant, but also etch the underlying material. Therefore, when selecting a cleaning method, all parameters should be investigated to maximize success.

Aqueous and Semiaqueous Cleaners

Aqueous and semiaqueous cleaning compounds are water-based and do not contain ODCs. Aqueous cleaners do not contain any VOCs, whereas semiaqueous versions typically do. In general, the higher the water content, the more likely the cleaner is to be less hazardous to human health and the environment, and the greater ease with which it may be regenerated or treated to remove contaminants.

Aqueous cleaners are a composite of water detergents and various additives formulated to be effective for the removal of organic and inorganic contaminants. The most commonly found characteristics and constituents of this class of cleaner are alkalinity, antifoaming agents, corrosion inhibitors, deflocculation, saponifiers, sequestering agents, and surfactants.

Alkalinity is a term used to express the capacity of a solution to accept hydrogen ions (H^+) or neutralize an acid. It is not a measure of the basicity of a solution, which is directly related to hydroxide-ion concentration.

Consider an example of a solution with pH 7 and a concentration of bicarbonate (HCO_3^-) equal to $1 \times 10^{-3}\ M$ (moles per litre). The two constituents capable of neutralizing hydrogen ions are hydroxide ions and bicarbonate. Therefore, the alkalinity of the solution will be equal to the sum of these two terms.

$$pH = -\log[H^+]$$

$$7.0 = -\log[H^+]$$

$$[H^+] = 1 \times 10^{-7}\ M \tag{3.11}$$

Because the product of hydrogen and hydroxide ions in water must equal 1×10^{-14},

$$[OH^-] = 1 \times 10^{-14}\ M / 1 \times 10^{-7}\ M$$

$$[OH^-] = 1 \times 10^{-7}\ M \tag{3.12}$$

Alkalinity is the sum of bicarbonate and hydroxide ions as follows:

$$Alkalinity = [OH^-] + [HCO_3^-] = 1 \times 10^{-7}\ M + 1 \times 10^{-3}\ M$$

$$Alkalinity = 1.000\ 1 \times 10^{-3}\ M \tag{3.13}$$

The pH of the solution has little bearing on alkalinity in this example. In fact, the bicarbonate exhibits a buffering effect on the solution. As hydrogen ions are added, they are neutralized by the bicarbonate, hence keeping the pH at 7. If this process is continued, the bicarbonate will eventually be used up and the pH will begin to drop.

The prior example demonstrates that a highly alkaline solution tends to exhibit stability. Although pH is not a measure of alkalinity, it does have an effect. Highly alkaline solutions typically have a high pH (greater than 12). Lower alkalinities are found around pH values of 9 to 11. The more alkaline a cleaning solution, the more likely it is to etch a wide variety of metals. This tendency is driven by the solubility of the metal in relation to pH. Therefore, with the exception of removing a small surface layer of oxides or hydroxides (which may frequently be the case before metal finishing processes), mild or low alkalinity solutions will typically be sufficient for removing contaminants.

Antifoaming agents and corrosion inhibitors are additives to aqueous cleaning solutions that are not directly related to the cleaning process. They are added to provide process improvements outside of the primary objective. The purpose of antifoaming agents is to reduce frothing within a cleaning bath that can cause interruptions in the flow of work. Corrosion inhibitors achieve their objective by coating or reacting with the surface of a clean metal part so that it will not form oxides, which is an effect that may be detrimental to the success of downstream processes or the final product.

Deflocculants are chemicals that are used to help keep contaminants dispersed within the cleaning solution, rather than agglomerating into larger particles and redepositing on cleaned surfaces or accumulating on the top or bottom of the process tank, depending on their density.

Saponifiers aid in the formation of soaps through chemical reactions with organic fatty acids such as vegetable oils and animal fats. A typical class of compounds used for this purpose includes organic amines such as ethanolamine, diethanolamine, and triethanolamine. Because these additives can exhibit volatility, evaporation from the cleaning solution with the passage of time may weaken cleaning strength. Therefore, proper maintenance of a cleaning bath containing saponifiers should include periodic addition of the compound to ensure efficiency.

For the most part, sequestering agents are chelating agents. They are used to prevent inorganic constituents, primarily calcium and magnesium, from precipitating out of solution. Sequestering agents such as sodium ethylenediamine tetraacetic acid and nitrilotriacetic acid (NTA) that are commonly used in metal plating baths are also used in aqueous cleaning solutions.

Surfactants act as detergents and are able to efficiently penetrate small spaces and get underneath deposited contaminants. The latter action is able to occur because of the lowered surface tension of the solution (water), which enables it to act as a "crow bar." There are many types of surfactants that exhibit cationic, anionic, or nonionic properties. Depending on the characteristics of the surfactant (either hydrophilic or lipophilic), specific objectives may be accomplished. Surfactants that are mostly hydrophilic, or water soluble, function well as detergents and "oil-in-water" emulsifiers. These properties tend to disperse oils, grease, and other organic compounds. Lipophilic surfactants, or those which are mostly soluble in oil, function best as wetting agents and "water-in-oil" emulsifiers. In situations where one medium (either oil or water) is present in much larger quantities than the other, emulsification is a desired approach. This permits small amounts of contaminants to be suspended. When relatively large amounts of each medium are present, emulsification is not the best solution. In this case, the natural immiscibility of the two liquids should perform the separation sufficiently.

Although it is a suitable substitute for traditional solvent cleaning methods in many applications, the operation of aqueous cleaning systems is intrinsically different and must be performed with several process changes. Whereas solvents primarily achieve the removal of contaminants through dissolution, aqueous cleaners must employ both chemical and physical means, the latter of which include spray heads, agitation, and ultrasonic measures.

Because of the presence of water and its inability to "flash-off" quickly, corrosion is an important issue. Corrosion is a significant drawback of the aqueous cleaning approach. Methods to reduce corrosion are geared toward the removal of water via accelerated evaporation rates or mechanical processes. One of the most effective techniques to remove water is the use of recirculated hot air. This technique operates much like an oven, except to prevent the buildup of moisture, fresh makeup air is introduced to keep the humidity low. Mechanical methods include pressurized air blown across a cleaned surface to physically remove water deposits.

Before implementing aqueous cleaning methods, it is important to evaluate the process and the parts in conjunction with the process. Substrates that are not subject to corrosion, such as plastics, may appear to be ideal. However, these same materials may degrade under alkaline conditions. Conversely, metal alloys that are susceptible to corrosion may be able to use aqueous cleaners as long as proper process controls are in place. In short, there is no single, best approach for all situations.

The management of wastes is also a consideration with aqueous cleaning methods. When spent, water-based cleaners contain varying amounts of organic and inorganic contaminants. Some aqueous cleaners may be suitable for discharge to sanitary sewers without prior treatment; however, this is not frequently the case. Treatment processes are dependent on the contaminants present. For metals, for example, precipitation is commonplace. For suspended solids and organic constituents, filtration may be amenable. In addition, where large amounts of oils are present, oil–water separators may be employed. Costs associated with waste treatment should also be part of the evaluation of an aqueous cleaning method before implementation.

A close relative to aqueous cleaners is the semiaqueous family of cleaners. This group, frequently referred to as an *emulsive cleaners*, includes mixtures of compounds that are both miscible and immiscible in water. Compounds that are compatible with water include alcohols, amines, and ketones. These are discussed in the Miscellaneous Organic Solvents section later in this chapter. These compounds are miscible in water because of their polar nature, which is a result of the ionic charges of the molecule or the functional group that is attached to the primary molecule. Immiscible solutions and combinations thereof include petroleum hydrocarbons, esters, terpenes, and glycol ethers. Figure 3.16 illustrates the different chemical structures and formulas associated with these compounds.

Family: Petroleum Hydrocarbons
Compound: Naphtha

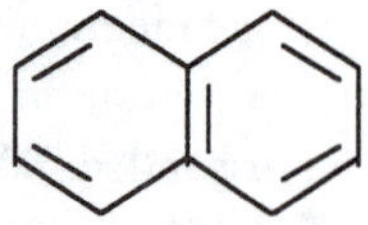

Family: Ethers
Compound: 2-ethoxyethanol

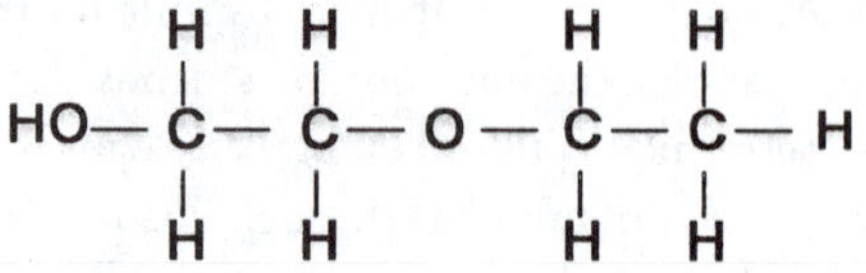

$$HO-\overset{\overset{\displaystyle H}{|}}{\underset{\underset{\displaystyle H}{|}}{C}}-\overset{\overset{\displaystyle H}{|}}{\underset{\underset{\displaystyle H}{|}}{C}}-O-\overset{\overset{\displaystyle H}{|}}{\underset{\underset{\displaystyle H}{|}}{C}}-\overset{\overset{\displaystyle H}{|}}{\underset{\underset{\displaystyle H}{|}}{C}}-H$$

Family: Esters
Compound: ethyl lactate

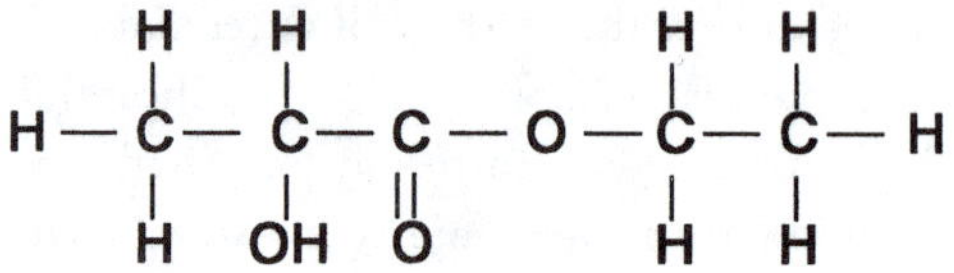

$$H-\overset{\overset{\displaystyle H}{|}}{\underset{\underset{\displaystyle H}{|}}{C}}-\overset{\overset{\displaystyle H}{|}}{\underset{\underset{\displaystyle OH}{|}}{C}}-\overset{\overset{\displaystyle H}{|}}{\underset{\underset{\displaystyle O}{\|}}{C}}-O-\overset{\overset{\displaystyle H}{|}}{\underset{\underset{\displaystyle H}{|}}{C}}-\overset{\overset{\displaystyle H}{|}}{\underset{\underset{\displaystyle H}{|}}{C}}-H$$

Family: Terpenes
Compound: d-limonene

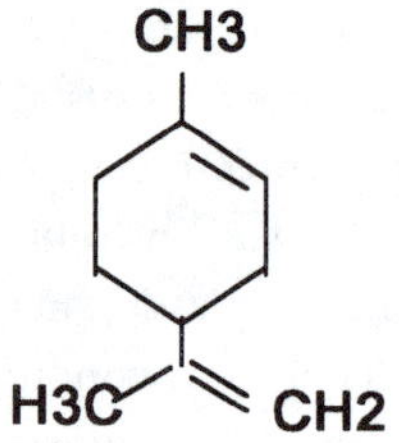

Figure 3.16   Chemical structures and formulas of common semiaqueous cleaning
compounds.

Semiaqueous cleaners are efficient at removing organic-based contaminants. The organic solvents in semiaqueous cleaners readily dissolve grease, oil, wax, and other similar compounds. A solvent's miscibility in water also determines how the parts should be rinsed. For solvents that are miscible, a simple water rinse will adequately purge the cleaning solution from all surfaces. Immiscible types, however, may require the use of another solvent, such as alcohol, to completely remove residues. In general, the semiaqueous approach provides better penetration during cleaning because of the lower surface tension of organic constituents and improves corrosion control over aqueous alternatives. As always, selection criteria are important because a semiaqueous approach may not be suitable for some plastics, elastomers, and rubbers.

As with most processes, semiaqueous cleaners must be used properly to be effective. Because VOCs are present, temperatures should be kept a minimum and low-vapor-pressure components should be used where feasible to limit emissions. These emissions should be viewed not only from an environmental standpoint, but from a worker health and safety standpoint as well. Some compounds used in semiaqueous cleaners, such as glycol ethers and terpenes, are known to cause a variety of teratogenic and carcinogenic effects. Other compounds may possess flashpoints so low that mists created from solution agitation or the drying of parts may result in an ignitable atmosphere. Common sense and proper attention can avoid unnecessary accidents.

Recovery methods for semiaqueous cleaners primarily involve distillation and membrane filtration. Where possible, these techniques should be used rather than traditional waste-water treatment approaches involving biodegradation. In situations where large waste streams are produced, a reasonable payback period for the purchase of recycling equipment may be realized. For those manufacturing operations where only small baths are purged periodically, isolation of semiaqueous cleaners may be advantageous to allow effective batch treatment or bulk disposal to an off-site facility.

Petroleum Hydrocarbons

Petroleum hydrocarbon cleaners, which have been in use for a number of years, have proven to be successful applications. The primary classifications of petroleum hydrocarbon cleaners are petroleum distillates and synthetic paraffinic hydrocarbons. Petroleum distillates are the most common and include compounds such as mineral spirits, kerosene, naphtha, stoddard solvent, and a military grade referred to as PD-680. These compounds consist of mixtures of numerous hydrocarbons and small amounts of aromatic constituents (benzene derivatives) that exhibit a range of boiling points. The paraffin family is considered a refined grade of petroleum hydrocarbon cleaner and contains alkanes with virtually no aromatic compounds. This tends to produce physical characteristics that lead to a more effective cleaner, but at higher costs.

Petroleum hydrocarbons are typically used in situations where water is not an acceptable ingredient because of its potential to initiate corrosion. The properties of petroleum hydrocarbons are similar to those of semiaqueous cleaners in that they are efficient at removing stubborn organic deposits and possess low surface tensions that allow for penetration into tight areas and under contaminants. As with semiaqueous cleaners, volatility is slow with petroleum hydrocarbons; thus, the drying process may require the blowing of pressurized air or convective methods.

The regulatory issues of petroleum hydrocarbons and semiaqueous cleaners are also similar. Flashpoint values can indicate the need for concern over combustible atmospheres, and volatility can be an issue in areas where ozone formation is a problem. In general, the health effects of primary concern are caused by aromatic constituents. For compounds that are synthesized (as in the paraffin family), benzene-derivative content is lower and, therefore, less of an issue.

Petroleum hydrocarbon cleaners can be advantageous in waste minimization programs because they do not produce wastewater. However, if they are not recovered through distillation techniques, the bulk of the wastes produced will have to be disposed of via various regulatory channels or commercial off-site treatment/recycling companies. Because this family of solvents is typically a composite of individually related compounds, each with varying flash-off rates, residues are likely to exist on part surfaces after the cleaning process is complete. In manufacturing operations where the predominant process involves the machining of various metal alloys, residues can be beneficial in protecting the parts from corrosion. In other applications, however, this film may be unacceptable and require rinsing in a solution capable of dissolving the thin layer. The latter procedure is not unlike rinsing procedures required for some semiaqueous cleaners that are immiscible in water.

Hydrochlorofluorocarbons

Because of the stratospheric ozone-depleting potential exhibited by many CFCs and rising concerns about the detrimental effects they pose to human health and the environment, a movement to regulate CFCs by banning or chemical replacement is in effect. By the year 2000, the use of CFCs and halons (that is, brominated compounds) is to be eliminated (Title VI–Stratospheric Ozone Protection [1990] Public Law 101-549).

Significant reduction in the use of these materials is required in the short term. As a result, waste minimization techniques that include recovery/recycling and emission reduction programs are becoming more prevalent. Solutions underway include source reduction programs to replace CFCs with HCFCs. The initiative is widespread in that it also affects the general public. An example of this is the replacement of traditional CFC air conditioners in automobiles with HCFC-compatible units. Hydrochlorofluorocarbons have 1% of the ozone-depleting potential of CFCs. Although the use of HCFCs represents a move in the right direction, these compounds are designated for phase-out by 2030.

The block of time allowed for phasing in the use of HCFCs and eliminating CFCs should allow for development of new and better chemical alternatives. However, an important question remains: Because the lag time for CFC emissions traveling into the stratosphere is approximately 10 years and their destructive capacity on arrival can remain for an additional 120 years, are the pollution prevention regulations and actions in place enough, or are they too late? Although only time will answer this question, the potential hazards posed by stratospheric ozone depletion warrant immediate attention.

Ongoing research is being conducted to develop chemical substitutes for ozone-depleting substances. A number of nearly azeotropic fluid mixtures that have saturation pressures similar to that of CFC-12 (dichlorodifluoromethane) while being only approximately 2% as damaging to the ozone layer are being evaluated. These mixtures exhibit low toxicity, inflammability, and greater compatibility with conventional lubricating oils than does HCFC-134a (1,1,1,2-tetrafluoroethane), which is currently the leading replacement for CFC-12. These new mixtures may be usable in commercial, automotive, and household refrigerators and air conditioners. Other operations subject to CFC replacement strategies include processes that involve electronics and metal cleaning, hospital sterilization, flexible and rigid foam manufacturing, fire extinguishers, and processes in the aerospace industry.

For degreasing operations, the most prevalent HCFC compounds to replace the predominant compound of the recent past, CFC-113, include HCFC-141b and HCFC-225. The physical characteristics that are important elements in cleaning, such as surface tension, boiling point, heat of vaporization, and density, indicate that little performance degradation can be expected as a result of replacing CFC-113 with HCFC alternatives. Indeed, virtually all substrates compatible with CFC cleaning are amenable with HCFCs as well. The only materials of concern include plastics such as acrylobutylstyrene, styrene, and acrylics.

The health and safety aspects of HCFC-141b and -225 are perhaps the best for all such compounds. Both are nonflammable and exhibit no flashpoint. Although they do possess a low level of toxicity that may pose some health risks, other HCFC compounds exhibit higher levels of toxicity and are therefore less desirable for use. Nevertheless, it should be reiterated that HCFCs are a temporary solution to the use of CFCs, whether for use in solvent cleaning and degreasing operations or for any other application.

Supercritical Fluids

Supercritical fluids are used in applications where precision cleaning is necessary and relatively low volumes of parts are processed. Typical uses of this technology include the cleaning of a range of substrates in connection with the following items: optical components, ceramics, electromechanical assemblies, porous metals, and critical parts used in an assortment of instrumentation and measurement devices. Supercritical fluids are efficient at removing organic contaminants as long as they are present in only small quantities. Deposits of inorganic materials, particulates, and significant residues of oil or grease are not compatible with this type of cleaning method.

Benefits derived from supercritical fluids are the result of their low viscosity and high diffusivity—characteristics that allow excellent solution penetration. To understand this, a discussion of supercritical fluids is presented. Supercritical fluids are found at conditions above critical temperatures and/or pressures. This is illustrated in Figure 3.17 using the traditional dome-shaped phase diagram for carbon dioxide (CO_2). The axes are molar volume versus pressure. The shape of the diagram for many compounds will be similar.

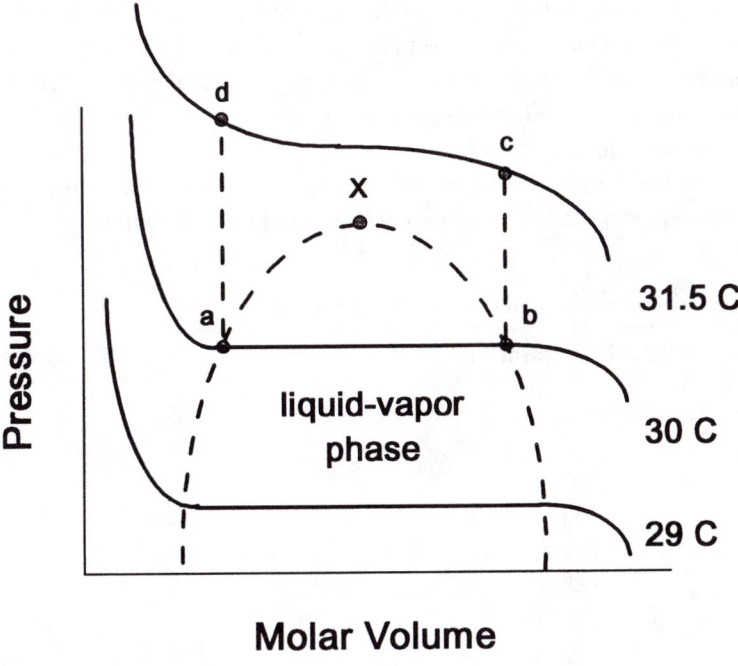

Figure 3.17 Phase diagram for carbon dioxide.

The critical point is defined as the temperature and pressure at which equal molar volumes of liquid and gas exist, thereby eliminating any distinction between the two phases. This state is labeled as point "x" on the previous diagram. Below this point or underneath the dome, vapor and liquid exist together in varying proportions. To the left of the dome, liquid exists. To the right, vapor exists. At temperatures or pressures above the critical point, "x," a supercritical fluid results.

Follow the isotherm for 30°C from its intersection with the dome from left to right ("a" to "b"). A liquid exists first ("a"), followed by an increasing volume in the vapor phase, until only vapor remains ("b"). This is best visualized by a cylinder of carbon dioxide liquid held at constant temperature and pressure, but expanded in volume. If the temperature is raised while keeping the volume constant ("c"), the pressure will need to be increased. Follow the 31.5°C isotherm from right to left. The pressure must be increased and the volume decreased to arrive at "d." The last step to complete a cycle and return to "a" is to decrease pressure and temperature while maintaining a constant volume. An interesting transition has occurred. The vapor phase at "b" somehow became a liquid at "a" without a distinct two-phase vapor–liquid system ever existing. This is because the system traveled through supercritical conditions. If the path taken from "b" to "a" were along the 30°C isotherm, the two phases of vapor condensing to liquid would be visible. The supercritical route achieves the same results, but the separation of the phases is not apparent because of its occurrence at a molecular level while in a single phase.

Table 3.9 provides a list of some common elements and compounds, their respective critical-point constants (T_c, P_c, V_c), and atmospheric boiling points. It is interesting to note that atmospheric pressure is always less than P_c. This accounts for the fact that as temperatures rise, solids melt to liquid and liquids evaporate to gas. It is also interesting to note that ambient temperature (25°C or 298 K) is less than the T_c value for most substances, except helium, hydrogen, neon, nitrogen, oxygen, and methane. Hence, these compounds, when pressurized, must be cooled to below their T_c values before a phase change (vapor to liquid) can take place.

One of the most common supercritical fluids used in cleaning operations is carbon dioxide because of its natural abundance in the atmosphere and nonpolluting characteris-

Table 3.9 Critical-point values for selected compounds.

Substance	T_b, K at 1 atm	T_c, K	P_c, atm	V_c, cc/mole
H_2	21	33.2	13	65
N_2	77	125	34	90
O_2	90	154	50.5	75
CH_4	109	191	46.5	
CO_2	195	304	72.8	95
HCl	188	325	83.2	
H_2S	212	373	90.4	
NH_3	240	405	113	72
H_2O	373	647	218	57

tics. Although some viewpoints contend that carbon dioxide contributes to global warming, because of the relatively small quantities generated from this application, it is generally not thought of as an issue. One of the characteristics of a supercritical fluid that may be tailored to create more efficient cleaning is solvent density: in general, the higher the density, the better the cleaning power.

The closer a supercritical fluid is to the liquid phase, the higher the density. Hence, density may be maximized by controlling the temperature of the fluid to within approximately 30°C in excess of T_c and exerting a pressure that will keep conditions of the substance to the left of V_c. This is commonly at a pressure equal to two or more times the P_c value. Maintaining liquidlike characteristics ensures good solvency characteristics. A generic process layout for a supercritical fluid cleaning system is shown in Figure 3.18.

Miscellaneous Organic Solvents

This classification of solvents, which has been in use for decades, is commonly used in a range of small-scale industrial applications. Cold solvent degreasing vats, hand-wipe cleaning, and enclosed spray washers are typical examples of organic solvents. Although none of these compounds exhibits ozone-depleting potential, all are considered VOCs unless specifically exempted by regulations. An example of a compound that has been exempted from VOC status is acetone. In addition, some solvents, such as methyl ethyl ketone and methyl isobutyl ketone, are considered HAPs, which are an entirely different classification of regulated compounds.

Depending on the size and location of operations that use VOCs, certain restrictions may exist. For example, in metropolitan areas classified as nonattainment areas by CAA,

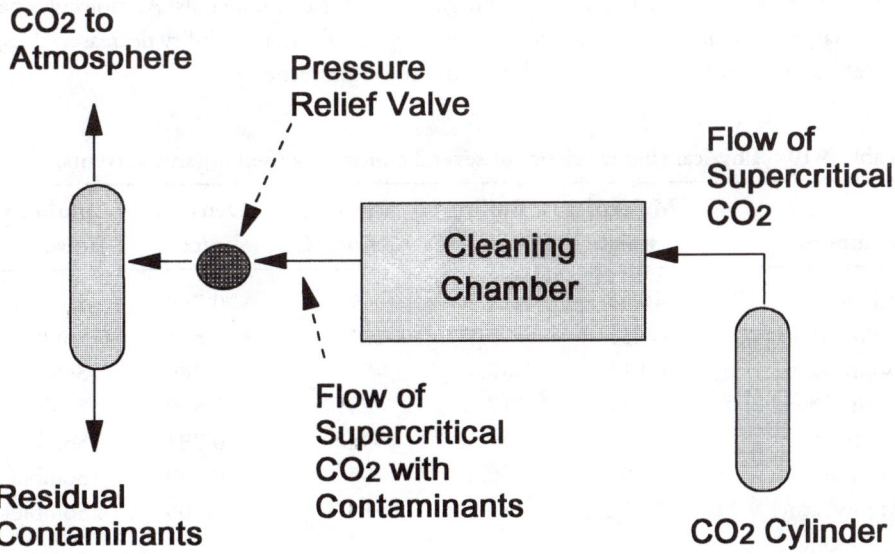

Figure 3.18 Process flow schematic for carbon dioxide supercritical fluid cleaning
        system.

VOC use may be highly regulated. Hazardous air pollutants, on the other hand, are not considered suitable substitutes for cleaning solvents and are regulated through a plethora of CAA NESHAPs, which govern specific sources rather than geographic regions. These compounds may also create hazardous wastes because of their RCRA classification as a "spent solvent" in the following waste codes:

- F003—The following spent non-halogenated spent solvents: xylene, acetone, ethyl acetate, ethyl-benzene, ethyl ether, methyl isobutyl ether, n-butyl alcohol, cyclohexanone, and methanol; all spent solvent mixtures/blends containing, before use, only the above non-halogenated solvents; and all spent solvent blends/mixtures containing, before use, one or more of the above non-halogenated solvents, and, a total of ten percent or more (by volume) of one or more of those solvents listed in F001, F002, F004, F005; and still bottoms from the recovery of these spent solvents and spent solvent mixtures.
- F005—The following spent non-halogenated solvents: toluene, methyl ethyl ketone, carbon disulfide, isobutanol, pyridine, benzene, 2-ethoxyethanol, and 2-nitropropane; all spent solvent blends/mixtures containing, before use, a total of ten percent or more (by volume) of one or more of the above non-halogenated solvents or those solvents listed in F001, F002, F004; and still bottoms from the recovery of these spent solvents and spent solvent mixtures [40 CFR 261.31(a)].

Some organic solvents that are useful as cleaning compounds are shown in Table 3.10. Most flash off quickly and act as good drying agents. Those listed under the alcohol and ketone family are generally water soluble, which is a property that allows for efficient cleaning characteristics over a range of inorganic and organic materials. As molecules become larger or molecular weights increase, the degree of water solubility decreases; thus, the ability to handle inorganic or polar compounds diminishes.

Table 3.10   Physical characteristics of several commonly used organic solvents.

Compound	Molecular weight	Boiling point, °C	Melting point, °C	Density, g/cc	Solubility in water
Ethanol	46.07	78.5	−117	0.789	Soluble
n-Propanol	60.11	97.4	−127	0.803	Soluble
Isopropanol	60.11	82.4	−90	0.786	Soluble
n-Butyl alcohol	74.12	117	−89	0.810	Slightly
Methanol	32.04	64.96	−94	0.791	Soluble
Acetone	58.08	56.2	−95	0.790	Soluble
Methyl ethyl ketone	72.12	79.6	−83.4	0.805	Soluble
Methyl isobutyl ketone	100.6	116	−84.7	0.798	Slightly

Some solvents may possess vapor pressures that are too high to be effective in typical cleaning applications. These solvents may also be a significant fire hazard. Others solvents may portray vapor pressures too low to flash off quickly and must therefore be hand-wiped with dry clean rags or assisted via pressurized air or oven heating. Each solvent will exhibit specific health, safety, and material compatibility problems. Toxicological data, for example, vary depending on a range of biological reactivity and exposure conditions such as lipid solubility, rate/depth of respiration, dermal exposure, cardiac output, blood:air ratio, and threshold limit values.

Most compounds in this category act as depressants to the human central nervous system. High-level human exposure to these compounds can result in disorientation, euphoria, dizziness, and confusion leading to unconsciousness, paralysis, convulsions, and even death. In the majority of cases, human recovery from central nervous system damage caused by these compounds is rapid and complete. Other solvents, such as petroleum hydrocarbons and chlorinated solvents, also possess many of these toxicological effects. Appropriate screening procedures should therefore be conducted before their use.

TECHNOLOGICAL DEVELOPMENTS

Wet Oxidation Cleaning

Wet oxidation is an aqueous-phase oxidation process conducted by "stripping" organic or oxidizable inorganic-containing liquids with a gaseous source of oxygen, typically air, and thermally oxidizing (that is, incinerating) the resulting effluent gases. Reactions may be executed in a batch or continuous-flow mode and typically operate at temperatures in the range of 150 to 325°C. In either event, liquid feed must be under high pressure and oxygen supplied via compressed air or a liquid oxygen vaporizer. A schematic of the wet oxidation process is shown in Figure 3.19.

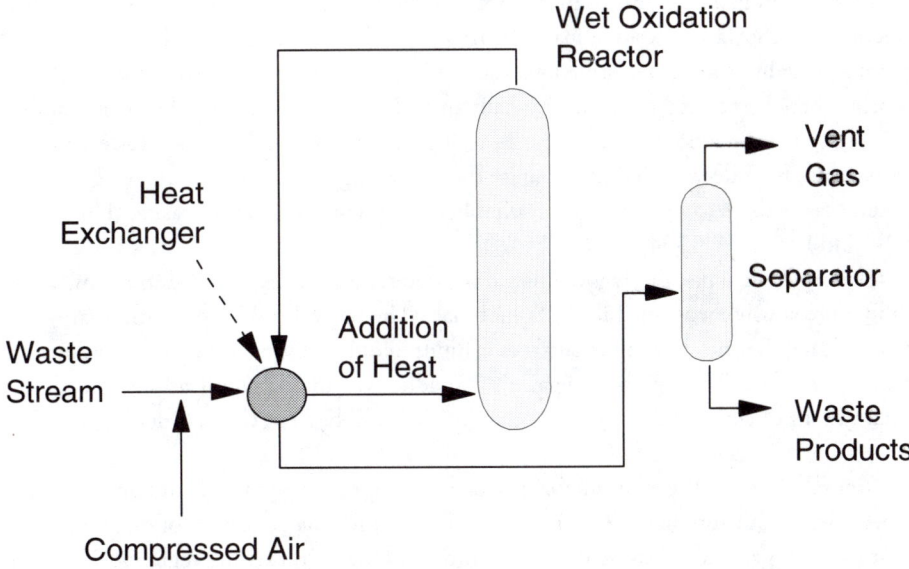

Figure 3.19   Typical wet oxidation process.

The application of wet oxidation to the cleaning process is new. In theory, organic contaminants from wet parts may be removed and disposed of in the same operation. A drawback of this, however, is that the oxidizing environment may remove substrate materials.

Absorbent Media Cleaning

Absorbent media cleaning is a process in which rags are used to wipe contaminated surfaces or bulk materials are used to "soak up" liquefied contaminants. No solvents are employed in these activities. The cleaning action is accomplished by the use of various fibers that trap contaminants in the weave of the cloth or, when used in bulk form, absorb organic materials (for example, oil and grease) into the internal matrix of the cleaning substance. The most suitable projects for implementing these methods involve substrates to which various solvents or aqueous alternatives are not amenable (that is, those that degrade on exposure to solvents or water).

Some of the most common substances used to perform dry media absorbent cleaning include polypropylene fibers and bulk materials such as cellulose pulp, starches, and silicates. Absorbent media cleaning is a simple process that has been used indirectly to some extent for years. An example of the process is the use of polypropylene bulk fibers, pads, pillows, and booms for the cleanup of oil spills or for use as drip pads. An older approach was to use clay cat litter, vermiculite, or peat-moss derivatives.

These materials were initially used for cleaning up spills from leaking equipment. However, the use of more in-process applications has evolved in recent years. An example of this evolution is the use of these same polypropylene wipers or starch-laden rags on surfaces where light oil contamination or fingerprints are present. Results have indicated that these methods can, in some cases, provide a cleaner surface than that provided by a traditional solvent wipe of methyl ethyl ketone.

The use of bulk absorbents in most applications is not a useful approach when applied directly to part surfaces. Residual particles frequently remain and cause additional cleanup procedures to be performed via vacuuming, sweeping, or similar methods. Bulk absorbents are best suited for their originally intended use in the realm of spill containment, equipment maintenance, and non-production-related applications. However, developments are underway that incorporate these cleaning materials into enclosed systems. In some cases, they may prove to be a suitable substitute for vapor degreasing (Doscher, 1991, and U.S. EPA, 1994).

Polypropylene wipes and rags are useful in production processes and in the same areas as those previously mentioned for bulk materials. The most sensible applications are those that require the cleaning of surfaces exhibiting only minimal oil contamination and those that are not employed before painting or adhesive bonding applications. An additional benefit of cloth media is that production cells can be kept clean, orderly, and efficient.

Disposal of spent absorbent media is dependent on the materials or contaminants absorbed. Because the original media is inert and is used for the collection of oil and grease, spent media is generally classified as nonhazardous. However, existing regulations prevent or prohibit the landfilling of biologically degradable absorbents. As a result, incineration of these materials is often a desirable option because of their high British thermal unit value.

Surface Coating

The surface-coating industry is as regulated and amenable to waste minimization strategies as the areas of metal finishing and solvent cleaning/degreasing. Coatings operations are primarily governed by CAA, but are also covered by statutes of other media such as RCRA and CWA because they generate various waste streams. The Emergency Planning and Community Right-to-Know Act (EPCRA) requires that releases and transfers of more than 600 toxic chemicals or chemical categories be reported annually. Although EPCRA applies to a number of operations in addition to those associated with surface coating, it is discussed here because nearly half of the toxic release inventory, or Form R, releases are accounted for as air emissions. Issues of RCRA and CWA have been discussed in prior sections of this chapter and will not be reiterated here. Instead, issues of CAA and EPCRA will be the primary focus. The information is provided in an attempt to keep the subject as clear and uncomplicated as possible.

The CAA, as enacted in the 1970s, restricted and controlled how the surface-coating industry performed its operations via the following statutory programs: air quality standards, state implementation plans, new source performance standards, air toxics, and new source review.

The CAA's air quality standards program specifies acceptable levels of air quality and defines objectives for achieving these levels through the implementation of a set of National Ambient Air Quality Standards. The National Ambient Air Quality Standards are based on scientific determinations of threshold levels for six pollutants below which no adverse effects will be experienced by humans or the environment. Areas or regions that exceed these threshold levels are classified as nonattainment areas and must initiate state implementation plans. The six criteria pollutants are ozone, carbon monoxide, particulate matter (less than 10 μm in diameter), sulfur dioxide, nitrogen dioxide, and lead.

The CAA's state implementation plans program defines the regulatory framework for comprehensive analysis of air quality in each region and establishes a strategy to achieve compliance with air quality standards and legally enforceable control requirements. The new source performance standards program requires the use of best demonstrated control technologies when major sources of air pollution are constructed or modified. The air toxics program allows for standards to limit emissions of toxic compounds, referred to as NESHAP. Seven HAPs were originally identified for the air toxics program: arsenic, asbestos, benzene, beryllium, mercury, radionuclides, and vinyl chloride.

Finally, the CAA's new source review program establishes approval guidelines for new, significant industrial growth projects located in attainment or nonattainment areas. Specifically, sources located in attainment areas (that is, sources that meet air quality standards) must meet special rules that prevent industrial growth from causing a significant deterioration to air quality. This program is titled Prevention of Significant Deterioration (PSD). Projects that fall under the category of PSD must implement best achievable control technology (BACT). Despite the implementation of BACT, emissions that remain must fall within an available increment of air quality degradation beyond which the area cannot exceed. Best available control technology, which is an approach identified by U.S. EPA as "the best state-of-the art control technology that could possibly be used," is determined on a case-by-case basis. New or modified sources must justify any variation of such an approach by demonstrating that it could not be used with respect to the actual proposed facility.

Sources located in nonattainment areas have more severe restrictions under CAA's new source review program. In addition to installing lowest achievable emission rate technology, nonattainment areas must provide offsets for any residual emissions attributable to the new or modified source. Lowest achievable emission rate technology is a control technology equal to or greater than BACT. Offsets must incorporate legally enforceable reductions in emissions from other sources above and beyond those that would otherwise be required. Offsets can be derived from the installation of advanced controls that result in reductions in emissions from existing sources or from the shutdown of sources. In most cases, offsets may be traded or banked for future use, although certain restrictions may apply. In short, areas classified as nonattainment must achieve a net air quality improvement.

It is not uncommon for an individual area to exhibit nonattainment status for one criteria pollutant and attainment status for others. Therefore, new or modified major sources of air pollution may be required to simultaneously comply with both nonattainment offsets and PSD requirements, but for different criteria pollutants.

The five CAA programs discussed have been expanded under the CAA Amendments of 1990. The most significant changes or additions as they relate to the surface-coating industry involve the following issues: non-attainment (Title I), air toxics (Title III), federal operating permits (Title V), and stratospheric ozone protection (Title VI).

Title I required a new classification of nonattainment areas and set deadlines for achieving attainment based on the severity of pollution. Reasonably available control technology (RACT) requirements were expanded and made more strict for existing and newer/smaller facilities. Control technique guidelines were defined as a method to provide generic definitions of RACT for specified industrial categories. State implementation plans were also required to be revised to ensure that the air quality standards would be achieved by specific deadlines. These requirements include reducing total emissions measured from a baseline by 15% over a 6-year period and 3% per year thereafter. In addition, the criteria of the new source review program were expanded to capture the effects of smaller sources located in nonattainment areas. Before the CAA Amendments of 1990, only sources emitting 100 tons per year of a criteria pollutant triggered new source review. After the 1990 amendments, the level of emissions defining the term *major* were changed. Emissions now range from 10 to 100 tons per year depending on the classification of the nonattainment category such as marginal, moderate, serious, severe, or extreme.

Title III redefined controlling emissions of HAPs. The National Ambient Air Quality Standards cover six criteria pollutants that were defined to monitor and control emissions in widespread areas. For toxic compounds emitted in relatively local and/or more concentrated quantities, CAA devised a set of standards for seven HAPs. In the CAA Amendments of 1990, this list of HAPs was expanded to include 189 compounds (the list of HAPs may be found in Appendix B).

In addition to a longer list of regulated toxic pollutants, U.S. EPA was instructed to develop tighter controls by implementing a new strategy that incorporated two phases. The first phase involves technology standards and requires facilities to install MACT. Maximum achievable control technology standards govern selected categories of industrial facilities that emit the newly defined list of air toxics. In some cases, these standards may require large investments in new control technology to prevent releases of toxic air pollutants. Once MACT standards have been promulgated and complied with, the second phase is to develop additional regulations to handle facilities that still emit concentrations of air toxics that may be harmful to exposed individuals.

One such MACT standard, the NESHAP for aerospace manufacturers and rework facilities, was introduced in the Solvent Cleaning and Degreasing section of this chapter. This regulation is also pertinent to portions of the surface-coating industry because it imposes coating use limitations and equipment restrictions. Specific operations covered include spray gun cleaning, primer application, topcoat application, depainting, and chemical milling masking operations. The most significant excerpts from 40 CFR 63 Subpart GG are summarized below.

> Spray Gun Cleaning—Each owner or operator of a new or existing spray gun cleaning operation subject to this subpart in which spray guns are used for the application of coatings or any other materials that require the spray guns to be cleaned shall use one or more of the following techniques, or their equivalent:
>
> 1. Enclosed systems. Clean the spray gun in an enclosed system that is closed at all times except when inserting and removing the spray gun. Cleaning shall consist of forcing solvent through the gun.
> 2. Nonatomized cleaning. Clean the spray gun by placing solvent in the pressure pot and forcing it through the gun with the atomizing cap in place. No atomizing air is to be used. Direct the solvent from the spray gun into a vat, drum, or other waste container that is closed when not in use.
> 3. Disassembled spray gun cleaning. Disassemble the spray gun and clean the components by hand in a vat, which shall remain closed when not in use. Alternatively, soak the components in a vat, which shall remain closed during the soaking period and when not inserting or removing components.
> 4. Atomizing cleaning. Clean the spray gun by forcing the solvent through the gun and direct the resulting atomized spray into a waste container that is fitted with a device designed to capture the atomized solvent emissions.
>
> Primer and Topcoat Application Operations—each owner or operator of a new or existing primer or topcoat application operation subject to this subpart shall comply with the requirements specified in the following paragraphs.

1. Uncontrolled coatings—organic HAP and VOC content levels. Organic HAP and VOC emissions from primers shall be limited to an organic HAP content level of no more than 350 g/l (2.9 lb/gal) of primer (less water) and a VOC content level of no more than 350 g/l (2.9 lb/gal) of primer (less water and exempt solvents) as applied. Organic HAP and VOC emissions from topcoats shall be limited to an organic HAP content level of no more than 420 g/l (3.5 lb/gal) of coating (less water) and a VOC content level of no more than 420 g/l (3.5 lb/gal) of coating (less water and exempt solvents) as applied.

2. Controlled coating—control system requirements. Each control system shall reduce the operation's organic HAP and VOC emissions to the atmosphere by 81% or greater, taking into account capture and destruction or removal efficiencies, as determined by specified regulatory procedures.

3. Compliance methods. Compliance with the organic HAP and VOC content limits specified shall be accomplished as follows either by themselves or in conjunction with one another:

  a. Use primers and topcoats with HAP and VOC content levels equal to or less than the limits specified (primers 2.9 lb/gal, topcoats 3.5 lb/gal).

  b. Use any combination of primers or topcoats such that the monthly volume-weighted average organic HAP and VOC content comply with the specified content limits, unless the permitting agency specified a shorter averaging period as part of an ambient ozone control program. Note: Averaging of primers together with topcoats is prohibited and is allowed only for uncontrolled primers or topcoats.

4. Application equipment. Each owner or operator of a new or existing primer or topcoat application operation subject to these requirements in which any of the coatings contain organic HAP or VOC shall comply with the following:

  a. All primers and topcoats shall be applied using one or more of the application techniques specified below (exemptions exist for specified situations):
     • flow/curtain coat application
     • dip coat application
     • roll coating
     • brush coating
     • cotton-tipped swab application
     • electrodeposition (dip) coating
     • high volume low pressure (HVLP) spraying
     • electrostatic spray application
     • other coating application methods which achieve emission reductions equivalent to HVLP or electrostatic spray application methods

5. Inorganic HAP emissions. Each owner or operator of a new or existing primer or topcoat application operation subject to these requirements in which any of the coatings that are spray applied contain inorganic HAP, shall comply with the following (exemptions exist for specified situations):

a. Apply coatings in a booth or hangar in which air flow is directed downward onto or across the part or assembly being coated and exhausted through one or more outlets.

b. Control the air stream from this operation as indicated below:

- for existing sources, pass the air through either a dry particulate filter systems or a waterwash system before exhausting it to the atmosphere
- waterwash booths shall remain in operation during all coating application operations
- dry filter booths shall include two-stage filter systems or the equivalent, as determined by the permitting agency
- for new sources, pass the air stream through either a two-stage dry particulate filter system or a waterwash systems before exhausting it to the atmosphere. If the primer or topcoat contains chromium or cadmium, control shall consist of either a three-stage filter system, HEPA filter system, or other equivalent control system, as approved by the permitting agency.

c. If a dry particulate filter system is used, it must be maintained in good working order. A differential pressure gauge must be installed across the filter bank and the pressure drop across the filter must be continuously monitored.

d. If a waterwash system is used, continuously monitor the water flow rate.

e. If the pressure drop across the dry particulate filter systems is outside limit(s) specified by the filter manufacturer or in locally prepared operating procedures, shut down the operation immediately and take corrective action. If the water path in the waterwash systems fails the visual continuity/flow characteristics check or the water flow rate exceeds the limit(s) specified by the booth manufacture's or locally prepared operating procedures, or the booth manufacturers or locally prepared maintenance procedures for the filter or waterwash systems have not been performed as scheduled, shut down the operation immediately and take corrective action. The operation shall not be resumed until the pressure drop or water flow rate is returned within specified limit(s).

From this overview, it is apparent that material and equipment restrictions may be placed on specific operations and that the manner in which the operations are conducted may also be defined. Although there are an equal number of requirements for depainting operations, the aerospace NESHAP excerpts provide a cursory view of the regulatory community's intentions and direction. Moreover, although this represents one regulation covering only a section of the surface-coating industry, it provides an indication of the existing and future trends manufacturers must follow.

The last two issues of the CAA Amendments of 1990 are Title V and Title VI. Title V, the federal operating permit program, was designed to address the need to precisely define the requirements applicable to each source and to enable effective enforcement by the consolidation of these requirements into a single document. The intention is that the Title V permit be similar in nature to the NPDES permit program under CWA. Although the Title V permit is a federal program, states are responsible for implementation of the federal requirements that establish detailed rules governing emissions from the source and related activities such as monitoring, recordkeeping, and reporting. The timeline for full implementation of the federal operating permit program varies from state to state. Some states already have a Title V permit program approved by U.S. EPA in place. States that do not have an effective program in place may lose the opportunity to shape their own programs because U.S. EPA will be required to take over in the wake of inadequate responses.

Title VI is a program designed to address growing concerns over the depletion of the earth's stratospheric ozone layer. With emerging evidence that CFCs and other chemicals were causing destruction to the stratospheric ozone layer, the CAA Amendments of 1990 established a schedule to phase out identified chemicals from production and sale. It was the first legislation to go beyond the restrictions defined within the Montreal protocol, which was discussed in the Solvent Cleaning and Degreasing section earlier this chapter.

In addition to CAA and its amendments, legislation pertinent to the surface-coating industry includes EPCRA. This act does not contain permitting requirements or operational restrictions. Rather, it is a reporting and recordkeeping statute. Its purpose is to establish a list of extremely hazardous substances, threshold planning quantities, and facility notification responsibilities necessary for the development and implementation of state and local emergency response plans. It also serves as a vehicle to inform communities and the general public of potentially hazardous operations and chemicals at a facility operating in their vicinity.

Reports required of facility operators by EPCRA cover issues of facility emergency coordinators, emergency release notices, material safety data sheet submissions to local and state emergency planning commissions, hazardous chemical inventory reports, toxic chemical releases (toxic release inventory, or Form R), and supplier notifications. Extremely hazardous substances, hazardous chemicals, toxic chemicals, and hazardous substances are compounds for which these reports must be filed. All the aforementioned chemical categories are defined by U.S. EPA with the exception of hazardous chemicals, which are defined by OSHA. The use of hazardous chemicals requires a manufacturer to create a material safety data sheet.

LOW-COST EMISSION CONTROL STRATEGIES

Minimizing costs of air pollution control is primarily achieved by the elimination of hazardous or toxic constituents in the materials used, or through the application of techniques and selected equipment options to help minimize the amount of contaminants that become airborne in the first place. This section addresses these concepts while focusing on low-cost strategies for implementation.

Spray Application Techniques

Spray application techniques for paint use three key mechanisms for atomization—compressed air, airless, and electrostatic methods. Although there are variations in each category, the approaches are elemental. In addition, some techniques blend the characteristics of separate groups.

HIGH-VOLUME, LOW-PRESSURE SPRAY GUNS. In past years, conventional air spray techniques have been popular approaches. Although air spray techniques are still used to some extent today, because of increasing concerns over air pollution and the generation of hazardous wastes, modifications to the basic technique are becoming more common. The most popular approach that still incorporates the basic principles of conventional air spray is referred to as high-volume, low-pressure (HVLP) atomization. The basic principle of HVLP spray guns is simple. Liquid is fed to the gun via a hose that is under pressure. Also connected to the gun is an air hose. As the trigger of the gun is depressed, liquid is passed through the nozzle while the compressed airstream atomizes and propels the paint while shaping the spray pattern. Depending on the pressures used in the atomizing air jets, the desired "fan" of paint may be produced.

High-volume, low-pressure spray guns operate at reduced pressures, typically less than 68.98×10^3 Pa (10 psi). This is considerably lower in comparison to conventional approaches that use air at pressures near 68.95×10^4 Pa (100 psi). This reduction in pressure reduces the degree to which atomization occurs and can improve transfer efficiencies nearly 60 to 70%. Conventional spray guns may only achieve transfer efficiencies close to 30 to 35%. The softer spray of HVLP spray guns allows the paint to strike a part with less energy, thereby sticking to the surfaces more readily and penetrating into cavities that would otherwise be missed. Increased efficiency in paint application yields other benefits such as lower paint usage, lower paint wastes, and, therefore, overall lower costs.

The usage of HVLP guns also has drawbacks. Because the spray is not as fine, the quality of the finish is not as good as with conventional guns. High-gloss coatings or additional surface polishing are sometimes applied to remedy this situation. This practice is especially commonplace in the automotive industry. Figure 3.20 illustrates a typical setup using HVLP guns and identifies the primary components involved in the process.

AIR-ASSISTED AIRLESS PROCESSES. Another pollution-prevention-compatible spray method is one that integrates both conventional air and airless technologies. Airless spraying is a technique that does not directly use compressed air to atomize the paint. Instead, hydraulic pressures are exerted on the fluid to atomize the paint as it leaves the nozzle.

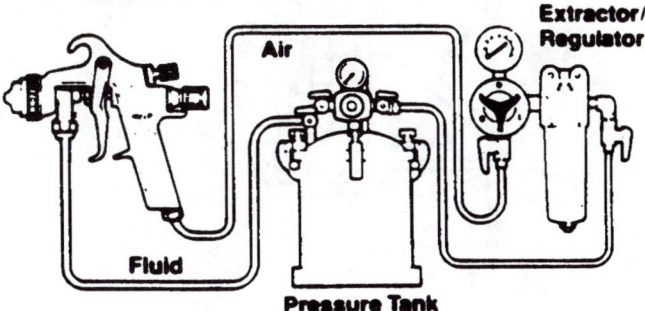

Figure 3.20   High-volume, low-pressure spray gun application system (illustration courtesy of Binks–Sames Corp.).

The airless method alone is an efficient process that exhibits high transfer rates and a minimum amount of overspray. A drawback of the method, however, is that the resulting surface finish is often rough and, although suitable for basic painting, is not amenable to the level of surface detail required for automotive or appliance applications.

The spray technique that integrates conventional air and airless technologies is referred to as *air-assisted airless*. This approach modifies airless methods by incorporating small amounts of compressed air to assist in the shaping and atomization process of the coating. Low-pressure configurations of this equipment yield HVLP characteristics and achieve similar transfer efficiencies. Hence, the technique can achieve compliance with the regulatory community. The main difference between an air-assisted airless approach and a conventional approach is the use of higher fluid pressures. Figure 3.21 illustrates the air-assisted airless technology.

ELECTROSTATIC SPRAYING. A spray method unlike those previously discussed is the electrostatic technique. Although the process of atomization is the same with electrostatic spraying, the manner in which the paint is "delivered" is not. The basic principle of electrostatic spraying is to negatively charge the paint particles so that they seek a grounded object—the part. This is accomplished by several approaches, the main principle being that the atomized particles are passed though a cloud of electrons produced by a high-voltage source administered through an electrode at the gun. Where traditional paint overspray is wasted, electrostatically applied material generally "finds" the part. Transfer efficiencies for this process are typically in excess of 65% and sometimes as high as 95%. Figure 3.22 illustrates the electrostatic process.

One of the requirements for electrostatic spraying is that the part be conductive. Therefore, composite parts are not compatible with this method. Another concern is that because various currents and voltages are present, electric shock is a potential hazard. Ad-

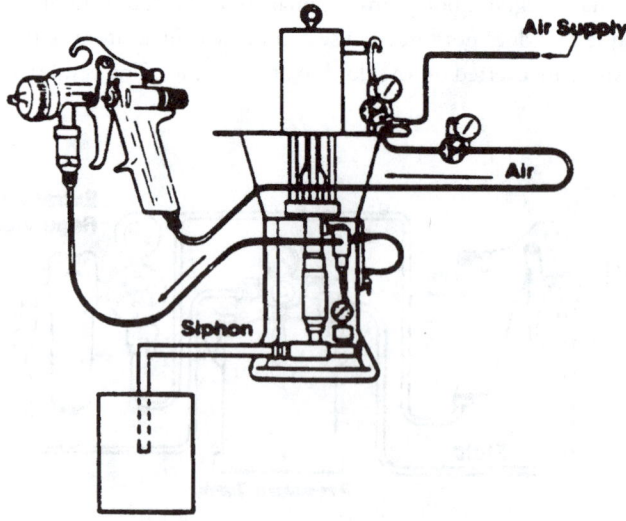

Figure 3.21   Air-assisted airless spraying technique (illustration courtesy of Binks–Sames Corp.).

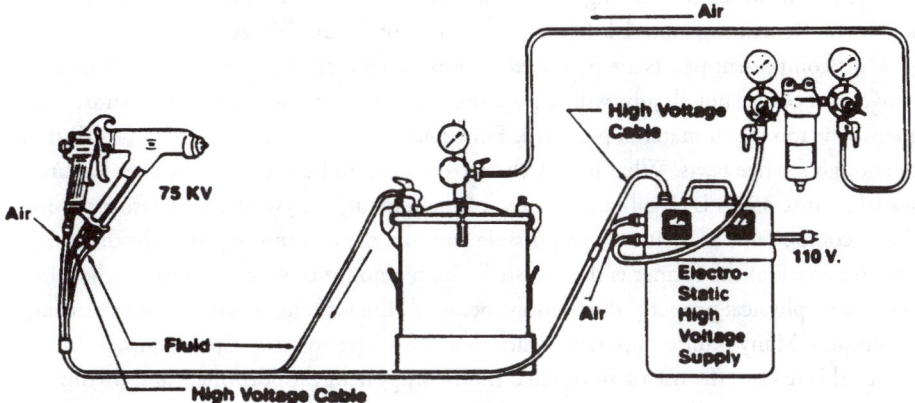

Figure 3.22   Electrostatic spray process (illustration courtesy of Binks–Sames Corp.).

equate safety measures should be considered and strictly enforced to minimize any risks of this nature.

Equipment Options

In addition to methods for depositing paints and coatings, there are a variety of equipment options to minimize air pollution and the generation of wastes. The CAA Amendments of 1990 have had a significant effect on pollution control and will likely continue to in the future. Control technologies that meet these regulatory requirements fall into three principal categories: process modifications, recovery, and oxidation.

Process modifications essentially control and/or minimize VOCs and HAPs by source chemical substitutions, process changes, and improvements in O&M procedures. These options are relatively easy to implement, are typically low cost, and greatly minimize the generation of air pollution and wastes from the onset. As such, examples of these technologies and others will be discussed.

AUTOMATED MULTICOMPONENT PROPORTIONERS AND MIXING. Many of today's coatings are formulated into kits that consist of two or more individual parts or containers (that is, part A, part B, and so on) that are to be mixed together before being applied to the material. This process is similar to the mixing of epoxy resins. Traditionally, the material constituents, which are referred to as resin and hardener, are mixed in specific ratios or proportions, such as 2 parts resin to 1 part hardener. Many surface-coating shops still use hand mixing. Hand mixing is quite amenable to multicomponent paints as long as each paint is contained individually in premeasured portions. With this technique, all that is required is to open each component, pour them into a suitable mixing chamber, and blend. Once the mixing is complete, the constituents begin to react. The reactions consist of crosslinking between various polymers contained in the resin and hardener matrix. These reactions determine the "pot life" of the material or the amount of time within which the coating must be applied before the reaction proceeds past a us-able form. Remaining materials left over past the pot life are wastes.

Multicomponent paints are prekitted to provide consistency from batch to batch. However, because not all jobs will require the same quantity of paint, there can fre-quently be too much material prepared. For example, a kit yielding 3.785 L (1 gal) may be enough for five parts. What if only three parts need to be painted? Two choices are available: mix 3.785 L (1 gal) and dispose of the remainder as waste, or measure the indi-vidual components as accurately as possible and mix the amount required for the task. The drawback of the former is the unusually high amount of waste. However, the latter may create physical property fluctuations because of inaccurate or imprecise measuring techniques. Many multicomponent coatings are sensitive to ratio adjustments.

For this reason, the use of automated multicomponent proportioners and mixing equipment is a technology that is becoming commonplace. Proportioning systems meter paint on a volumetric basis ensuring that mixing ratios are consistently maintained. The mixing portion of the equipment blends the metered portions before arrival at the spray gun. These systems are "on demand" in that as the trigger of a gun is displaced, the paint is metered, mixed, and conveyed to the nozzle. The only wastes generated are those in the mixing chamber and the hose joining the spray guns to the mixing equipment. Periodic flushing with a solvent is required to ensure that no paint "kicks" while in the system. Wastes generated are significantly lower than those resulting from hand-mixing opera-tions.

Proportioning is accomplished primarily by two processes: convergent stream and in-tegration. Convergent stream is a process in which two streams flow together, most typi-cally into a static mixing tube. Integration is different in that one steam is metered into the other and then passed through a mixing device. The advantages of the latter are that fluctuations in paint demand do not affect the process. Therefore, convergent stream pro-portioners are more ideally suited for applications in which there is virtually a constant flow of material. The methods for performing proportioning may be mechanical or electromechanical, although mixing actions are traditionally accomplished through the use of a static mixing tube. This ensures complete and uniform properties. Figure 3.23 il-lustrates the two approaches and a basic schematic for automated proportioners and mix-ing equipment.

The overall benefits from the integration of the aforementioned equipment are reduc-tions in raw materials usage, the minimization of wastes generated, and quality improve-ments through product or coating consistency. Recordkeeping for environmental pur-poses may also be streamlined by integrating peripheral instrumentation to document the daily activities of the equipment. The resulting savings can often pay for the initial costs within a relatively short period of time.

ENCLOSED GUN CLEANERS. To a large degree, solvent emissions are generated through the application of paint but are also generated secondarily through cleaning op-erations. To ensure accurate and uniform coating, every paint shop should keep equip-ment used for the process calibrated and clean. This requires purging pressure pots, spray guns, hoses, and other items with solvent up to several times a day so that they can be

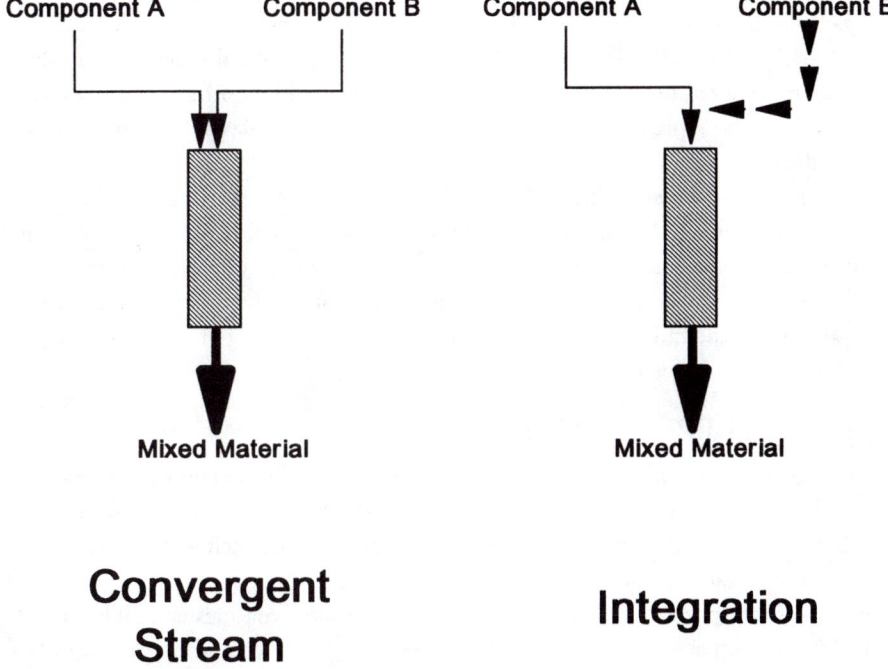

Figure 3.23 Automated proportioner and mixing.

clean for the next type of paint to be used. In the enclosed gun cleaning process, solvent is routed through the guns and dissolves and removes dried paint as it travels. However, rather than spraying the solvent into the atmosphere, it is contained within a cleaning chamber. This simple process is a regulatory requirement in some areas, particularly in ozone nonattainment areas. Solvent may be reused more than a dozen times, thereby greatly reducing solvent usage and emissions. Enclosed spray gun cleaners reuse a given quantity of solvent within the closed container until the solvent becomes so contaminated that it can no longer serve its function adequately.

RECOVERY TECHNIQUES. Recovery is typically accomplished by distillation, condensation, absorption, and adsorption to remove or collect VOCs and HAPs from an airstream or liquid waste. The most common technologies used in this family include solvent-recovery equipment and carbon adsorption beds. Each technology may be implemented individually or in tandem and will exhibit variations in cost and applicability from process to process. As these options are not specific to the surface-coating industry and the reduction of pollutant emissions, they are not elaborated here. An overview of recovery techniques is provided in the Source Reduction section in Chapter 1.

OXIDATION. The final category of pollution control is oxidation. Oxidation " . . . offers the greatest potential for meeting both VOC and HAP compliance requirements.

Nearly 75% of the emissions targeted by the CAA [Amendments of 1990] can be controlled effectively through one or more of the oxidation technologies available" (Pennington, 1994). Typical applications of these technologies include the following: rotary-kiln incineration, fluidized-bed reactors, liquid injection incineration, infrared incineration, catalytic combustion, plasma arc combustion, pyrolysis, wet oxidation, and supercritical water oxidation.

The aforementioned applications are variations on the same theme. Each performs the destruction of VOCs and HAPs through thermal incineration or oxidation, which is the conversion of organic compounds to carbon dioxide and water at high temperatures. Although they can achieve compliance requirements, they are frequently the most expensive option to implement and are considered "end-of-pipe" solutions. For this reason, these approaches are not discussed.

CHEMICAL REPLACEMENT

Coatings are specifically designed to provide a thin film of protection or decoration to a given object. Depending on the purpose of the film, a host of chemicals may be employed. The constituents may be referred to in broad categories such as binders, pigments, thinners, and miscellaneous additives.

Binders are the primary vehicle by which most traditional coatings exist. They are the resin system or polymer to which all other components are added and may consist of single-component or multicomponent systems such as epoxies, urethanes, acrylics, and polyesters. Each resin system is selected based on the requirements of the product. For example, epoxy-based coatings generally exhibit good strength and chemical resistance, whereas acrylics have little resistance to water and solvents.

Pigments are insoluble constituents that are added to a resin system to modify or enhance physical properties or to tailor visual characteristics. Pigments altering physical properties can be employed for environmental protection, conductivity, or reinforcement. Common materials used for corrosion resistance include certain metals such as chromium, zinc, and aluminum. Nickel, copper, and silver provide electrical conductivity. The most common application of pigments, however, is as a colorant. Examples of pigments used as colorants are provided in Table 3.11.

Care should be taken during pigment selection because the light stability of a colorant may be limited. For this reason, several supplemental products have been developed and are continuing to evolve that incorporate silica. Silica coatings improve heat and light stability compared to uncoated grades. Another important factor to consider is that if a product is intended to be reclaimed, recycled, or permanently disposed of in some fashion, the constituents should be reviewed for their potential to create hazardous wastes. Therefore, inorganic pigments such as chromium oxides, cadmium, and lead oxides should be considered for replacement wherever possible.

Additives are compounds employed to lend specific in-process characteristics or enhancements to a product. Constituents of this nature ease application processes, adjust cure times or drying rates, and improve physical attributes such as ultraviolet (UV) or heat resistance. Additives are typically low-molecular-weight compounds and are a minor constituent within a coating matrix.

Table 3.11   Pigments used as colorants.

Pigment	Colors
Titanium dioxides, zinc sulfides	White
Iron oxides	Red, yellow, brown, black
Lead chromates	Yellow, orange
Cadmium	Red, yellow, orange, maroon
Chromium oxides	Green
Carbon black	Black

Most coating ingredients are not harmful to the environment. The exceptions to this include solvents used as thinners or strippers and heavy metals used for corrosion protection or coloring. These solvents and heavy metals are targets for regulation and, therefore, a focus in most waste minimization programs.

Solvents and Water-Based Paints

Until recently, most paints used solvents such as methyl ethyl ketone, toluene, xylene, and others to thin the coating before application. The purpose for thinning was to tailor application viscosity before spraying to achieve good atomization and a smooth finish. However, the problem with this approach today is that these solvents are now regulated as VOCs or HAPs. Their emissions are regulated in quantities via permits in hourly, monthly, and/or yearly rates. At times, this can inhibit production. Although implementation of various thermal oxidation or incineration equipment can rectify the issue, the associated high costs are frequently a deterrent to this line of action.

To avoid these pitfalls, numerous efforts have been initiated by coatings manufacturers to find alternatives to the use of solvents. In some cases, regulated solvents are being replaced with nonregulated or more environmentally friendly compounds. An example of this approach is the reformulation of paints to incorporate the recently exempted solvent, acetone. Acetone is no longer a VOC, nor is it a HAP. Therefore, for all practical purposes, it is similar to water. The most common ideologies are to replace solvent-based paints with water-based alternatives or to increase the amount of solids while lowering the VOC content.

HIGH-SOLIDS LOW VOLATILE ORGANIC COMPOUND BASED PAINTS. Two predominant resin systems used in the formulation of high-solids low-VOC based paints include the epoxy and polyurethane families. Each is used for applications compatible with its properties. Where epoxies generally exhibit flexibility, chemical resistance, and adhesion to a variety of substrates such as wood, glass, metals, plastics, and others, they are commonly employed as primers. Because they are relatively resistant to UV light, polyurethanes are a frequent choice for top coats.

Most epoxy-based high-solids, low-VOC coatings are two-component systems. The determination of whether it is an air or high-temperature bake system will dictate the ne-

cessity for mixing the components before packaging. High-temperature epoxies are generally mixed before packaging because they do not begin the crosslinking process until temperatures range from 65 to 175°C (150 to 350°F). Therefore, these systems appear to be
a single component when received by the end user.

High-solids epoxy alternatives are generally well below regulated limits for VOC content in coating, which is typically near 0.4 kg/L (3.5 lb/gal). Before the development of
these low-VOC coatings, is was not uncommon to see VOC content exceeding 0.6 to 0.7
kg/L (5 to 6 lb/gal). Therefore, by simply substituting materials rather than using costly
abatement and control technologies, a noticeable reduction can be achieved.

Polyurethane versions are also popular resin systems. This coating family is a two-
component system whereby polyurethane is formed by the reaction of various polymers,
such as acrylics and polyesters (each of which contain hydroxyl groups), with a polyiso-
cyanate (versions of toluene diisocyanate, hexamethylene diisocyanate, and others). Single-component systems also exist that use atmospheric moisture as a catalyst for curing,
although they are not as widely used because of the potential for moisture contamination
during storage and variations in cure time caused by climatic conditions.

Polyurethanes are also capable of meeting regulatory VOC limitations. As previously
mentioned, they are a good choice for use as top coats because of their UV tolerance,
abrasion resistance, and hardness. In addition to these qualities, curing is easily accomplished at temperatures ranging from subzero to elevated levels. Aliphatic, acrylic, and
polyester urethanes may be selected based on exhibited variations to these characteristics.

The primary drawbacks to high-solids, low-VOC coatings that differentiate them
from traditional multicomponent systems are that pot life may be shortened and the ease
with which they may be applied to a uniform thickness may be compromised on complex
surfaces. However, integration of these coatings with application equipment such as
HVLP guns can provide an excellent base for VOC and HAP emission-reduction programs while minimizing capital costs.

WATER-BASED PAINTS. Another method for achieving regulatory compliance within
coating operations is to use water-based low-VOC materials. The measurement criteria
typically used to determine regulatory compliance is a unit based on pounds per gallon
(kilograms per litre) of VOC, as applied. An example of this concept is provided in the
following subsection.

CALCULATION OF VOLATILE ORGANIC CHEMICAL CONTENT. The regulatory
limit for the VOC content of top coats is designated as 3.5 lb/gal. A vendor has introduced a new two-part polyurethane coating that appears to be ideal for surface-coating
applications within a specified portion of a facility. The mix ratio of part A to part B is 4
to 1, and the VOC content of each, less water and exempt solvents, is 3.3 lb/gal and 1.4
lb/gal, respectively. This information is verified on the coating's material safety data
sheet. Will this coating meet the regulatory requirement?

First, a table should be set up to record coating data. Because all the VOCs contained
within the coating will flash off, the entire amount is considered to be emitted into the
environment. The last column in Table 3.12 is the product of the volume of each con-

Table 3.12 Coating data example.

Coating components	Volume, gal[a]	VOC content, lb/gal[b]	Emissions, lb[c]
Part A	4	3.3	13.2
Part B	1	1.4	1.4
Total	5		14.6

[a]gal × 3.785 = L.
[b]lb/gal × 0.119 8 = kg/L.
[c]lb × 0.453 6 = kg.

stituent used and the corresponding VOC content. The VOC content for the mixed paint may be determined by adding the total emissions and dividing by the resulting total paint volume.

The calculation for total VOC content is

$$VOC = 14.6 \text{ lb}/5 \text{ gal} = 2.92 \text{ lb/gal} \qquad (3.14)$$

This coating meets the VOC-content limitation. What about a situation in which additional thinner must be added to properly atomize the paint (Table 3.13)? This may be required periodically because of varying environmental conditions such as temperature and humidity. The prior exercise should be repeated where 1 gal of mixed paint must be thinned by 2 pints of solvent (the density of the solvent is 6.8 lb/gal and 1 pint equals 0.25 gal.).

The calculation for total VOC content is

$$VOC = 4.62 \text{ lb}/1.25 \text{ gal} = 3.70 \text{ lb/gal} \qquad (3.15)$$

On days when thinning is required, there is a high probability that the coating will be noncompliant. As a general rule, thinning should be avoided whenever possible. Thinning defeats the purpose of using low-VOC paints.

As a rule, water-reducible coatings satisfy VOC-content restrictions. They are excellent solutions for minimizing air pollution. This class of coatings is available in resin sys-

Table 3.13 Coating data example with thinner added.

Coating components	Volume, gal[a]	VOC content, lb/gal[b]	Emissions, lb[c]
Part A	0.80	3.3	2.64
Part B	0.20	1.4	0.28
Thinner	0.25	6.8	1.70
Total	1.25		4.62

[a]gal × 3.785 = L.
[b]lb/gal × 0.119 8 = kg/L.
[c]lb × 0.453 6 = kg.

tems incorporating epoxies, polyurethanes, acrylics, or most any system that is readily available in traditional solvent-based forms. A minor setback to the inclusion of high amounts of water is that a small amount of durability and chemical resistance may be sacrificed. In applications involving construction equipment, electrical enclosures, miscellaneous tools, and the like, this is probably of no consequence. However, in some structural applications, it may be a concern. In any event, proper care should be exercised during the selection process of any new production-related material before implementation to ensure compatibility with the use of the product. Water-based formulations are progressing to the point that corrosion-inhibiting primers are now widely used throughout the airline and aerospace industry. Options probably exist for the integration of water-based alternatives in several applications.

An in-process parameter that differentiates water-based coatings from their solvent counterparts is curing or drying time. Because water does not flash off as quickly as solvents, humidity and low temperature will only serve to aggravate this condition. Therefore, it is not uncommon to use air-forced or low-temperature oven drying processes to accelerate the curing or handling time of the paint.

A coating application method used for solvent-based paints in most cases should prove to be adequate for water-based paints as well. However, with the presence of water comes the potential for corrosion. This is particularly true in cases in which exposure to steel and other metal components is likely. In some instances, plastics or rubbers may react or degrade as well. Therefore, it is a good idea to isolate water-based systems with equipment that incorporates 316 stainless steel and other water-tolerant materials.

If systems are to toggle between solvent- and water-based alternatives, care should be taken to ensure that there are no material incompatibilities between the constituents. In some instances, lines may become clogged because of the mixing of water with solvent-based components. Another important consideration is the mixing of water-based paints with polyurethane systems. Incompatibilities of isocyanates with water can cause exothermic reactions that may be dangerous to the health and safety of employees and the facility. Electrostatic application methods, which employ electricity, should also be carefully examined, as water is a good conductor.

Overall, water-based coatings are an excellent solution for reducing air pollution and minimizing hazardous wastes. Their use will undoubtedly continue to become more prevalent as the regulatory environment in respect to the emission of VOCs falls under tighter controls.

Heavy Metals

Heavy metals such as chromium, lead, cadmium, and others are used in coatings to impart specific physical attributes or even color enhancement. Most uses of these metals for cosmetic purposes have been phased out over recent years. However, substitutes for corrosion protection are not as simple. Chromium and lead are the primary constituents in many corrosion-inhibiting primers, especially within airline, aerospace, and military applications. Although strides have been made to curtail VOC emissions through the minimization of solvent content in these coatings, HAP content and the emission of particulate matter is an entirely different issue.

To control the emission of heavy metal particulates, paint arrestors or filter media and water wash curtains are the primary methods. However, these cannot ensure 100% removal before release of the water into the environment. In fact, the most harmful particulates (that is, those less than 10 μm in diameter) are those that pass through these control measures. Many dry filter systems are based on removal efficiency as calculated on a weight basis. Therefore, large particles are retained and constitute the majority of the mass controlled. Particulates that are much smaller in size and, therefore, lower in mass are less controlled.

Public health criteria are derived from exposure to these smaller particles. Hence, emission limitations exist. This corresponds to the amount of paint containing heavy metals—many of which are suspect carcinogens—that may be applied within a given amount of time and is manifested by hourly rates (that is, cubic metres per second or gallons per hour). Although filter media are constantly being developed to improve particulate removal efficiencies, this approach to filtration is essentially an end-of-pipe solution. Reduction or elimination of the heavy metal constituents is true source reduction. Another detrimental side effect to the use of heavy metals in coatings is that solid wastes produced are likely to be hazardous. Therefore, additional regulation and higher costs are incurred.

For reasons of public health, waste disposal and liability, production rate limitations, and cost, the search for alternatives to the use of heavy metals is highly desirable. Current research is underway by many paint manufacturers and end users to minimize or eliminate the use of heavy metals. One such study is being conducted through the Rotorcraft Industry Technology Association to determine the feasibility of replacing heavy metals with nonhazardous constituents while still maintaining the requirements defined within military specifications for corrosion resistance, adhesion, flexibility, and other parameters.

Stripping Options

The majority of chemical stripping methods have employed the use of methylene chloride, phenols, and various corrosive mixtures. Alternatives to these options are constantly under development. While some methods use more environmentally friendly chemicals, others achieve their objectives through physical processes. The most recent chemical formulations derived as substitutes to chlorinated and phenolic ingredients are alkaline in nature and may be performed in either hot or cold bath environments.

A concern with corrosive strippers, however, is that the substrate may be subject to attack in addition to the coating. Fortunately, most epoxies and new formulations of waterborne and high solids coating alternatives are fairly resistant to alkaline solutions. Heating the solution may also help to increase the speed and efficiency of alkaline stripping methods. Development along this line has resulted in two-phase solutions, the lower phase of which may be a highly alkaline concentrate, and the upper layer a water-based or alkaline-insoluble constituent. Boiling points minimize emissions and improve the effective life of the solution. It is not uncommon for solvents such as ketones, glycols, or amines to be incorporated into this approach. Most alternative chemical strippers are not as efficient or effective as their predecessors.

The physical options that have been developed have proven to be a successful new approach to the problem. They may not be used in all applications, but if selected carefully, they have proven to be faster and less expensive than time-consuming chemical methods. The two most commonly used physical stripping methods are plastic media blasting and dry ice blasting.

PLASTIC MEDIA BLASTING. Plastic media blasting (PMB) is a fast, efficient, and cost-effective method for the removal of organic coatings from primarily metallic substrates. Although it is similar to sandblasting, PMB has a much lower hardness and will greatly minimize effects to the substrate material. The hardness of PMB is such that it will displace deposited organic coatings while leaving the metal substrate relatively untouched. Compatible metals include those that are relatively soft such as aluminum, brass, and copper. Some plastic substrates are also compatible with the PMB process; however, proper care and investigation are warranted. This is possible by selecting a media that has a lower hardness than that of the substrate.

The original use of PMB was fostered through the U.S. Department of Defense in the aerospace industry as a proactive approach to the elimination or replacement of toxic chemical strippers such as methylene chloride. Upon development and implementation, the method was found to be environmentally friendly and economically attractive. Many applications of PMB have seen significant reductions in labor costs. This technology has continued to grow during the last decade and appears as though it will continue to do so in the future. Parameters to consider range from substrate compatibility issues, to proper plastic media selection, to a host of operational and equipment variables.

Several issues exist because of the impingement of the plastic particles on a substrate. These issues involve the partial removal of material, crack propagation, fatigue life reduction, and other concerns. Although the pressure at which the PMB is applied is low (typically less than 2.8×10^5 Pa [40 psi]), damage can result. Therefore, PMB is available in a variety of polymers such as polyesters, acrylics, urea, and melamines. Figure 3.24 provides an illustration of the usefulness and general applicability of the materials (Organic Finishing, 1995).

In addition to eliminating or reducing the necessity for toxic stripping methods and providing economic benefits as a result of the reduction of labor costs, PMB exhibits other environmentally proactive options through recycling of the spent material. This is possible because PMB is traditionally a thermoplastic material. Thermoplastics differ from thermosets in that they can be reformed or melted down for reprocessing again and again. Thermosets, once cured or crosslinked, do not present this option. Recycling programs have been set up by PMB vendors to lease the material to users subject to return of the spent product for reprocessing into bath fixtures, pen holders, and other products. Therefore, the implementation of PMB saves time and money and reduces wastes and long-term liabilities.

Although imperfections exist, PMB can be an attractive option for a number of coating removal applications. Improper selection of media type and grit size can result in ineffective coating removal or damage to substrates via pitting, warping, or even metal fatigue. Therefore, adequate hardness testing of all constituents involved should be

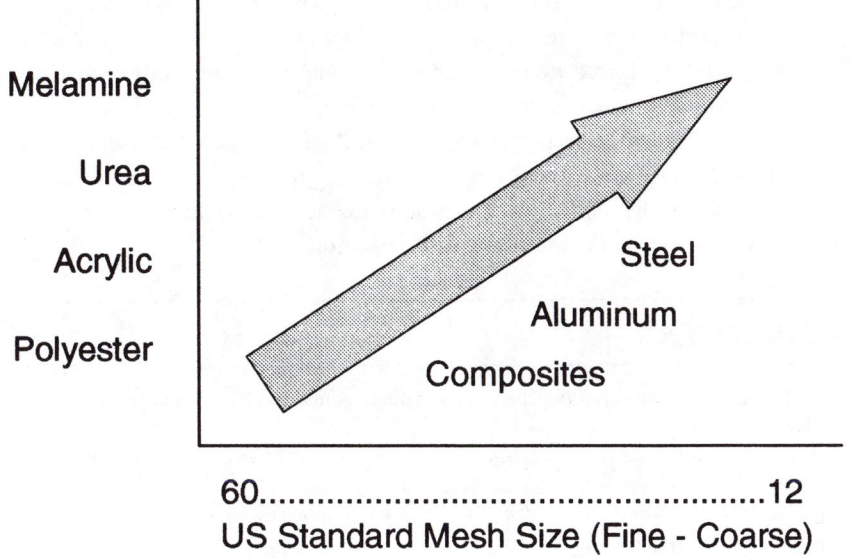

Figure 3.24   **Plastic media blasting selection criteria.**

performed. Operational problems may also be present because of the work environment or raw material contaminants. However, with specifications to identify acceptable PMB parameters and diligence in controlling the process, PMB can be a viable alternative to chemical stripping methods.

DRY ICE BLASTING. In the quest for environmentally compatible alternatives to chemical paint stripping, dry ice or carbon dioxide blasting techniques have developed. Originally investigated by the Lockheed Corporation in the early 1980s as a surface preparation or cleaning process, this method has received increasing attention and has evolved into a relatively successful coating removal process (Organic Finishing, 1995).

The dry ice blasting process consists of converting liquid carbon dioxide into a carbon dioxide "snow." Once in a particulate form, it is compressed into pellets and projected through nozzles at a high velocity, whereby the impingement forces are capable of removing select contaminants. The forces exerted by carbon dioxide pellets differ significantly from sand or PMB methods. Carbon dioxide pellets achieve their objectives by penetrating surface contaminants, dislodging them from underneath, and sweeping away the residues via the continuous flow of carbon dioxide. Another physical stripping mechanism is achieved from the outside moving inward, with the transfer of energy to the substrata being considerable higher. This sweeping action does not ensure that all residues are removed from the part. In fact, the contaminants may merely be relocated to another area. Therefore, operator skill and experience play a role in the process.

For these reasons, carbon dioxide blasting is most efficient as a cleaning method for the removal of manufacturing process residues rather than as an organic coating stripping tool. However, a consistent advantage to this method is that the pellets disappear on contact and do not damage substrate materials. In addition, the method is highly compatible with a wide variety of substrates and generates a minimal amount of wastes (that is, only contaminant residues remain). Carbon dioxide blasting is a "dry process" alternative with the same benefits as those realized though sand and PMB techniques. No electrical hazards exist and minimal surface preparation and masking procedures are required. However, one drawback of the method is that carbon dioxide pellets exhibit slower coating removal rates compared to other physical stripping options.

REFERENCES

Dammer, H.J. (1980) Non-Cyanide Chemical and Electrochemical Metal Stripping. *Galvanotechnik* (Ger.), **71**, 29.

Doscher, P.A. (1991) Grease and Oil Removal Using Absorbent Media. *Proc. 2nd Annu. Int. Workshop Solvent Substitution*, Phoenix, Ariz.

Freeman, H.M. (1989) *Standard Handbook of Hazardous Waste Treatment and Disposal.* McGraw-Hill, Inc., New York, N.Y.

Hammel, B. (1982) Metal Stripping. *Oberflaeche-Surf.* (Ger.), **23**, 215.

Hartinger, L. (1986) Effluent and Heavy Metal Discharge Reduction in the Metal-Working Industries Using Ion Exchange. *Korresp. Abwass.* (Ger.), **33**, 895.

Hartinger, L. (1994) *Handbook of Effluent Treatment and Recycling for the Metal Finishing Industry.* 2nd Ed., Finishing Publications LTD/ASM International Stevenage, Herts, U.K., 405/724.

Kauczor, H.W. (1969) Solid Ion Exchangers to Concentrate Metal Containing Solutions. *ERZMETALL 22*, B19.

Oehme, C.H. (1971) Experiences with Ion-Exchangers in Rinsewater Recycling Loops. *Oberflaeche-Surf.* (Ger.), **12**, 105.

Organic Finishing Guidebook and Directory Issue (1995). *Met. Finish.*, **93**, 5A, Elsevier Science, Inc., New York, N.Y., 352.

Pennington, R.L. (1994) VOC, HAP Control Systems Compared. *Environ. Protection*, **5**, 9, 56.

U.S. Environmental Protection Agency (1991) *Stratospheric Ozone: The Problem and the Strategy*, EPA Journal, Washington, D.C., 35.

U.S. Environmental Protection Agency (1992) *Modifications to Reduce Drag Out at A Printed Circuit Board Manufacturer.* EPA-600/R-92-114, Office Res. Dev., Washington, D.C.

U.S. Environmental Protection Agency (1994) *Guide to Cleaner Technologies: Alternatives to Chlorinated Solvents for Cleaning and Degreasing.* EPA/600/R-93/016, Office Res. Dev., Washington, D.C.

SUGGESTED READINGS

Amdur, M.O., *et al.* (1991) *Casarett and Doull's Toxicology, The Basic Science of Poisons.* 4th Ed., McGraw-Hill, Inc., New York, N.Y.

Capaccio, R.S. (1996) Metals Control. *Ind. Wastewater*, **4**, 1, 21.

Chemical Rubber Company (1970) *Handbook of Chemistry and Physics.* 51st Ed., Cleveland, Ohio.

Childers, D.G. (1997) Environmental Economics in the Aerospace Industry. *Proc. 18th Annu. Am. Electroplaters and Surf. Finish./EPA Pollut. Prevention & Control Conf.*, Orlando, Fla.

Davies, V.R. (1995) Troubleshooting Ion Exchange Systems. *Ind. Wastewater*, **3**, 5, 32.

Dickerson, R.E., *et al.* (1979) *Chemical Principles.* 3rd Ed., The Benjamin/Cummings Publishing Company, Inc., Menlo Park, Calif.

Manahan, S.E. (1991) *Environmental Chemistry*. 5th Ed., Lewis Publishers, Inc., Chelsea, Mich.

Meier, J.F., *et al.* (1995) Pollution Prevention in Organic Finishing Processes. *Proc. 88th Annu. Meeting Exhibition Air Waste Manage. Assoc.*, San Antonio, Tex.

Quarles, J., and Lewis, W.H. (1990) *The NEW Clean Air Act, A Guide to The Clean Air Program as Amended in 1990*. Morgan, Lewis & Bockius, Washington, D.C.

Title VI— Stratospheric Ozone Protection (1990) Public Law 101-549.

U.S. Environmental Protection Agency (1995) *Implementation Strategy for the Clean Air Act Amendments of 1990 (1995 Update)*. Off. Air Radiat., EPA-410/K-95-001, Washington, D.C.

U.S. Environmental Protection Agency (1992) The Evaluation of an Advanced Reverse Osmosis System at the Sunnyvale, California, Hewlett Packard Facility. In *Pollution Prevention Case Studies Compendium*. EPA-600/R-92-046, Risk Reduction Eng. Lab., Office Res. Dev., Washington, D.C.

U.S. Environmental Protection Agency (1992) *Waste Minimization Assessment for a Manufacturer of Printed Circuit Boards*. EPA-600/M-91-022, Office Res. Dev., Washington, D.C.

National Aeronautics and Space Administration (1991) Earth's Vanishing Ozone. *NASA Tech Briefs*, Washington, D.C., 114.

57 Federal Register 135 (1992) Protection of Stratospheric Ozone.

4 Engineering Decision Analysis and Economics

The implementation of waste minimization strategies requires that considerable attention be given to cost versus benefit. Although some environmental projects may appear to be the "right thing to do," the project must be financially sound and prove to be worthwhile from a business standpoint. Engineering decision analysis and economics is a method or process whereby alternatives are compared based on their worth from a cost perspective. Numerous arguments that environmental issues are a cost of doing business are prevalent. However, in all probability, environmental drivers are frequently opportunities for reducing operating costs and increasing profitability. Various financial topics will be presented in this chapter. These tools should be an integral part of any waste minimization project. Up to this point, discussions have focused primarily on technical issues. Examples used in this chapter are tailored to illustrate how important contributions can be made by using economic decision analysis.

Individual corporate business strategies will affect the direction of analyses performed. In most cases, long-term goals will compete with short-term goals. Profitability may be based on survival versus the long-term health and stability of a company. Nevertheless, as long as the parameters of a problem are defined, engineering decision analysis and economics is the primary driver for the implementation of any new process.

Tools of Engineering Economics

The old saying, "time is money," is at the heart of all financial decisions. This concept is translated into a single financial term, *interest*. Interest is essentially the earning power of money. It is also frequently used to make charges for services rendered such as administrative costs for processing loans, compensation for the risk of an investment, and, perhaps most importantly, the loss of the use of the funds during the life of a loan. The following subsections address the basic concepts of interest. Further discussions incorporate cash flow and the time value of money.

INTEREST FACTORS AND DEPRECIATION

Simple interest is the interest earned in relation to a defined amount of capital involved in a loan over a given amount of time. This quantity is given by the following equation:

$$I = P(i/n) \tag{4.1}$$

Where

I = simple interest,
P = principal or capital involved in the loan,
i = interest rate over specified period of time, and
n = number of interest periods.

Because the principal or amount of the loan is fixed and the interest charges are constant for each time period, the total charge for interest becomes:

$$F = P + I = P + P(i/n) = P(1 + i/n) \tag{4.2}$$

where F is the future value of money.

It is important to note that, in the case of simple interest, no interest charges are incurred for interest accrued over time. The principal remains fixed as does the rate at which the interest is accumulated.

Compound interest is a concept that incorporates the effect of unpaid interest reflected over time. Interest charges are accumulated with the principal, which creates a cascading effect over time. For example, if the simple interest rate of a $1 000 loan is defined as 8% annually, the annual interest earned on the loan is

$$I = P(i/n) = (\$1\ 000)(0.08) = \$80 \tag{4.3}$$

However, if the $80 in interest is not paid at the end of the year, then the unpaid portion is added to the initial principal amount yielding $1 000 plus $80, or $1 080. This may be more easily understood from the perspective of a savings account. Interest earned is compiled on top of the principal amount allowing accelerated growth of the fund rather than a fixed amount of interest earned annually. Compound interest may be expressed as

$$F = P(1 + i)n \tag{4.4}$$

For example, if $1 000 is deposited in a savings account for 3 years at an annual interest rate of 6.5%, the total balance at the end of the third year will be

$$F = 1\ 000(1 + 0.065)^3$$
$$F = \$1\ 207.95$$

Interest rates discussed thus far have been quoted on an annual basis, which is typically the rate at which they are compounded. However, interest may also be compounded monthly or quarterly. A method for expressing interest compounded over such time periods is referred to as the *nominal interest rate*. More frequent compounding periods yield higher accumulated interest. An annual interest rate of 6.5% compounded monthly yields a monthly interest rate of 0.54% (6.5% divided by 12). Using Equation 4.4 for compound interest, where n equals 36 months (3 years) and a principal value of $1 000, the following calculation can be made:

$$F = 1\ 000(1 + 0.005\ 4)^{36}$$

$$F = \$1\ 213.95$$

Notice the higher value of money at the end of 3 years when interest is compounded monthly versus annually. Again, the more frequent the compounding period, the larger the amount of interest that will accrue over time.

Another term used to express interest is the *effective interest rate*. Effective interest is a method for providing a direct comparison of interest rates that compound at different intervals. In the two previous examples, a balance of $1 000 accrued interest at 6.5% compounded annually and 6.5% compounded monthly. Without actually calculating the future value of money at the end of the third year, no direct comparison was possible. However, by determining the effective interest rate on an annual basis using the following equation, this confusion may be avoided:

$$\text{Effective interest rate} = i_e = (1 + r/m)^m - 1 \tag{4.5}$$

Where

r = nominal interest rate, and
m = number of compounding periods per year.

Therefore, a nominal interest rate of 6.5% compounded monthly yields an effective interest rate of

$$i_e = (1 + 0.065/12)^{12} - 1$$

$$i_e = 0.067 = 6.7\%$$

This figure may now be directly compared to the 6.5% compounded annually without having to conduct any further calculations. A rate of 6.5% compounded annually will incur less interest charges than 6.5% compounded monthly (6.7% compounded annually).

In comparing nominal interest rates, consider the example of a loan to be taken out on the purchase of a new paint booth that costs $20 000. Lender A is willing to finance the project at 8% compounded monthly, whereas lender B has agreed to do the same at a rate of 10% compounded quarterly. Which loan will incur the lowest amount of interest if the life of the loan is 5 years?

The loan to select is the one that exhibits the lowest effective interest rate. Therefore,

$$\text{Lender A} = i_e = (1 + 0.08/12)^{12} - 1 = 0.083 \text{ or } 8.3\%$$

$$\text{Lender B} = i_e = (1 + 0.10/4)^4 - 1 = 0.104 \text{ or } 10.4\%$$

The loan with the least amount of interest charges is that from lender A at 8% compounded monthly.

Where interest is often referred to as the "power of money," another financial concept is depreciation. *Depreciation*, defined as a decrease in worth, is a method for recovering invested capital funds used for income-generating assets. This is a particularly important tool when waste minimization projects are initiated by the purchase of new or specialized equipment. Because new equipment is traditionally worth more than old equipment, depreciation charges are a deductible item from gross income when determining overall tax liabilities. It is in this manner that money spent on capital equipment is recuperated.

There are several methods for quantifying depreciation charges over time. These include straight line, sum of digits, declining balance, and sinking fund.

Depending on the method selected, depreciation charges may be accelerated to allow the asset to be written off earlier. Sum of digits and declining balance are two such approaches. Where these approaches may allow invested money to be recovered more quickly, the acceleration decreases the time value of taxes. The reasons for depreciating an asset vary. In industry, those reasons are attributed to physical deterioration or the ability to keep pace with expanding operations. Another reason may be attributed to the technological obsolescence of the asset. Time value will be covered in the Methods for Project Analysis section of this chapter. Regardless of the method selected, the total taxes are the same. Figure 4.1 illustrates a comparison of the most commonly used depreciation methods and their relative rates for recovering invested capital funds. It should be noted that the values plotted are book values (BVs). Book values are the difference between the original cost of an asset minus the accumulated depreciation.

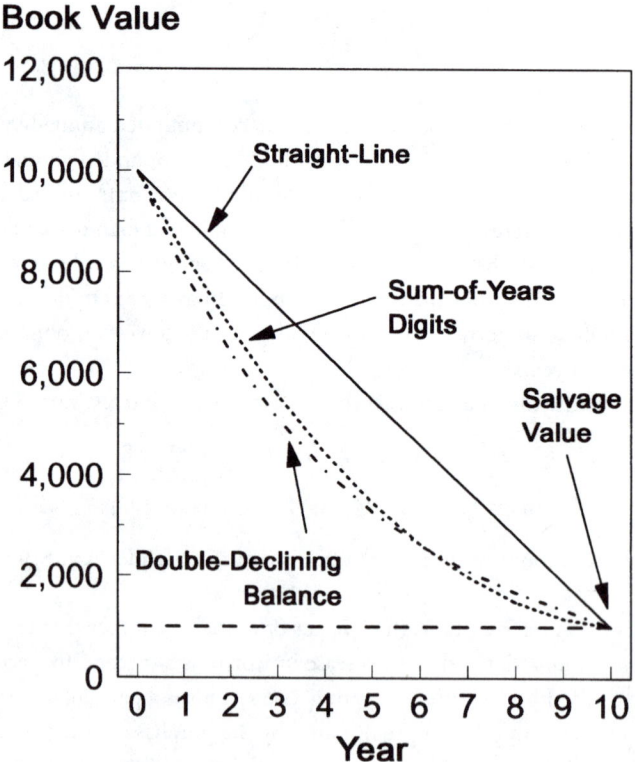

Figure 4.1   Comparison of depreciation methods.

Straight-Line Depreciation

Straight-line depreciation is the simplest to calculate and, therefore, the most frequently used method. Depreciation over time with this concept is constant. It is the slope of the line between the initial cost of the asset and the residual or salvage value.

$$D = (P - S)/N \tag{4.6}$$

Where

D = annual depreciation charges;
P = initial cost of asset, purchase price;
S = salvage value; and
N = life of asset, years.

The BV at the end of each year may be calculated using the following equation:

$$BV_n = P - n(P - S)/N \tag{4.7}$$

where n is the number of years depreciated from time of initial purchase.

Sum-of-Digits Depreciation

The sum-of-digits depreciation method accelerates the charges toward the initial purchase of the asset. The sum-of-digits calculation is so named because of the procedure used for arriving at a depreciation value. The depreciation amount annually is derived from the ratio of the remaining years of life of the asset (N minus n plus 1) to the sum of the years' digits for the life of the asset (for example, 1 plus 2 plus 3 plus . . . N) multiplied by the purchase price minus the salvage value (P minus S). Note that the sum of digits for the life of the asset may be represented by the following formula:

$$Sum\text{-}of\text{-}years\ digits = N(N + 1)/2 \tag{4.8}$$

The annual depreciation value for the sum-of-years digits and the remaining BV may be determined from the following equations (the variables are as previously defined):

$$D_n = 2(N - n + 1)(P - S)/[N(N + 1)] \tag{4.9}$$

$$BV_n = 2[1 + 2 + 3 + \ldots (N - n)](P - S)/[N(N + 1)] + S \tag{4.10}$$

The BV expression may be further simplified as the following equation shows:

$$BV_n = (N - n)(N - n + 1)(P - S)/[N(N + 1)] + S \tag{4.11}$$

Declining-Balance Depreciation

Like the sum-of-digits method, the declining-balance method also accelerates the depreciation of an asset toward the early years of use. The depreciation rate used to determine these amounts from year to year is given by the following expression:

$$\text{Declining balance depreciation rate} = 1 - (S/P)^{1/N} \qquad (4.12)$$

For this method, the salvage value of the asset must be greater than zero. Otherwise, Equation 4.12 will be meaningless. This rate is then used to arrive at a BV, or remaining undepreciated balance, which, in turn, is used to arrive at a depreciation charge. To determine the current year's depreciation charge, the BV for the prior year must be calculated first:

$$BV_n = P(1 - \text{depreciation rate})^n = P(S/P)^{n/N} \qquad (4.13)$$

$$D_n = BV_{n-1}[1 - (S/P)^{1/N}] \qquad (4.14)$$

The most commonly used declining-balance method is the double declining-balance method. This procedure does not use the S/P ratio; rather, it uses the following formula to determine the depreciation rate:

$$\text{Double declining-balance depreciation rate} = 2/N \qquad (4.15)$$

As may be seen, this method allows the depreciation rate to be twice that of the straight-line method. The BV and depreciation charges are still calculated in the same manner as basic declining balance methods, but when rearranged become the following:

$$BV_n = P(1 - 2/N)^n \qquad (4.16)$$

$$D_n = BV_{n-1}(2/N) \qquad (4.17)$$

When using the double declining-balance method, it is important to keep an eye on the salvage value (the salvage value is not used in the previous equations). Therefore, when the BV becomes equal to or less than the salvage value, depreciation of the asset must cease.

Sinking-Fund Method

The sinking-fund method for quantifying depreciation charges is different from the approaches previously discussed in that it amortizes the value of an asset slowly during the initial years, with an accelerated depreciation occurring in the later years. This method is also less popular than the other methods because of the manner in which the depreciated values are handled. For example, traditional depreciation charges are reflected on company books for tax purposes as working capital, but the sinking fund is normally handled externally in the form of a reserve that is established by payments physically made outside of the company. The size of payments made to the fund is determined by the estimated future replacement cost of the asset, taking into consideration the interest that is earned on the funds accumulated within the reserve over the economic life of the existing asset. The formula required to perform sinking-fund calculations takes into account the time value of money, which has not been covered in this section. However, the following equations are provided and will be referenced at a later point.

Depreciation charges, or deposits made to a reserve for the purchase of a future asset to replace an existing one, are given by the following equation:

$$\text{Depreciation payment} = (P - S)(A/F, i, N) \qquad (4.18)$$

The quantity (A/F, i, N) is a factor that, when multiplied by the value P minus S, yields a constant annual charge for depreciation that takes into account the time value of money. The variable, A, stands for annuity, whereas F is the future value. The quantity, N, is, as previously defined, the number of years of life for the asset. A point of confusion is that as the payments (depreciation charges) are accumulated in an account, they are earning interest. Therefore, the depreciation fund over time is reflected by

$$\text{Depreciation fund} = D_n = (P - S)(A/F, i, N)(F/A, i, n) \qquad (4.19)$$

In Equation 4.19, the quantity (F/A, i, n) is a factor that takes into consideration the interest earned on the account. Here, it is important to note the difference between N and n. The quantity n pertains to the year in which information is desired; it is not the life of the asset. These topics will be discussed in the following section and are provided here for future reference only because their likelihood for use is minimal. In closing, the BV for the sinking-fund method is found by the following equation:

$$\text{BV}(n) = P - (P - S)(A/F, i, N)(F/A, i, n) \qquad (4.20)$$

Or

$$\text{BV}(n) = P - D_n$$

INFLATION AND TAXES

Inflation causes a rise in prices while decreasing the purchasing power of a given quantity of money. Inflation rates are typically determined by the Consumer Price Index, which takes into account numerous factors such as goods purchased, geographies, demographics, and annual household income.

Although inflation is not always accounted for in the evaluation of economic decisions, it should always be considered. If the inflation rate during a specified period of time is thought to be only 2 to 3%, it may have little to no effect in the overall scheme. However, if the rate approaches double digits, it will have ramifications that will need to be addressed. For a constant inflation rate, expressed as f, the cost of an asset will increase in relation to its present cost as shown in the following equation:

$$F = P(i + f)^n \qquad (4.21)$$

All variables are as previously defined. However, the future worth of money must also be accounted for, and its devaluation is reflected by the following equation:

$$F = P/(1 + f)^n \qquad (4.22)$$

This expression, when the investment is made at a constant interest rate, i, is modified and becomes the following equation:

$$F = P[(1 + i)/(1 + f)]^n \qquad (4.23)$$

When interest is earned, it is subject to taxes. As long as the interest period is the same as that for the tax period the analysis may be simplified. For example, if the interest earned for a given period is iP, the amount of taxes subject to these earnings is $(iP)t$, where t is the tax rate. Therefore, the net return after taxes is equal to the interest earned minus the taxes paid, iP minus $(iP)t$. When this is taken into account in regard to the future value of money, the equation for future worth after 1 year will become

$$F = P + (1 - t)iP = P[1 + (1 - t)i] \qquad (4.24)$$

If the effects of inflation are incorporated, then the future worth of the net interest earned after taxes for that year will be

$$F = P\{[1 + (1 - t)i]/(1 + f)\} \qquad (4.25)$$

Equation 4.25 may be rewritten as

$$F = P\{1 + [(1 - t)i - f]/(1 + f)\} \qquad (4.26)$$

Therefore, the overall interest rate, i_o, which reflects interest, taxes, and inflation, may be summarized as follows:

$$i_o = [(1 - t)i - f]/(1 + f) \qquad (4.27)$$

Hence, the equation for evaluating future worth can be expressed as

$$F = P(1 + i_o)^n \qquad (4.28)$$

This takes into account effects that may be experienced over a period of years.

Inflation and Taxes Scenarios

Given a present sum of $5 000, calculate its future worth for each of the following scenarios: scenario 1—taking into account only interest and time ($i = 7\%$, $n = 5$); scenario 2—taking into account interest, time, and inflation ($i = 7\%$, $n = 5$, $f = 4\%$); and scenario 3—taking into account interest, time, inflation, and taxes ($i = 7\%$, $n = 5$, $f = 4\%$, $t = 30\%$). The solution for scenario 1 is as follows:

$$F = P(1 + i)^n \qquad (4.29)$$

$$F = \$5\,000(1 + 0.07)^5$$

$$F = \$7\,013$$

The solution for scenario 2 is as follows:

$$F = P[(1 + i)/(1 + f)]^n \qquad (4.30)$$

$$F = \$5\,000[(1 + 0.07)/(1 + 0.04)]^5$$

$$F = \$5\,764$$

The solution for scenario 3 is as follows:

$$i_o = [(1 - t)i - f]/(1 + f) \qquad (4.31)$$

$$i_o = [(1 - 0.30)0.07 - 0.04]/(1 + 0.04)$$

$$i_o = 0.008\,65$$

$$F = P(1 + i_o)^n$$

$$F = \$5\ 000(1 + 0.008\ 65)^5$$

$$F = \$5\ 220$$

Note the effect of inflation and taxes on interest earned. As long as the time intervals for interest, inflation, and taxes are the same (that is, compounded annually), the calculations can be greatly simplified. If the time intervals are not the same, however, this procedure may still be followed, but the calculations must be performed separately for each interval on a sequential basis.

CASH FLOW

Changes in the time value of money are based on numerous factors. To make intelligent economic decisions, all information must be reduced to simplified cash flow patterns. These patterns, once formed, may be further analyzed to provide a direct comparison of economic options at specific points in time. In many cases, these analyses will yield a single number for each option that provides a basis for decision making. The latter of these tools is discussed in the following section and ties in the concepts of annual, present, and future worth.

First, however, cash flow concepts must be defined. Cash flow diagrams are tools by which transactions over time may be reflected or mapped. Establishing the time frame of a problem is typically addressed up front. This parameter is represented by the horizontal scale in Figure 4.2 and is broken down into units that typically reflect transactions on an annual basis. The actual transactions are denoted by vertical lines or arrows. When received, they are positive and point upward from the x-axis. When disbursed, they are negative and point below the x-axis. The size or length of the vertical lines is not important, but the quantity associated with it should be reflected in the manner as shown on

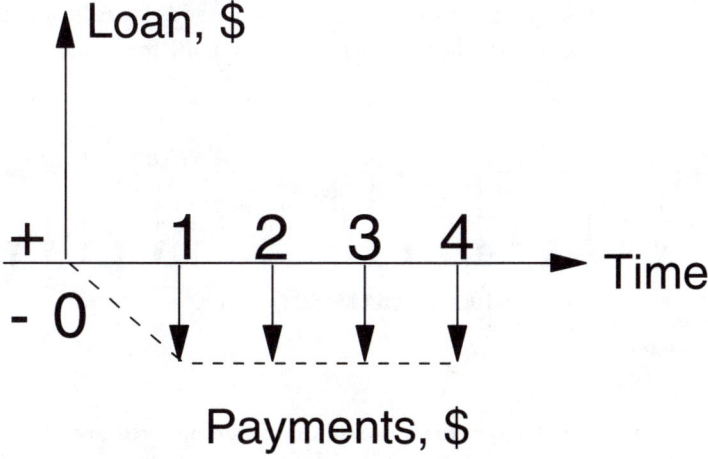

Figure 4.2   Cash flow diagram.

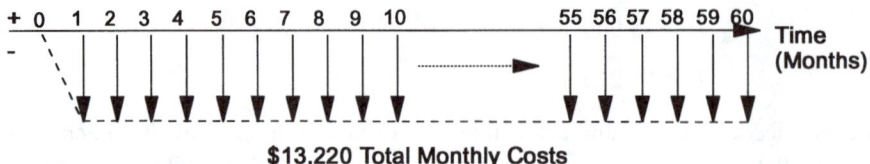

$13,220 Total Monthly Costs

Figure 4.3   Cash flow diagram for existing paint-mixing process.

the graph. Figure 4.2 is a cash flow diagram for the purchase of an asset from the borrower's perspective.

In most industrial applications, constructing a cash flow diagram is not as simple because the diagram will have to reflect more information such as depreciation, salvage values, effects of increased efficiency, regulatory permit costs, and maintenance. This mapping of information clarifies thought processes and simplifies the calculations necessary for arriving at a logical decision. A frequent scenario encountered in waste minimization projects is the need to compare the "way we have been doing things" to the "spending of money to make improvements." While some companies are better than others at making capital improvements, most view waste minimization activities as a cost rather than an opportunity for savings and improved productivity. Therefore, the cash flow diagram is not only a tool for streamlining calculations, but a pictorial for management on which decisions may be based. Figures 4.3 and 4.4 illustrate these concepts. An example of a decision analysis using the calculation of net present worth for these two potential courses of action is presented in the following section. Information outlining multicomponent paint mixing equipment was discussed in the Surface Coating section in Chapter 3.

The information on which to base a decision (that is, the differences and effects of each alternative) is efficiently summarized in the diagrams presented in Figures 4.3 and 4.4. When integrated with calculations that summarize numerous transactions into a single present or future worth value, engineering decision analysis can become a powerful tool for change. The upcoming sections elaborate on methods for reducing the data of a cash flow diagram and evaluate the information to determine the benefits and viability of each option.

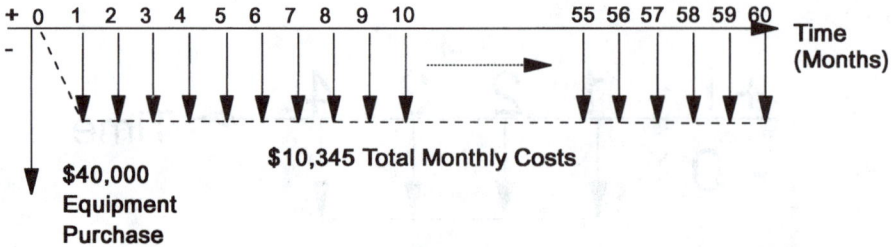

$10,345 Total Monthly Costs

$40,000
Equipment
Purchase

Figure 4.4   Cash flow diagram for waste minimization opportunity—
            multicomponent

Table 4.1   Interest factors for discrete compounding.

Factor name	Converts	Symbol
Single-payment compound amount	Present worth to future worth (P to F)	$(F/P, i\%, n)$
Single-payment present worth	Future worth to present worth (F to P)	$(P/F, i\%, n)$
Uniform-series sinking fund	Future worth to annuity amount (F to A)	$(A/F, i\%, n)$
Uniform-series compound amount	Annuity amount to future worth (A to F)	$(F/A, i\%, n)$
Capital recovery factor	Present worth to annuity amount (P to A)	$(A/P, i\%, n)$
Uniform-series present worth	Annuity amount to present worth (A to P)	$(P/A, i\%, n)$

Methods for Project Analysis

A few of the basic tools of engineering economics have been discussed to provide a framework for strategic plans in relation to financial expenditures. Waste minimization projects are traditionally evaluated in a scientific light to determine their viability in specific production environments. Without considering financial impacts, however, the plans are worthless. Therefore, it is important that efforts be dedicated to define financial alternatives as business is geared toward manufacturing products at a profit. The following methods of financial project analysis are time-proven tools. Every engineering business decision should take them into consideration.

PRESENT, ANNUAL, AND FUTURE WORTH

This text covers six basic interest factors for discrete compounding. Each factor is denoted by a symbol in the form of $(F/P, i\%, n)$. The first portion of the symbol, F/P, refers to the desire to obtain a future value, F, given the present value, P. For example, if a deposit of \$1 000 is made to a savings account today, what will be its value in 5 years if the interest rate is 7%?

$$F = \$1\ 000(F/P, 7, 5) \tag{4.32}$$

Here, $(F/P, 7, 5)$ is a factor multiplied by the present value, P, to arrive at a future value, F. In this example, $(F/P, 7, 5)$ is equal to 1.402 3, which yields a future value of \$1 402.30. This methodology is consistent throughout all of the symbols used in this text. A summary of the types of factors and their notations is given in Table 4.1.

A brief discussion is provided for each type of factor. As each factor is discussed, their uses and benefits will become apparent. However, it may be useful to remember that the following relationships exist. (Other relationships may also be derived; however, to minimize confusion, they will not be elaborated on at this point.) Detailed derivations are not covered because this discussion assumes that a cursory knowledge of these concepts already exists. References for additional information are provided at the end of this chapter.

$$(F/P, i, n) = 1/(P/F, i, n) \tag{4.33}$$

$$(A/F, i, n) = 1/(F/A, i, n) \tag{4.34}$$

$$(A/P, i, n) = 1/(P/A, i, n) \tag{4.35}$$

Compound-Amount Factor

The factor for determining the future value of a present value or single payment is referred to as the *compound-amount factor*, $(F/P, i, n)$. This factor was covered earlier as compound interest (see Equation 4.4) and is reiterated in the following equation:

$$(F/P, i, n) = (1 + i)^n \tag{4.36}$$

Present-Worth Factor

Conversely, the present worth of a future sum is given by the following equation:

$$(P/F, i, n) = 1/[(1 + i)^n] \tag{4.37}$$

Sinking-Fund Factor

Sinking-fund factors, $(A/F, i, n)$, are established to determine the size of a uniform series of payments necessary to accumulate a desired future amount. This can be conceptualized in terms of a person wishing to retire with $1 million in a bank account. What is the amount of money that must be deposited annually for 20 years in a savings account that earns 8% (compounded annually) to accumulate $1 million?

$$A = F(A/F, i, n) \tag{4.38}$$

$$(A/F, i, n) = i/[(1 + i)^n - 1]$$
$$(A/F, i, n) = 0.08/[(1 + 0.08)^{20} - 1]$$
$$(A/F, i, n) = 0.021\ 85$$

Therefore,

$$A = \$1\ 000\ 000(0.021\ 85)$$
$$A = \$21\ 850$$

Series Compound-Amount Factor

Series compound-amount factors, which are the reciprocal of sinking-fund factors, determine the future value of a series of annual payments. Consider the sum of $3 000 that can be saved each year and deposited annually in a savings account. What will be the size of the retirement fund if the interest rate and time period are the same as in the previous example?

$$F = A(F/A, i, n) \tag{4.39}$$

$$(F/A, i, n) = [(1 + i)^n - 1]/i$$

$$(F/A, i, n) = [(1 + 0.08)^{20} - 1]/ 0.08$$

$$(F/A, i, n) = 45.76$$

Therefore,

$$F = \$3\ 000(45.76)$$

$$F = \$137\ 280$$

Hopefully, the interest rate that is earned on the deposited funds will be larger than this amount and the time period for the deposits will be prolonged.

Capital-Recovery Factor

A commonly used interest factor in the financing of capital expenditures is the capital-recovery factor. The capital-recovery factor, $(A/P, i, n)$, is used to determine the annual payments that are necessary to pay back a quantity of money borrowed. This factor is often used in financing arrangements involving the purchase of a home or an automobile. The equation for calculating capital-recovery factors is

$$(A/P, i, n) = i(1 + i)^n/[(1 + i)^n - 1] \tag{4.40}$$

The procedure to determine the amount of the payments on a $100 000 loan that is to be paid back over a period of 10 years at 9.5% interest is as follows:

$$A = P(A/P, 9.5, 10) \tag{4.41}$$

$$A = \$100\ 000\ \{0.095(1 + 0.095)^{10}/[(1 + 0.095)^{10} - 1]\}$$

$$A = \$100\ 000(0.159\ 3)$$

$$A = \$15\ 930$$

Series Present-Worth Factor

The final factor to be covered is the series present-worth factor. This factor, $(P/A, i, n)$, is the reciprocal of the capital-recovery factor, $(A/P, i, n)$. Rather than determining the size of the payments during a specified period of time that are required to pay back a sum of money, this method is useful in bringing back to the present a value that is equivalent to a number of annuities made over time. The series present-worth factor and the basic present-worth factor are important tools when comparing cash flow diagrams. By using present-worth and series present-worth factors, transactions of the future—whether occurring annually or only once during the period of time the projects are analyzed—can return these values back to a single present-worth value so that the options may be directly compared. The equation for determining the series present-worth factor is as follows:

$$(P/A, i, n) = [(1 + i)^n - 1]/[i(1 + i)^n] \tag{4.42}$$

To demonstrate the usefulness of this concept, the following two options will be examined. Which of the options is the most financially attractive:

1. A contract to receive annual payments of $5 000 for 5 years at an interest rate of 7%, or

2. A contract to receive annual payments of $3 000 for 10 years at an interest rate of 5%?

The present worth of option 1 is

$$PW = A(P/A, 7, 5)$$

$$PW = \$5\ 000\{[(1 + 0.07)^5 - 1]/[0.07(1 + 0.07)^5\}$$

$$PW = \$5\ 000(4.100)$$

$$PW = \$20\ 500$$

It is important to note that the total present worth is less than the sum of the five annual payments, which equals $25 000. This is because of the time value of money. The future value of each $5 000 payment is worth less as time passes. This may be easier to understand when thought of in terms of inflation.

The present worth of option 2 is

$$PW = A(P/A, 5, 10)$$

$$PW = \$3\ 000\{[(1 + 0.050)^{10} - 1]/[0.05(1 + 0.05)^{10}]\}$$

$$PW = \$3\ 000(7.722)$$

$$PW = \$23\ 166$$

From this analysis, option 2 appears to be more financially attractive. Having covered the basic concepts of project analysis, these topics will now be expounded so their interpretation can provide a meaningful decision path.

NET PRESENT WORTH AND RATE OF RETURN

A proven method for reducing numerous financial transactions as time passes is to reduce these values to a single net present-worth value. This allows direct comparisons to be made for a number of business options. Although this may appear to be a simple procedure, it can become complicated by subjective factors. To perform a net present-worth comparison, the following information is required: the amount of dollars involved, the time period for which these transactions occur, and the interest rates that apply to the conditions. Numerous other less-tangible factors come into play with this information. Are the transactions pre- or post-tax dollars? Regardless of the viability of the project, are the funds available at the time period for which they are required? Are the interest rates selected a reflection of reasonable real-life scenarios? Are the projected cash flows accurate? Is inflation a factor, and if so, what is its value?

Present-worth analyses depend on variations in the financial parameters. This, however, should not discourage these investigations. Rather, it should serve to demonstrate the importance that should be given to the topic. As present-worth comparisons are made, it becomes apparent that although one option has a higher present-worth value, it does not mean that it is the option of choice. At what point does one option become more viable than another? The rate-of-return concept is a primary issue involved in determining this. Rate-of-return requirements are often selected or set within a corporation's business practices to serve as a "go/no-go" gauge for proceeding with a particular project.

Before elaborating on the issues, a more lengthy discussion of net present-worth valuations is in order. Here, Figures 4.3 and 4.4 are revisited. Is the proposed waste minimization project as viable a solution from a financial perspective as it is presumed to be from a production or environmental standpoint?

Net Present Worth for a Waste Minimization Proposal

A paint shop in a manufacturing facility currently mixes paint in 18.925-L (5-gal) batches. After several hours (typically no more than 2 to 4), any unused paint must be disposed of as waste because of the expired pot life of the material. As a result, cleaning procedures must also be conducted frequently. This practice leads to the generation of large amounts of hazardous waste and, thus, high disposal costs.

To minimize the generation of this waste stream, various options were investigated. The most promising was found to be an automated multicomponent proportioning and mixing system (detailed discussions of this process and equipment are provided in the Surface Coating section in Chapter 3). Implementation of this system can virtually eliminate the generation of wastes because the mixing of individual paint components occurs on demand by the triggering of the spray guns. Solvent flushing of the paint lines also produces minimal wastes. The cost of the equipment is $40 000.

Table 4.2 lists pertinent information for conducting a net present-worth comparison

Table 4.2   Waste minimization proposal cost and material data.

Option	Materials, gal/month[a]	Cost, $/month
Batch paint mixing		
Raw materials	500	12 800
Wastes generated	60	210
Cleaning solvents	60	210
Total cost (monthly)		13 220
Multicomponent mixing equipment		
Raw materials	400	10 240
Wastes generated	10	35
Cleaning solvents	20	70
Total cost, monthly		10 345

[a]gal/month × 3.785 = L/month.

of the two options. Refer to Figures 4.3 and 4.4 for the cash flow diagrams for each option. Determine the net present worth for each option for a 5-year period at an annual interest rate of 10%. Is the new system worth implementing?

The net present worth for the batch paint mixing option is

$$PW = A(P/A, 10\%/12, 60)$$

$$PW = \$13\ 220\{[(1 + 0.008\ 3)^{60} - 1]/[0.008\ 3(1 + 0.008\ 3)^{60}]\}$$

$$PW = \$13\ 220(47.11)$$

$$PW = \$622\ 794$$

The net present worth for the multi-component paint mixing option is

$$PW = \$40\ 000 + A(P/A, 10\%/12, 60)$$

$$PW = \$40\ 000 + \$10\ 345\{[(1 + 0.008\ 3)^{60} - 1]/[0.008\ 3(1 + 0.008\ 3)^{60}]\}$$

$$PW = \$40\ 000 + \$10\ 345(47.11)$$

$$PW = \$527\ 353$$

The multicomponent paint mixing system is a less-expensive option (by \$95 441) for the 5-year period than the current method used for mixing paint. Not only will this option yield savings, but it may also improve process quality through the achievement of higher levels of control and consistency.

This example provides a simple illustration of how useful net present-worth comparisons can be. In this particular example, the time period during which the evaluation occurred was equal for both options. In many situations, however, this is not possible. Comparisons should be made on an "apples-to-apples" basis, not an "apples-to-oranges" basis. Comparisons, in some cases, may be made assuming infinite asset lives, although this is not a traditional practice in manufacturing environments. Therefore, if multiple financial options do not have the same lives, which is often the case, then a procedure must be implemented that can carefully take the conditions into account.

One method for evaluating projects that have different lives is to select a period of time that is a common multiple of the various project lives. For example, if three projects each having a life expectancy of 2-, 5-, and 10-years are to be calculated, then the least-common multiple to evaluate these options is over a 10-year period. In this case, the 2-year option would occur five times and the 5-year option would occur twice.

Another method for providing meaningful comparisons is to conduct an analysis based on the expected time period during which a set of options is likely to exist. This is typically a time period that is less than the life expectancy of the assets involved. In the case of a manufacturer that comes out with a new product line every 2 years, the tooling may have an expected life of 10 years, although it realistically may only be needed for 2 years. As is the case with personal computers, for example, technology may be progressing so fast that although a computer purchased in 1995 will be operational for a number of years, it may become obsolete in as little as 3 to 4 years. Therefore, setting a specified time period, rather than evaluating options over expected lives, can be practical. An example illustrating these two concepts is presented in the following subsection.

Net Present Worth Evaluations Versus Assets of Unequal Lives

Consider an electroplating shop that is trying to determine which one of two waste minimization options to implement. Each option attempts to reduce the amount of rinsewater used at the facility. Annual wastewater treatment costs are nearly $50 000 (this figure also reflects the purchase of water). Determine the net present worth of the options using common multiple-of-asset lives of 20 years. In addition, evaluate these options during the short term at 5 years. The acceptable rate of return set by the company is 15%.

Option 1 is to add several additional rinse tanks to set up a countercurrent rinsing scheme. This option will cost $10 000 in capital equipment (that is, tanks, piping, and controls) and approximately $5 000 in labor. The life expectancy of the equipment is 20 years with no salvage value. Maintenance costs are negligible. Water usage reductions are estimated to be approximately 80%.

Option 2 is to install an ion-exchange system to recycle used rinsewater. This equipment is much more expensive, nearly $45 000. However, the equipment is estimated to reduce water usage by 95%. Life expectancy for the ion-exchange system is 10 years, and its salvage value is estimated to be $5 000. Maintenance costs are $1 500 per year.

Assuming common multiple-of-asset lives of 20 years, option 1 will have an initial outlay of $15 000 for capital equipment and labor. The yearly savings incurred will result from a reduction in wastewater treatment costs (80% of $50 000), nearly $40 000 annually. No salvage value or maintenance costs exist. The cash flow diagram for this option is illustrated in Figure 4.5.

Assuming common multiple-of-asset lives of 20 years, Option 2 will have an initial outlay of $45 000 for capital equipment and installation charges. Yearly wastewater treatment cost savings will be $47 500 (95% of $50 000). Maintenance costs are $1 500 annually. The salvage value is $5 000. The cash flow diagram for this option is illustrated in Figure 4.6. Note that the cash flow pattern is repeated twice and does not reflect the potential increase in cost of the ion-exchange equipment 10 years in the feature. When conducting analyses of this nature, it is important to remember that asset replacement values are held at the same value as the initial costs. This allows the equivalent service of each

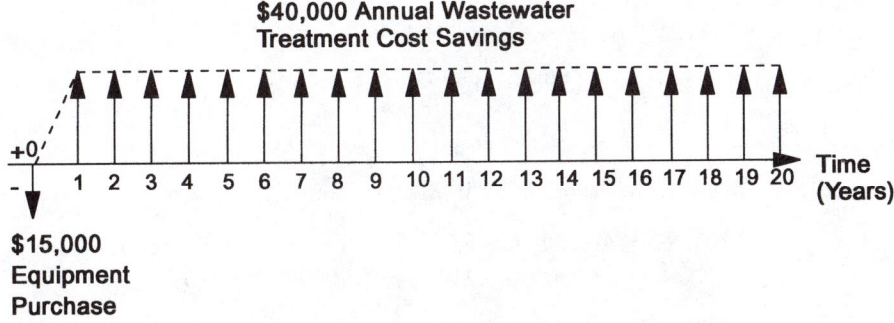

Figure 4.5   Cash flow diagram for option 1 (20-year period).

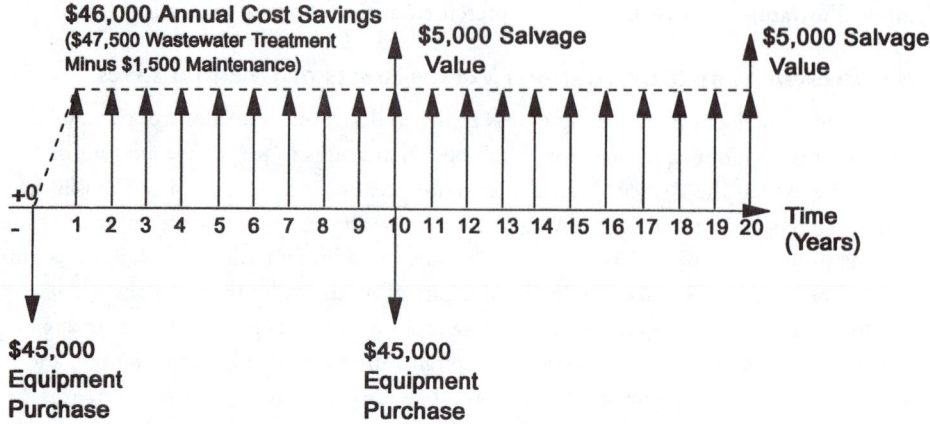

Figure 4.6   Cash flow diagram for option 2 (20-year period).

project to be compared over a period of lives divisible by the common multiple identified.

The net present-worth (PW) calculations for the cash flow diagrams in Figures 4.5 and 4.6 are as follows:

For option 1,

$$PW(1) = -\$15\,000 + \$40\,000(P/A, 15, 20)$$

$$PW(1) = -\$15\,000 + \$40\,000\{[(1 + 0.15)^{20} - 1]/[0.15(1 + 0.15)^{20}]\}$$

$$PW(1) = -\$15\,000 + \$40\,000(6.259)$$

$$PW(1) = \$235\,360$$

For option 2,

$$PW(2) = -\$45\,000 + \$46\,000(P/A, 15, 20) - \$45\,000(P/F, 15, 10)$$
$$+ \$5\,000(P/F, 15, 10) + \$5\,000(P/F, 15, 20)$$

Where

$(P/A, 15, 20) = 6.259$ (from previous calculation);

$(P/F, 15, 10) = 1/[(1 + 0.15)^{10}] = 0.247\,2$; and

$(P/F, 15, 20) = 1/[(1 + 0.15)^{20}] = 0.061\,1$.

Substitute the following in the original $PW(2)$ equation:

$$PW(2) = -\$45\,000 + \$46\,000(6.259) - \$45\,000(0.247\,2)$$
$$+ \$5\,000(0.247\,2) + \$5\,000(0.061\,1)$$

$$PW(2) = \$233\,332$$

The calculations are now performed assuming the period of use is set at 5 years. For option 1,

$$PW(1) = -\$15\ 000 + \$40\ 000(P/A, 15, 5)$$

$$PW(1) = -\$15\ 000 + \$40\ 000\{[(1 + 0.15)^5 - 1]/[0.15(1 + 0.15)^5]\}$$

$$PW(1) = -\$15\ 000 + \$40\ 000(3.352)$$

$$PW(1) = \$119\ 080$$

For option 2,

$$PW(2) = -\$45\ 000 + \$46\ 000(P/A, 15, 5) + \$5\ 000(P/F, 15, 5)$$

$$PW(2) = -\$45\ 000 + \$46\ 000\{[(1 + 0.15)^5 - 1]/[0.15(1 + 0.15)^5]\}$$
$$+ \$5\ 000[1/(1 + 0.15)^5]$$

$$PW(2) = -\$45\ 000 + \$46\ 000(3.352) + \$5\ 000(0.497\ 2)$$

$$PW(2) = \$111\ 678$$

In summary, the net present worth of options 1 and 2 using common multiple-of-asset lives at 20 years is as follows:

$$PW(1) = \$235\ 360$$

$$PW(2) = \$233\ 332$$

The net present worth of options 1 and 2 with a period of use set at 5 years is as follows:

$$PW(1) = \$119\ 080$$

$$PW(2) = \$111\ 678$$

Each of the analytical methods used indicates that the countercurrent rinsing project, option 1, yields the highest net present-worth value, although not by much. Because the net present-worth values of options 1 and 2 were close, other factors may come into play in making the final decision on which waste minimization option to execute. Nevertheless, net present-worth analyses are useful in comparing single numbers with multiple sets of transactions.

This problem can also be performed using net annual-worth values for each option. The methodology is the same. However, this approach would have been accomplished by spreading the relative capital costs and salvage values over the life of each option. Conducting net annual-worth comparisons is useful in cases in which annual cash flows may be fixed or limited to small fluctuations.

In the previous problem, the rate of return was selected at 15%. *Rate of return* is commonly associated with the expressions "return on investment" or "internal rate of return." When performing net present-worth calculations, the rate of return is essentially the interest rate used in determining the interest factors. The *minimum acceptable rate of return* is the lowest level of interest at which an alternative is considered attractive or viable with internal company standards. For a waste minimization project to appear attractive when compared to existing operations, a company may select a minimum acceptable rate of return equal to 15 or 20%. This means that the cost of investing funds will be at a rate of 15 to 20%. Although a waste minimization project may appear to be the "right thing to do" from an environmental standpoint, business decisions are based on profit margins. If analyzed and planned properly, projects of this nature should not be an investment burden. Waste minimization and process efficiency proposals thought of in these terms from the onset will yield a higher probability of success.

Another frequently used method for assessing the viability of a project is one that determines payback periods. A payback period is the segment of time it takes an amount of invested capital to be recovered. For example, if $20 000 is used to purchase a piece of equipment that will save a company $5 000 per year, then the payback period is

Payback period = Capital investment/Annual savings (or net annual profits)   (4.43)

Payback period = $20 000/$5 000 = 4 years

It is important to note that this method does not reflect the time value of money, nor does it allow for the determination of the rate of return. Therefore, its use should be scrutinized when making economic decisions.

It is also important to remember that a rate of return that yields zero net present worth means that the project will merely break even. Minimum acceptable rate of return is the acceptable profitability of a project. These terms and the methods for determining their values prioritize how capital funds are distributed within corporations. Without these analyses, credibility may be jeopardized.

Minimum Acceptable Rate of Return

Consider the following cost-reduction proposal. A piece of equipment, which will last for 5 years, may be purchased for $50 000. This equipment can produce 1 000 more units of a particular product per year at no additional internal cost. Each product can be sold for $20 (disregard the salvage value of the equipment). Should the company proceed with the transaction?

First, a cash flow diagram should be constructed (Figure 4.7). The net savings per year are equal to the benefits from the production of additional units, or $20 000 (1 000 times $20).

The equation reflecting these relationships is as follows:

$$PW = -\$50\ 000 + \$20\ 000(P/A, i, 5) \tag{4.44}$$

The rate of return at which the net present worth is equal to zero is the break-even point. To find this value, several iterations must be performed. A sample calculation is illustrated below for a rate of return equal to 10%. In addition, Table 4.3 lists the results for a range of rate-of-return values.

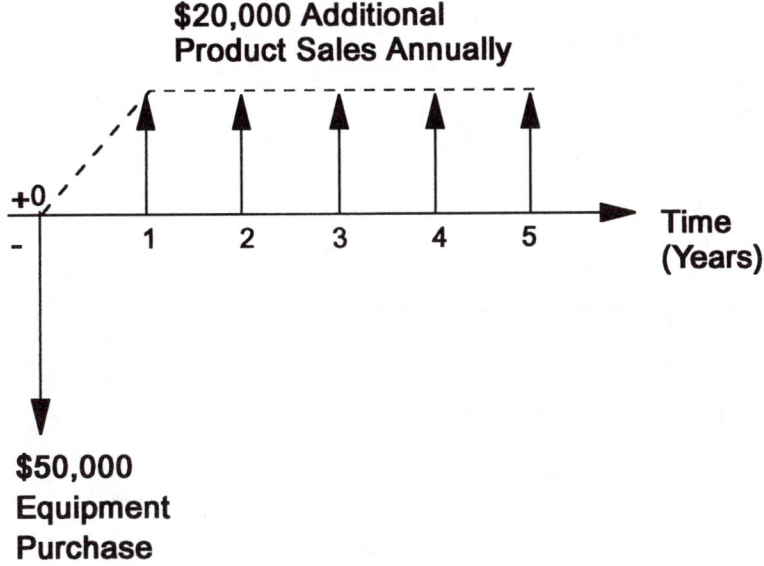

Figure 4.7 Cash flow diagram for minimum acceptable rate of return.

$$PW = -\$50\ 000 + \$20\ 000\{[(1 + 0.10)^5 - 1]/[0.10(1 + 0.10)^5]\}$$

$$PW = -\$50\ 000 + \$20\ 000(3.791)$$

$$PW = \$25\ 820$$

Depending on the rate of return selected, the net present value of the cost-reduction proposal will vary. By plotting these values, the true rate of return is illustrated; the lower the rate of return, the higher the net present value and, therefore, the higher the profits. (Note the minimum rate of return and the shape of the curve in Figure 4.8.)

Project Comparison Using Rate of Return

In the earlier section that discussed net present-worth evaluations versus assets of un-equal lives, a profitability analysis was conducted to compare two potential courses of ac-

Table 4.3   Rate-of-return values.

Rate of return (i)	Net present value, $
10	25 820
15	17 043
20	9 812
25	3 786
30	−1 289

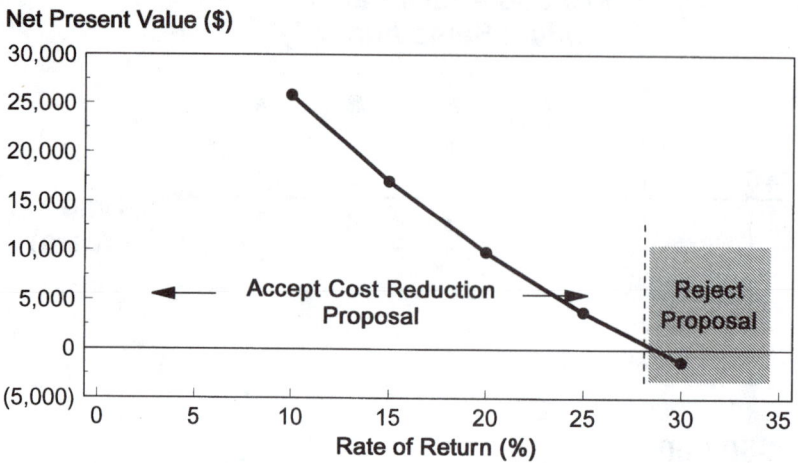

Figure 4.8   Plot of rate of return versus net present value.

tion. In performing the analysis, a minimum acceptable rate of return was selected at 15%. In the section that discussed minimum acceptable rate of return, a simple rate-of-return analysis was performed. This exercise integrates both problems, illustrating the importance of the minimum acceptable rate of return as an important tool in the financial decision-making process.

First, an analysis to determine the break-even point for each option outlined in the Net Present Worth Evaluations Versus Assets of Unequal Lives section will be conducted. The net present value is set equal to zero. This equates the initial cost of capital with the return on investment. To simplify the illustration, only the method of analysis conducted for the 5-year period will be examined.

$$PW(1) = -\$15\,000 + \$40\,000(P/A, i, 5) = 0$$

$$PW(2) = -\$45\,000 + \$46\,000(P/A, i, 5) + \$5\,000(P/F, i, 5) = 0$$

The interest rate is denoted by i. Several iterations have been performed using various interest rates. The calculations are the same as those performed in the aforementioned subsection. The results are listed in Table 4.4.

In this analysis, there is no break-even point even at an interest rate as high at 40%. This is a strong indicator that both of the projects provide an extremely high return on investment. The benefits of each option will have merits of their own. However, even though no net present value of zero was determined, additional information was uncovered. As the rate of return changes, so may the option of choice. At low rate-of-return values, option 2 will prove to be slightly better than option 1. At rates of return exceeding 5%, option 1 is the superior choice.

Table 4.4   Rate of return and net present value for options 1 and 2.

Rate of return (i)	Net present value, $	
	Option 1	Option 2
5	158 180	158 075
10	136 632	132 481
15	119 080	111 678
25	92 571	80 345
40	66 407	49 547

BENEFIT/COST RATIO

The benefit/cost ratio is a method for determining or evaluating the benefits of a project's implementation versus its associated costs. It is a frequently used tool in public or municipal projects and is sometimes used for private-sector decision-making processes. The equation for a benefit/cost ratio is as follows:

$$\text{Benefit/cost ratio} = (B - D)/C \tag{4.45}$$

Where

B = the projects perceived benefits,
D = the drawbacks, and
C = the cost of the project.

The numerator is defined as the overall benefit of the project. It should also take into account any drawbacks of the project. The denominator is the direct costs associated with implementation of the proposed project. These costs can include a range of expenditures, from the purchase of capital assets to operation and maintenance costs. The benefits of implementing a countercurrent rinsing scheme at a plating facility, for example, are improved rinsing efficiency and a reduction in water usage. A drawback may include the downtime of the process lines because of the installation of the addition tanks. If a project affects surrounding wildlife, river usage, or other major environmental aspects, potential drawbacks can be difficult to estimate and, in some cases, may become quite large.

Benefit/cost ratios must exceed a value of at least 1 to be considered viable. In many cases, the ratio may be set at 2 or at a value determined internally by a company or organization. Benefit/cost analyses may also be performed using net present-worth or annual-worth comparisons. As long as the figures used in the analysis are consistent, results should be the same. To illustrate the usefulness of this tool, the following example is provided.

Benefit/Cost Analysis

In this problem, the example in the Net Present Worth Evaluations Versus Assets of Unequal Lives section will again be revisited, and a benefit/cost analysis will be con-

Table 4.5   Benefit/cost analysis summary.

Proposed project	Capital costs (net present value), $	Maintenance costs (net present value), $	Total costs (net present value), $	Benefits, $	Benefits/ costs ratio
Counter-current rinsing	15 000	0	15 000	134 088 (net present value of annual savings)	8.94
Ion exchange	42 514 (incorporates salvage value)	5 030 (net present value of annual costs)	47 544	159 229 (net present value of annual savings)	3.35

ducted on the countercurrent rinsing scheme versus the ion-exchange program. A summary of the benefit/cost analysis for the proposed projects is provided in Table 4.5. Only the method of analysis pertaining to the 5-year evaluation period will be examined. Downtime for the implementation of either project is neglected in this analysis as it is considered a drawback.

Where the net present value of the countercurrent rinsing project, option 1, was not much better than that for the ion-exchange program, option 2, the benefit/cost analysis indicates that the countercurrent rinsing program is more desirable, regardless of the fact that both options meet most reasonable minimum benefit/cost ratios that can be defined.

Sensitivity and Risk Analysis

What if some of the estimated parameters used in the prior examples were in error or slightly adjusted? If a specific parameter is changed, does the entire analysis cascade or does it remain relatively stable? These questions are frequently asked and should be considered in any financial decision-making process.

Sensitivity analysis is a method for determining whether initial estimates are reliable. If a particular parameter changes and the overall result of the analysis has wide swings in its values, then the project may be labeled sensitive. If the changing parameters result in only small fluctuations, then the project is stable and labeled insensitive. These changing parameter values can include the underestimation or overestimation of project maintenance costs, the misjudgment in the cost of capital assets, or a host of other assumptions. In essence, a sensitivity analysis is one in which prior computations are repeated using parameters modified from initial assumptions. The resulting values are compared and determine the project's overall sensitivity or the sensitivity to only specific parameters. The more varied the parameters and the more permutations performed, the more complex is the analysis.

Risk analyses, which are often performed in conjunction with sensitivity analyses, go a step further in that probabilities are assigned to various parameters based on the likelihood of fluctuation and the occurrence of losses. This type of analysis attempts to quantify uncertainties. Typical assessments for risk assimilate the following tasks:

- Estimating the range of possible outcomes that may affect the profitability of the proposed projects,
- Calculating the range of profitability caused by the possible outcomes,
- Estimating the probability that these potential outcomes will occur, and
- Determining the expected value of the project by adding up the individual probability and profit products to yield a weighted average of the proposal's overall profitability.

Up to this point, assumptions have been made regarding specific proposals, and financial parameters have been calculated in relation to those assumptions. In a profitability analysis, this yields net present-worth or annual-worth values. These values, however, were based on one set of assumptions. A risk analysis takes into account the probability that several outcomes are possible.

In a number of examples used throughout this chapter, analyses were conducted comparing a countercurrent rinsing scheme to an ion-exchange program. Consider for a moment only the ion-exchange program. The analysis of this option assumed that the ion exchange would work at a specific efficiency to yield a 95% reduction in the amount of water used. What if this assumption were incorrect? Furthermore, what if the efficiency resulted in only a 50% reduction in the amount of water used because of the various operational parameters such as ion-exchange resin fouling? In this case, a risk analysis can be useful in reflecting an "expected value" that takes into account the probability that a range in efficiencies will occur. The example in the following subsection demonstrates this concept.

RISK ANALYSIS OF THE ION-EXCHANGE PROGRAM

Assume that the following probabilities are associated with the efficiency levels of the ion-exchange equipment proposed in the Net Present Worth Evaluations Versus Assets of Unequal Lives section (all other factors are as previously stated, the useful life of the asset is 5 years, and the rate of return is equal to 15%):

1. Forty percent probability that water usage will be reduced by 95%,
2. Fifty-five percent probability that water usage will be reduced by 50%, and
3. Five percent probability that water usage will be reduced by 97%.

First, the present-worth values for each scenario must be determined. The net present worth of assumption 1 is as follows:

$$PW(1) = -\$45\ 000 + 0.95[\$50\ 000(P/A,\ 15,\ 5)] + \$1\ 500(P/A,\ 15,\ 5)$$
$$+ \$5\ 000(P/F,\ 15,\ 5)$$

Here, the first term in the equation is the capital cost of the equipment, the second is the reduction in water usage (0.95 times 50 000 gpd), the third is the maintenance costs, and the fourth is the salvage value of the equipment. To complete the calculation,

$$PW(1) = -\$45\ 000 + \$47\ 500(3.352) + \$1\ 500(3.352) + \$5\ 000(0.497\ 2)$$

$$PW(1) = \$121\ 734$$

The net present worth of assumption 2 is as follows:

$$PW(2) = -\$45\ 000 + 0.50[\$50\ 000(P/A,\ 15,\ 5)] + \$1\ 500(P/A,\ 15,\ 5)$$
$$+ \$5\ 000(P/F,\ 15,\ 5)$$

$$PW(2) = -\$45\ 000 + \$25\ 000(3.352) + \$1\ 500(3.352) + \$5\ 000(0.497\ 2)$$

$$PW(2) = \$46\ 314$$

The net present worth of assumption 3 is as follows:

$$PW(3) = -\$45\ 000 + 0.97[\$50\ 000(P/A,\ 15,\ 5)] + \$1\ 500(P/A,\ 15,\ 5)$$
$$+ \$5\ 000(P/F,\ 15,\ 5)$$

$$PW(3) = -\$45\ 000 + \$48\ 500(3.352) + \$1\ 500(3.352) + \$5\ 000(0.497\ 2)$$

$$PW(3) = \$125\ 086$$

With the various net present-worth values calculated, the expected net present worth may be determined. These values take into account the associated probabilities of each scenario and are summarized in Table 4.6.

The expected or composite value is not entirely different from the $111 678 determined for option 2 in the Net Present Worth Evaluations Versus Assets of Unequal Lives section, although it is more conservative and, perhaps, more realistic. It is important to note that the sum of all probabilities must equal one or the outcome will be meaningless.

Table 4.6   Ion-exchange proposal.

Item	Efficiency	Net present worth, $	Probability	Expected value, $
1	0.95	121 734	0.40	48 694
2	0.50	46 314	0.55	25 473
3	0.97	125 086	0.05	6 254
Total expected value				80 421

Hidden Costs

The investigation of the feasibility of waste minimization projects often yields a range of financial parameters that, when compiled, give an indication of a project's viability. For the decision-making process to have any credence, however, the figures must be reflected accurately. This means that pertinent parameters must be identified and that estimated values must be carefully evaluated.

Typical financial parameters analyzed in business decisions include items relating to the costs of capital assets, operation and maintenance, product yields, labor, and a host of other factors. In analyzing the viability of waste minimization projects, however, many items that may be termed as cost avoidance parameters are overlooked. These parameters are sometimes thought to have negligible effects.

In the case of the countercurrent rinsing scheme that has been discussed throughout this chapter, the cost of capital equipment (that is, tanks, instrumentation, piping, and so on), installation labor, maintenance, and savings resulting from a reduction in water usage come to mind immediately. In many cases, however, only a crude analysis is performed. For example, if the cost of buying rinsewater is only pennies per litre, why bother with the project at all? Unless management is properly educated on the subject, this may be the prevalent attitude. The answer, however, does not focus on the cost of rinsewater; rather, it focuses on the cost of rinsewater and its subsequent treatment. Subsequent treatment involves the purchase of treatment chemicals, energy costs for running the treatment facility, increased maintenance costs for the higher volume of water used, increased accumulation of wastewater treatment sludge (which may be caused by water hardness and not from the process line), and increased labor costs. Hence, the cost of water now becomes three to four times more than the figure that was initially used.

Identifying these "hidden costs" is difficult because some of them may be hard to quantify. This may be caused by incomplete information or lack of accumulated data. Nevertheless, the figures must be estimated if the financial analysis of a proposed project is to be of value. This approach is not only pertinent to waste minimization projects, but to a plethora of manufacturing processes for start-up and modification activities. Just as there are life-cycle costs associated with products, there are life-cycle costs associated with processes. How often is the cost of process wastes used in decision making to implement a specific manufacturing process? If the costs are thought to be small, they are generally not included. As the prior example demonstrates, financial decisions involving manufacturing should contain a line item that takes into consideration waste generation costs.

In assessing the viability of a waste minimization project or the implementation of a new process, the following parameters should be reviewed:

- Quantity and type of waste generated per unit input to the process;
- Waste disposal costs, which may include items such as waste handling, packaging, inspection, transportation costs, and treatment costs (that is, chemicals, energy, labor, subcontractors, and so on);

- Legal costs, which may include items such as regulatory costs (environmental, health, and safety oversight at the manufacturing facility), permit costs, liability costs (spills, accidents, remediation, and so on), and insurance costs; and
- Miscellaneous costs, which may include items such as training of personnel and public perception of company activities.

SUGGESTED READINGS

Kurtz, M. (1984) *Handbook of Engineering Economics.* McGraw-Hill, Inc., New York, N.Y.

National Council of Examiners for Engineering and Surveying (1995) *Fundamentals of Engineering (FE) Reference Handbook.* Clemson, S.C.

Riggs, J.L. (1982) *Engineering Economics.* 2nd Ed., McGraw-Hill, Inc., New York, N.Y.

Thuesen, G.J., and Fabrycky, W.J. (1989) *Engineering Economy.* 7th Ed., Prentice Hall, Englewood Cliffs, N.J.

5 Standards Driving Environmental Economics

In addition to technical and analytical methodologies, there is a third, more holistic approach toward waste minimization—management systems. Management systems have grown at an accelerated rate during the last decade, which has resulted in the formation of both quality and environmental versions. Although each version may be composed of unique elements and used for distinctly different purposes, their products are nearly the same. In addition, each system consists of phases associated with planning, executing, measuring/assessing, and improving.

The philosophy of management systems is based on the concept of continuous improvement. Continuous improvement is a business process rather than a manufacturing process. The difference between the two is that a business process produces no physical or tangible products. Management systems are vehicles by which measurements, controls, efficiencies, and results are assessed for the operation of a business as a whole. They represent an important step toward achieving quality, profitability, and environmental stewardship in today's arena of global competition.

The most well-known and predominantly used management systems are those formulated through the International Organization for Standardization (ISO). Located in Geneva, Switzerland, ISO attempts to address quality, environmental, and other issues and their effects on global competition through the development of various standards. The organization is made up of more than 100 countries, from highly technical and industrial nations to those that are considerably less economically developed. The U.S. representative to ISO is the American National Standards Institute (ANSI). This institute is responsible for issuing technical and safety standards for a number of processes, tests, and specifications governing a wide range of products and services in the U.S. Other representatives of ISO include the British Standards Institution of the United Kingdom, the Standards Council of Canada, and similar organizations from regions and countries such as Europe, Australia, China, Japan, Mexico, India, and Latin America.

Numerous international technical and safety standards have been in existence for decades. Examples of standards in daily life that may be taken for granted include automobile display symbols, film speeds, paper sizes, computer programs, and more. Each is specified by codes, definitions, test methods, weights, and other parameters. Today, these standards, rather than being technical or safety oriented, have evolved to take into account how businesses may be measured in their operations. Thus, management systems have been created. These systems are designed to take into account the performance of an operation or business, rather than the characteristics of a product. The first of these organizational standards to be implemented was ISO 9000, the standard for international quality management. The organizational standard for environmental management, ISO 14000, was implemented not long after.

The development of ISO standards is primarily consensus-based rather than formed as a result of centralized activities. Parties from private business and government and nongovernmental entities pool their thoughts at roundtable sessions. These sessions result in a draft document, which, after extensive review, evolves into final versions and eventually achieves consensus through a voting process involving all member countries. The intent of the process is to ensure that all interested parties can provide input into the formation of a set of standards at the forefront. Although industry is a key player in this course of events, technical committees are formed that include representatives from diverse entities such as consumer groups, research, legal council, testing laboratories, government, media organizations, and others.

After a set of standards has been drafted, the draft is circulated among committee members for final input and technical revisions; the public is also invited to submit comments on the draft standards. To become an official standard, a draft standard must receive a two-thirds vote from the ISO members that worked on its development and a 75% vote of approval from all eligible ISO members. When this occurs, the standard is considered final and published as an official international standard. Published standards are reviewed every 5 years. This ensures that any information, requirements, or technologies that have become obsolete are detected, and that actions are initiated to remedy the deficiencies.

The U.S. first began to recognize ISO with the passage of the ISO 9000 series of standards. Companies all over the world are now certifying their quality management systems to these standards. The drive to seek registration to ISO standards is primarily customer driven, where the term *customer* can mean anything ranging from the consumer (that is, the public) to other companies that require subcontractors and suppliers to, in turn, certify their products.

The effects that the ISO 14000 standards will have on businesses are not as clear. The progression of companies pursuing certification by these standards is slower than that of ISO 9000 because of a current "wait-and-see" attitude.

Nevertheless, these two international standards—and the manner in which they will ultimately shape the way business is conducted domestically and internationally—will continue to unfold. In addition, their role in the future will undoubtedly continue to grow.

ISO 9000 Standards

The late 1970s and early 1980s marked the beginning of an era in which quality emerged as a priority in the relationship between customer and seller. Many standards were being developed to implement quality systems for a wide range of industries, military applications, and commercial activities. While some of these documents were contractual in nature between a major producer and its subcontractors, others were used only as reference or guidance materials.

Although these early efforts were necessary and a natural evolution of quality systems, a problem with them was that there was no consensus. Application of the principles could not be consistently made on an international basis. For these reasons, the ISO 9000 series of standards was developed. One of the most significant drivers in this effort was that the marketplace for many small and large companies alike had become global. This was most prevalent in Europe, where distances between countries are small, but diversity is great.

The European Community has embraced the quality revolution. Pressure has been applied to companies worldwide who desire to trade or compete within Europe. A predominant effort in this area has been the implementation of third-party registration for certifying and auditing a company's quality system for compliance. This idealogy is also becoming commonplace with other international standards, such as the ISO 14000 series. Third-party auditing and certification is a means by which companies may have their quality system documentation and implementation strategies reviewed against the requirements of applicable ISO 9000 series elements. If the information reviewed meets specific criteria, a third-party auditor or registrar may grant certification. The use of a third-party in the process not only aids in the evaluation of quality systems, but also gives additional credibility because of the outside review of specific requirements.

OVERVIEW OF CORE REQUIREMENTS

The term *quality* connotes different meanings. According to ISO 8402 (which identifies the vocabulary associated with the international quality standards), quality is defined as "the totality of characteristics of an entity that bear on its ability to satisfy stated and implied needs" (ISO, 1994). Other uses of the term revolve around conformance to requirements and a quantification of excellence. The problem with these latter uses is that they may often portray opposing images. For example, conformance to requirements may imply that a product is built right the first time, creating a situation in which operation and maintenance costs are low. This satisfies the needs of the customer by implying that quality will cost less with the passage of time. The other end of the spectrum, however, is that quality should cost more so the product or service provided is a cut above the competition. These perceived differences give rise to the term *grade*, which is used to differentiate technical variations. For example, an economy car is a different grade of automobile than a luxury car or limousine. While both may exhibit quality by conformance to specifications or requirements, each is in a different class.

Although quality is often thought of as a characteristic that pertains to physical products, it can refer to multiple generic product categories such as hardware, software, processed materials, and services. Accepted definitions of these terms are found in Table 5.1 (ISO, 1993).

Another set of terms that may appear confusing are those that commonly refer to individual processes within the general topic of quality management such as quality control, quality assurance, quality improvement, quality policy, quality planning, and others. The term *quality management* is essentially all of these functions and is driven by an overall

Table 5.1   Generic product categories.

Category	Definition or product example
Hardware	Manufactured parts, pieces, assemblies (automobiles, computers, refrigerators)
Software	Computer software that may consist of information, concepts, transactions, or procedures (Windows 95™/Microsoft™ products, Lotus™ software)
Processed materials	Intermediate of final products consisting of solid, liquids, gases, or any combination thereof (bulk chemicals, pressurized gas cylinders)
Services	Intangible products that are not physical and are activity based (marketing, selling, delivering, training)

management structure as depicted in Figure 5.1. To provide a basic level of understanding of these topics, brief definitions of each quality element are presented below (ISO, 1994):

- Quality policy—the embodiment of upper management within an organization to the quality initiative. It states the overall direction and visualization of the program.

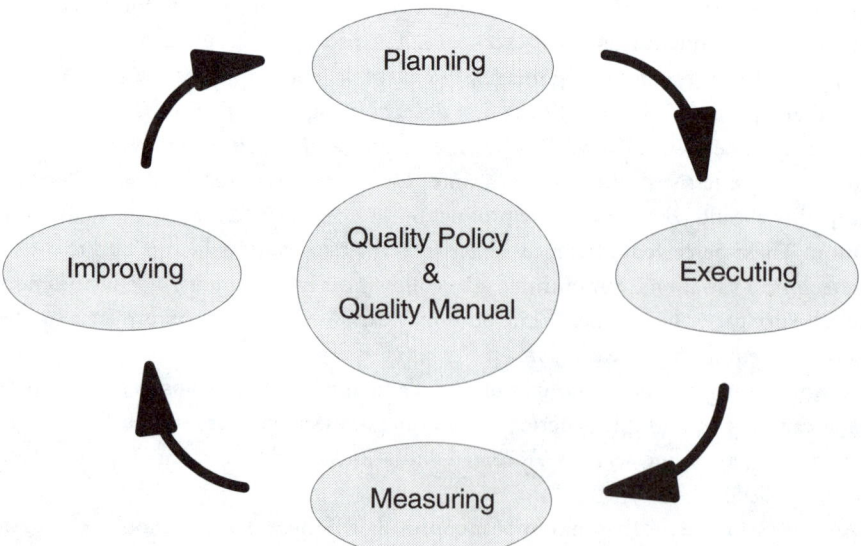

Figure 5.1   Quality management structure.

- Quality planning—those activities that provide the framework for setting objectives and requirements of the quality program and its mode of application. This includes product planning to identify specific quality criteria and management functions such as organizing, scheduling, and planning for continuous improvement.
- Quality control—operational techniques and activities used to monitor processes and correct unsatisfactory performance within all phases of the quality program. This may include efforts such as statistical process control and result in cost reductions or economic benefits.
- Quality assurance—activities geared toward providing internal and external confidence that quality requirements are being met. Internal assurance is focused on a company's management structure. External assurance is directed at customer satisfaction.
- Quality systems—all of the elements required to successfully implement quality management such as organizational hierarchy, responsibilities, procedures, processes, and resources.
- Quality improvement—the cycle by which continuous improvement to the overall quality process is made. These efforts benefit both internal and external customers of an organization while increasing the efficiency and effectiveness of the quality program.
- Quality manual—although this is not a specific function of a quality program, it is the embodiment of a company's quality policy and documents the manner in which quality will be achieved. The quality manual contains specific measures to be taken to ensure quality. Implementation of these procedures entails the aforementioned efforts and a potential list of many others.

A strong emphasis along these lines was initiated in the late 1980s under the name of total quality management (TQM). Total quality management was one of the first significant initiatives by U.S. companies that focused on quality in business commitments with the idea that long-term success in this area was crucial to a company's survival both domestically and globally. To reap the benefits of TQM, the principles should be embraced by all factions of a company. A common belief was that energetic implementation of TQM would yield business prosperity and additional benefits to the workforce and society in general.

However, because TQM was often perceived as a voluntary procedure, its overall effectiveness may have been marginal. The ISO 9000 series is not perceived as being optional. Rather, it is viewed as more of a requirement of doing business.

Quality is an important survival tool in today's highly competitive worldwide marketplace. Although the implementation of ISO 9000 may appear to be a high-cost initiative for competition on a global basis, quality competition creates better products, more highly satisfied customers, and more efficient/consistent manufacturing processes and service operations. The latter of these benefits frequently translates into increased profits for a company because of reduced overall operating costs.

QUALITY SYSTEMS

Quality systems can be viewed from two perspectives. The most encompassing perspective results from a purchaser's overall satisfaction of the quality of the product or service offered. The second pertains to a purchaser's confidence in a supplier's quality systems. A supplier must directly satisfy the producer's quality requirements, whereas the producer must satisfy all of the expectations of the consumer.

The ISO standards applicable to quality management systems are broken down into several documents, the primary of which are

- ISO 9000—Quality Management and Quality Assurance Standards: Guidelines for Selection and Use;
- ISO 9001—Quality Systems: Model for Quality Assurance in Design/Development, Production, Installation, and Servicing;
- ISO 9002—Quality Systems: Model for Quality Assurance in Production, Installation, and Servicing;
- ISO 9003—Quality Systems: Model for Quality Assurance in Final Inspection and Test; and
- ISO 9004—Quality Management and Quality System Elements.

The ISO 9001, 9002, and 9003 documents cover quality system elements to varying degrees. Certification is made using one of these three standards. The ISO 9001 document covers items of design and development, production, installation, and servicing. The ISO 9002 and 9003 documents are essentially subsets of ISO 9001 in that only a portion of these elements is required, some of which are less stringent. For example, ISO 9002 does not include elements for design and development activities, while ISO 9003 covers only those criteria pertaining to final inspection and test. Quality systems discussed herein will focus primarily on ISO 9001 to introduce the full range of quality system topics.

Other documents that exist for various quality functions, such as general guidance, auditing, equipment, and statistical methods, include

- ISO 8402—Quality—Vocabulary;
- ISO 10011—Guidelines for Quality Auditing Systems;
- ISO 10012—Quality Assurance Requirements for Measuring Equipment;
- ISO 10013—Guidelines for Developing Quality Manuals;
- ISO/TR13425—Guidelines for the Selection of Statistical Methods in Standardization and Specification; and
- ISO Handbook 3—Statistical Methods.

Twenty quality system requirements must be met to achieve registration for ISO 9001 per the 1994 revisions. Table 5.2 lists these elements by the standard's paragraph section numbers and compares them to those of ISO 9002. A brief discussion of each element follows.

Table 5.2   Quality system requirements for ISO 9001 and 9002.

	ISO 9001 (1994)	ISO 9002 (1994)
4.1	Management responsibility	4.1
4.2	Quality system	4.2
4.3	Contract review	4.3
4.4	Design control	na[a]
4.5	Document and data control	4.5
4.6	Purchasing	4.6
4.7	Control of customer-supplied product	4.7
4.8	Product identification and traceability	4.8
4.9	Process control	4.9
4.10	Inspection and testing	4.10
4.11	Control of inspection, measuring, and test equipment	4.11
4.12	Inspection and test status	4.12
4.13	Control of nonconforming product	4.13
4.14	Corrective and preventive action	4.14
4.15	Handling, storage, packaging, preservation, and delivery	4.15
4.16	Control of quality records	4.16
4.17	Internal quality audits	4.17
4.18	Training	4.18
4.19	Servicing	4.19
4.20	Statistical techniques	4.20

[a]Not applicable.

Management Responsibility

Effective management responsibility requires that the following elements be established: the quality policy, organizational definition, and the review cycle. Without achieving these elements, there will not be a formative structure, sense of purpose, or backing to achieve meaningful results.

The quality policy is a concise general statement depicting a company's views and intended purpose for achieving quality. The quality policy should include internal organizational goals that take into consideration the needs of the customer. Effective statements must be understood by all levels of management and workers within a company. As such, it is possible for the quality policy to be a single sentence that incorporates these aspects.

Management should also define an effective organizational structure for addressing quality. This should lay out the responsibility and authority of company representatives and how their interaction will be achieved. A portion of this may be accomplished through the formation of a detailed organizational chart. Individuals should be designated from management to implement and maintain the quality program and compliance with ISO 9001, 9002, or 9003, as applicable. Adequate resources should also be provided. These may be in the form of capital or expendable funds, materials, equipment, personnel, and time.

A management review cycle should also be established. This process should be conducted at specified intervals (typically every 6 months) to ensure that the quality system continues to effectively implement and meet the requirements defined by ISO. Internal audits are an efficient means of performing this function and include reviews of information such as customer complaints/satisfaction, quality data, manufacturing rates/yields, and other issues. These findings should be well documented and used for initiating improvements in the overall quality system.

Quality System

The quality system consists of documentation in the form of a quality manual. The quality manual incorporates procedures that effectively document and implement the requirements of ISO 9001 and the company's quality policy. Procedures define the work to be performed and provide instructions on how various tasks should be carried out. The level of detail varies according to the complexity of the job, the method or skills needed, and, in all probability, the training required. In addition, quality plans should be prepared that, at a minimum, address the following issues: maintenance of quality records, identification of criteria used to accept or reject each product or element of a process, identification and acquisition of resources to achieve the required level of quality, verification that the documentation matches the processes and design specifications for the product produced, continuous improvement cycle for quality procedures and processes, and continuous improvement cycle for the measurement systems used.

Contract Review

Documented procedures should be in place to ensure that agreements made between a supplier and a customer are effectively tracked and implemented. These procedures entail performing reviews to verify a contract before it is agreed on, establishing that the supplier can meet the contract, and ensuring that the contract accurately represents a customer's requirements. Part of this effort should also include establishing an effective system for contract amendments or changes, that, when made, flow through a supplier's internal chain of command and ensure that the product is produced to the latest requirements. As with all elements of a quality system, records of all contract reviews should be well documented and maintained.

Design Control

This phase, or portion, of an international standard is applicable only to ISO 9001. It is important that control and verification processes be an integral part of design efforts to ensure that all specified requirements of a product can be met. After a product has been conceptualized, a plan should be formulated to accomplish the objectives. Part of the planning effort should incorporate typical project management tools that reflect project evaluations and reviews, schedules, budgets, and personnel. However, it is also important that organizational/technical lines between departments and individuals be set up so that information may be transferred or communicated easily. Cooperation and synergy are important elements for the design process to be timely, accurate, and effective at meeting the internal operations of a company and satisfying customer requirements.

Another important part of the design process is defining input requirements. Items covering statutory, regulatory, structural, maintenance, and other issues must be clear and concise so that the product may be evaluated alongside these parameters in an objective fashion. Unclear or conflicting input may result in a product that fails some requirements.

Design output is equally as important as design input. The result of the design of a product or service should be a match with customer criteria. Output should meet input requirements and be demonstrable via various acceptance criteria (that is, measurements to meet blueprint parameters, mechanical testing to verify physical attributes, dynamic evaluations to confirm operational characteristics). Characteristics should also be identified that are pertinent to the overall operation and safe use of the product.

Periodic and systematic reviews of the design process are necessary to identify problems and initiate solutions. This is generally conducted throughout the process, rather than at its completion. In doing so, small errors may be detected early on in a process, thereby minimizing any cascading effects that may result in higher design costs or a product that does not meet customer requirements or exhibits attributes that are detrimental to product reliability, dependability, and safety. An element of these reviews is design verification. Design verification is defined as those efforts geared toward the review, inspection, testing, or other initiatives that establish and document whether various items are conforming to specified requirements.

Conversely, design validation is a process used to ensure that the product, rather than in-process design efforts, meets customer specifications and needs. Examples of validation include the following efforts: beta tests for newly developed software, which serve to uncover glitches in the operations of the product prior to its widespread release; field testing of automobiles, aircraft, and watercraft, which may reveal handling problems, operational deficiencies, and ergonomic details; and food and beverage taste tests, which determine the likes and dislikes of specific markets as well as specific product characteristics that are beneficial or detrimental to a potential customer base.

Part of any design process is the process of change. This may be implemented before initial completion of a project or after many years of successful production. Regardless, design changes or modifications must go through a systematic chain of events to ensure that the desired results will be obtained. All changes should be adequately identified, documented, reviewed, and approved by authorized personnel before their implementation. During these efforts, considerations should be made regarding how the changes will affect the product in addition to individual elements or processes.

Document and Data Control

Documented procedures should be established to identify methods of control for all documents and data associated with the quality system. This includes internal documents and potential external ones such as customer drawings, specifications, and standards. Appropriate control measures should be set up to ensure that only reviewed and approved materials are issued. The following elements define a baseline approach to document and data control: applicable documents that are required to perform specific functions as related to the quality system are available at all locations within the company; all documents relating to the quality system are maintained to ensure that only up-to-date versions are permitted for use, and all obsolete or outdated documents are either removed or in some fashion prevented from inadvertent use; and document retention policies are specified and performed to meet legal and/or other company requirements. All document or data changes should be reviewed by the individuals most familiar with the original versions. Changes that are approved should be noted so that the history of the document may be maintained.

Purchasing

Purchasing provisions within a quality system are necessary to ensure that purchased products meet any and all specified requirements. For example, if the purchase of sulfuric acid for an internal manufacturing process requires a technical grade rather than an industrial grade, procedures within the purchasing system must guarantee that only the technical grade is supplied. This may be accomplished by specifying the amount of control a company may want to have over its suppliers. Selecting suppliers based on their track record for meeting product criteria and the suitability of their own internal quality systems is a conventional approach. Another reliable method for determining the reliability of suppliers is their status in regard to ISO 9000 registration. Therefore, approved lists of suppliers should be created to document the quality system elements at work within a purchasing department.

Provisions for the verification of quality in purchased products and in a supplier's quality system should be included in a purchasing system. This effort may be performed most efficiently at a supplier's facility and should be executed on a periodic basis. Typical information captured within purchasing documents may include class and grade of products; pertinent technical references (that is, drawings, specifications, processes, personnel, and other relevant requirements); key identifiers such as names, titles, and serial numbers; and applicable references to quality systems standards.

Control of Customer-Supplied Product

In situations where a customer supplies materials and/or equipment to be incorporated into the final product of a company, procedures should be established and maintained that ensure these items will be taken care of and, if damaged, reported to the customer. All efforts associated with these activities should be documented.

Product Identification and Traceability

In some cases, procedures for identifying products on completion and during manufacturing and assembly operations may be appropriate. Information such as applicable and pertinent specifications, drawings, batch numbers, and individual part numbers are frequently used for this process. These practices improve traceability and help enure overall product quality.

Process Control

Every supplier, whether manufacturing products or performing services, should identify, plan, and execute their processes in a consistent, orderly fashion. Documentation of these elements and specific criteria is crucial. Conditions that should be controlled include the following: identification and documentation of procedures that are pertinent to ensuring and maintaining product quality; use of equipment suited for specific tasks and a working environment conducive to occupational health and safety standards; compliance with all applicable governmental statues and regulations, industrial standards, and internal quality protocols; oversight and control of all processes and product characteristics; approval of all equipment and processes used during the course of manufacturing products or providing services; definition of workmanship standards via written documents or representative hardware; maintenance of equipment to ensure continued capabilities in manufacturing and services; and training and skills necessary for individuals to perform special processes (that is, painting, welding, chemical processing) and the qualifications required for the equipment and processes used.

Inspection and Testing

Procedures and supporting documentation must exist to conduct proper inspection and testing. This process is conducted throughout the sequence of operations necessary for manufacturing a product or providing services. Requirements for receiving materials and supplies should be specified and the methods for verifying their conformance established. If quality issues are addressed and problems are uncovered up front, they will be less costly to remedy later on. Therefore, suppliers should be screened on their ability to provide items that meet prescribed specifications before initiating purchasing activities with them. In some cases, suppliers that have been approved for certain materials may not necessitate incoming inspections before their products are used.

In-process inspections and testing requirements should also be in place and be fully documented. Although these will vary by manufacturers and their products, the purpose of inspections and requirements is to track the progress of a product or service to ensure that key operations are performed to meet specifications. All inspection and testing requirements of the quality plan should be executed and their results fully recorded. This includes tasks performed on process equipment as well as the actual product.

Final inspection and testing should also be conducted per the quality plan and should incorporate enough detail to ensure that the product or service has conformed to applicable requirements or specifications. A portion of these efforts involves verifying that all incoming and in-process inspection and testing requirements have been met, conducted, and fully documented. Products should not be shipped to a customer until all of the identified final inspection procedures have been conducted and documented.

When conducting inspection and testing procedures, it is important that records be maintained and retained to provide evidence that they were indeed conducted. Records should include data or information regarding pass/fail criteria. The findings of each product should be recorded regardless of the outcome. In addition, individuals responsible for approving shipment of a product should be employed.

Control of Inspection, Measuring, and Testing Equipment

Equipment that is used to verify that a product meets specified requirements should be controlled. The procedures defining this process should document the control elements, calibration techniques, and maintenance requirements for each item of inspection, measuring, and test equipment. To be effective, each piece of equipment should be able to provide a sufficient level of accuracy, allow the degree of measurement uncertainty to be known, and be used in a manner consistent with its intended purpose and capabilities.

Calibration checks should be performed at suitable and specified frequencies to ensure proper operation. Records of these activities should also be maintained. Calibration plans may include a host of items ranging from small instruments and personally owned tools to sophisticated process control equipment. All inspection, measurement, and testing equipment should be identified with a sticker or other suitable marker to indicate calibration status.

Physical controls may also need to be implemented to ensure document and software reliability. Software is an issue that is not always adequately addressed. If a version of software can be inadvertently or intentionally modified without proper authorization, chances are that security or other measures are not sufficient. Therefore, records documenting these checks and balances should also be maintained.

Inspection and Testing Status

After a product has been inspected or tested, the results of the test, or the product's status, should be identified on or with the product as it moves throughout the production environment. This should occur regardless of whether the product passed inspection, is pending results, or failed. This element should also apply to incoming materials, subassemblies, and finished goods. Documentation should be such that if an item is separated from a grouping or batch of products, its test status may still be identified.

Control of Nonconforming Product

If a product is determined to be in nonconformance with requirements at any stage of manufacturing, procedures should be sufficient to ensure that it is not inadvertently used. Nonconforming products should be properly identified, documented, and segregated from other "in-work" items to the extent practicable.

Responsibilities for the review and disposition of a nonconforming product should be defined. This may include identifying specific individuals to perform the following activities: acceptance with or without repair and customer approval as required, rework to achieve the original requirements or specifications, reclassification to alternative uses, and rejection or scrapping.

It is important to note the difference between the functions of the terms *repair* and *rework*. Reworking a product can achieve the originally defined specifications of that product, whereas repairing a product cannot. Regardless of the method selected, it is important that those making the decisions are competent in the process and have the approval to do so. As with all elements of ISO 9001, the procedures and actions taken should be fully documented during the process and/or at the conclusion of each activity.

Corrective and Preventive Action

The occurrence of nonconformities precipitates the need for corrective and preventive actions. Procedures that outline the methodologies for accomplishing these actions should be documented. As the continuous improvement cycle is adapted or changed, procedural documentation should be updated to reflect these activities.

Corrective action is different from preventive measures in that the former is intended to solve existing problems and the latter is used to foresee and prevent problems from occurring. Corrective actions typically include the following: proactive responses to customer complaints regarding product nonconformities; investigative efforts to determine the cause of nonconformities in products, processes, or failures in the overall quality management system elements, and subsequent documentation recording the findings; elimination of the source of nonconformities; and initiation and follow through of a particular action.

Preventive measures follow a similar process line. However, their focus is to use a host of information from sources such as production processes, customer complaints, quality records, and audit findings to detect collapses in the system before their escalation into nonconformities. Problems that have been detected may serve as indicators for other, perhaps larger problems that may occur at a later time. Therefore, the preventive action cycle is continuous and requires dedicated efforts by management.

Handling, Storage, Packaging, Preservations, and Delivery

At all stages of manufacturing and delivery of a product, procedures should be in place to ensure adequate protection. This includes receiving raw materials, handling and storage of intermediate subassemblies, packaging and preservation measures, transportation, and issues surrounding the final delivery and/or installation of the product.

Handling procedures should be sufficient to prevent damage or deterioration of the product, each of which will differ depending on the type of product involved. For example, some products may be susceptible to temperature variation, moisture contamination, corrosive elements, or physical/mechanical/impact damage. As such, storage areas to control environmental conditions or segregation of raw materials or products before a manufacturing process or final delivery may be required. Elements that address the movement of in-process products or materials to and from these storage facilities and documentation indicating that all handling and storage criteria are being met should be included in these procedures.

Part of the handling, storage, and delivery process precipitates the need for suitable packaging. Packaging is dictated by a number of requirements that may be driven by customers and regulatory agencies (that is, U.S. Department of Transportation), although packaging primarily refers to items such as boxes, drums, tanks, and other alternatives.

The preservation of the product from inception to final delivery is of primary concern. It is not until final delivery that these activities cease. Again, procedures and documentation tracking these issues should be in place to verify their performance.

Control of Quality Records

Quality records are referenced in almost every section of ISO 9001. They are a key element to successfully demonstrating compliance with the standard. Therefore, as with all the processes discussed in this section, procedures that identify quality records and the manner in which they are collected, processed, stored, and maintained should be in place.

These records ultimately demonstrate conformance to specified requirements and establish whether a quality management system has been effectively implemented. Inadequate control of quality records may be a sign that the quality system has not been executed or properly maintained.

The legibility of all records is important, and retention times for each should be specified. The ISO 9001 series of standards requires that specific quality records be maintained from each of the following topic areas:

- Management responsibility;
- Contract review;
- Design control;
- Purchasing;
- Control of customer-supplied product;
- Product identification and traceability;
- Process control;
- Inspection and testing;
- Control of inspection, measuring, and testing equipment;
- Control of nonconforming product;
- Corrective and preventive action;
- Internal quality audits; and
- Training.

Internal Quality Audits

To verify the existence and performance of a documented quality management system, internal audits should be conducted. This activity also serves as a mechanism for continuous improvement through periodic reviews of the findings. Procedures and documentation should be in place defining the specific actions to be taken, the frequency with which they are to be conducted, the manner in which the results will be evaluated for compliance to the ISO 9001 standard, an overall assessment of the system's effectiveness, and the identification of action plans to correct deficiencies.

The scheduling and frequency of internal audits should be tailored to match the relative importance of activities as they pertain to the quality of the product. Auditors selected should be competent in the subject matter and possess objectivity. Selecting auditors typically results in the formation of audit teams made up of individuals who are separate from the departments or functions being audited. An auditor's role includes observing work practices, examining quality records, and identifying deficiencies that are found.

It is important that all deficiencies detected be resolved in a timely fashion. As such, the management of the areas being audited should be completely informed of all findings. In addition, summaries of all audits conducted should be routed to executive management for review to ensure the effectiveness of the quality management system. Guidance for quality audits is provided in the ISO 10011 standard.

Training

To ensure that all personnel are qualified to perform the specific tasks assigned to them that affect quality, records should be maintained identifying and tracking the applicable education, training, and experience of personnel. The basic elements of training in ISO 9001 must include procedures and documentation that identify and/or determine the training that is required for each person whose work affects the quality of products or services offered; mechanisms by which such personnel are qualified to perform these activities based on their education, training, or experience; and records documenting that the prior two items are being executed to ensure that only workers qualified to perform a designated task are actually doing so.

Servicing

Servicing a product, which is an activity typically performed after a product has been sold or installed, is not a requirement in all cases. Servicing may be specified as an agreement within a sales contract. Where it is a requirement, written procedures should be in place that define the actions to be performed. In addition, a mechanism for verifying and reporting that the actions taken for the services provided meet the specified requirements for customer support should be included in the documents.

Statistical Techniques

The need to implement statistical procedures is a requirement that is derived within a manufacturer's organization. Identifying the processes that will use statistical procedures and the specific techniques to be performed should be sufficient to establish, control, and verify process capabilities and product characteristics. In all processes in which statistical process control is used, procedures should be in place that document how to carry out the activity and validate the technique used for the designated application.

ISO 14000 Standards

The ISO 14000 set of standards is broken down into two groups—organizational standards and product standards. The organizational standards are

- ISO 14001—Environmental Management Systems (published September 1996);
- ISO 14010—Environmental Auditing, General Principles (published October 1996);
- ISO 14011—Environmental Auditing, Auditing of Environmental Management Systems (published October 1996); and
- ISO 14012—Environmental Auditing, Qualification Criteria for Environmental Auditors (published October 1996).

The product standards are

- ISO 14024—Environmental Labeling,
- ISO 14040—Life Cycle Assessment, and
- ISO 14060—Inclusion of Environmental Aspects in Product Standards.

This new wave of environmental management approaches is one in which environmental issues are identified and addressed through proactive means driven by business concerns rather than regulatory fears. In the late 1970s and early 1980s, public concerns generated by industrial accidents in Love Canal (near Niagara Falls, New York) and Bhopal, India, where hundreds of people were injured or killed by chemical releases, provided the impetus for the current regulatory framework in the U.S. for dealing with a wide range of environmental concerns. The traditional regulatory approach used to ensure that companies were protecting the environment was based around the concept of "command and control." As this regulatory structure has evolved, its prescriptive nature has proven to be somewhat costly for both private business and government. It has proven to be inefficient in that the speed at which government can react is considerably slower than that required of businesses to keep pace in today's highly competitive world market. Figure 5.2 illustrates the path corporate management should follow for sustainable development (Childers, 1997).

The environmental issues that companies face today are no longer centered only on public concerns. Indeed, environmental issues surrounding common business principles have come into play. In a world market where competition is high and extremely fast-

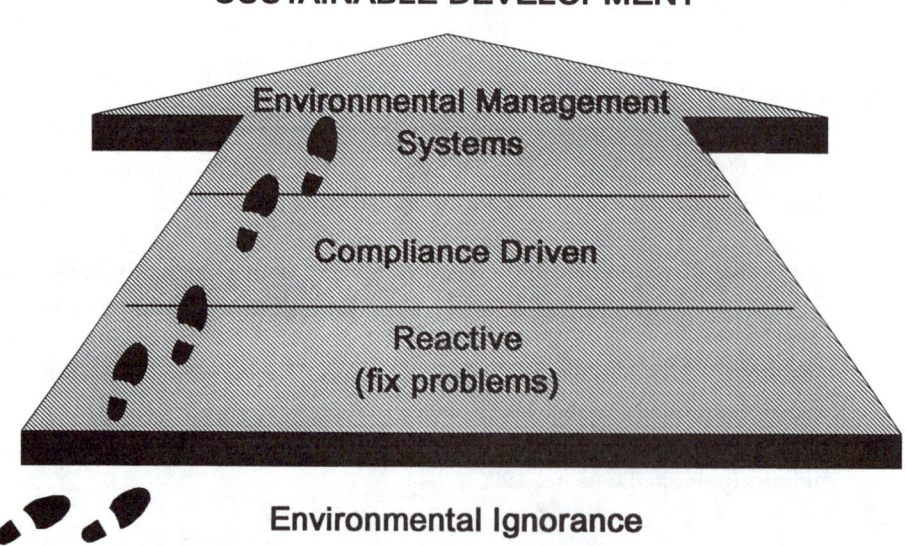

Figure 5.2   Environmental management system evolution.

paced, competitive fairness is a primary concern. Companies operating in undeveloped countries, where environmental regulation and enforcement are relatively nonexistent, put companies operating in developed countries at a significant disadvantage. As rival companies seek certification, this distinction becomes another differentiating factor in the sale of products and services. Consumers want to know how environmentally safe the products they buy are. This also pertains to larger corporations seeking to do business with subcontractors and suppliers, as demonstrated with the set of ISO 9000 standards. The three largest U.S. automobile manufacturers are a prime example of a case in which all suppliers must be certified to certain quality standards or risk losing business.

Because companies that have poor environmental management practices have a higher risk of environmental accidents, insurance companies are being more selective in deciding whom to cover. Environmental accidents are costly for insurance companies. Therefore, from an investment perspective alone, insurance companies often look for methods by which businesses may be evaluated as to their risk and proactive stance in dealing with protecting the environment.

Regulatory relief is also a possibility. As a company becomes more proactive in its approach to environmental issues, the leniency of governmental entities may follow. In some cases, the U.S. Environmental Protection Agency (U.S. EPA) is making strides to create a more compatible regulatory atmosphere where command and control is lessened for options that create a feeling of teamwork between government, industry, and the public. The effects of this compatible regulatory atmosphere include a more streamlined permitting process, fewer on-site inspections, reduced penalties, and reduced reporting requirements. This process is part of U.S. EPA's Project Excellence and Leadership, commonly referred to as Project XL, which is designed to provide regulatory relief to companies with sound initiatives and ideas for improving environmental performance.

Another important reason for implementing ISO 14000 is that it is based on efficiency and continuous improvement. With these factors in place, bad business practices may be identified and eliminated, spills and material wastes may be reduced, and violations and future liabilities may be avoided.

The potential benefits of successful implementation of an environmental management system include cost control and process efficiency enhancements, industry and government relationship improvements, regulatory violation and liability reductions, insurance risk minimizations, customer image improvements and potentially higher market share, and customer and public relations enhancements.

ENVIRONMENTAL MANAGEMENT SYSTEMS

An environmental management system (EMS) is a vehicle by which companies can integrate proactive measures into their environmental programs to upgrade them from compliance-related activities. An environmental management system is not a stand-alone system. It can be worked in with other management requirements to provide environmental gains and help achieve economic goals. Although ISO 14001 does not state specific performance criteria, it does require that companies or organizations create policies and objectives that take into account regulatory requirements and issues that may pose significant environmental impacts. An EMS is also a continuous improvement process

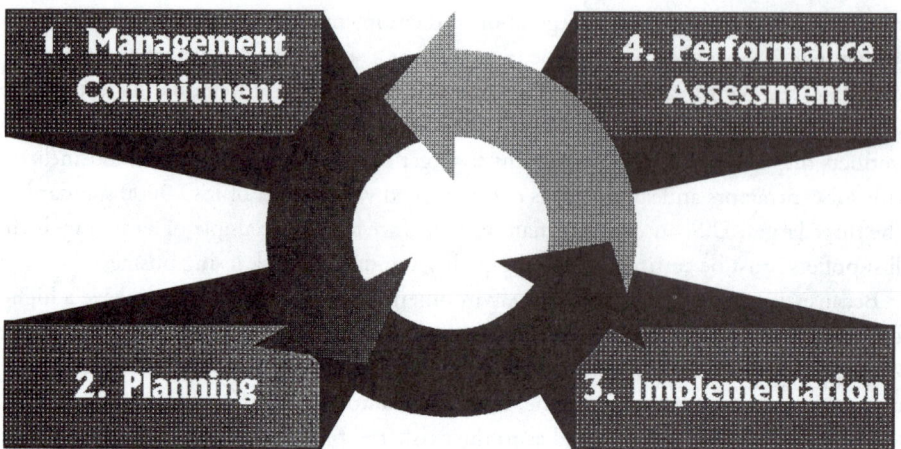

Figure 5.3   Environmental management system continuous improvement cycle.

that should be periodically reviewed for measuring performance against selected targets and identifying new goals as business priorities evolve and change. A diagram of this concept is presented in Figure 5.3. An EMS should incorporate the following attributes:

- Top management support illustrated through the development of environmental policies outlining issues related to pollution prevention, commitment to regulatory compliance, continual improvement with respect to specific goals and targets, documentation, training, and responsibilities;
- Environmental planning to identify aspects and legal issues and to ensure they are kept up to date and considered in an organization's activities, processes, products, and services;
- Environmental objectives and goals, the management process for attaining these goals, and adequate resources to achieve these goals;
- The definition of roles and responsibilities to ensure and promote protection of the environment to include training, internal/external communication, documentation control, operational control, and emergency preparedness and response; and
- A management system to audit and review the EMS to determine progress, initiate corrective actions, and identify opportunities for continuous improvement.

As previously stated, for a program of this type to be effective, top management support is required as it is their responsibility to define the company's policy on environmental management. This policy should include the company's commitment to the continuous improvement process and the prevention of pollution, compliance with regulatory and other internal requirements, and methods for setting goals and reviewing progress for meeting these objectives. In addition, it is important that provisions be outlined in an environmental management policy and that the information be disseminated to all employees in a company. An effective method for achieving this is employee training.

Policies and procedures contained in an environmental management policy should be made available to the public. This factor is sometimes more difficult to grasp because some companies may feel these documents should be private. This viewpoint, however, is contrary to the concept of an EMS, which is designed to be open and proactive. Therefore, open communication is one of the primary building blocks to achieving these objectives.

With management commitment to the EMS process in place, planning becomes the next important step. This process includes the definition of specific regulatory and internal requirements, identification of a company's environmental aspects (that is, those activities, products, or services within a company that may contact or affect the environment), and the establishment of specific objectives and goals to be accomplished. The planning phase should also identify how these targets are to be met and the documentation scheme that will track their progress.

Effective roles and responsibilities should be established to implement plans successfully. These should be supported by management through the supply of adequate resources. These resources may include personnel, technology, equipment, or financing. To ensure this process is coherent, the same message should be delivered to all employees. Again, this may be accomplished through a training program.

Initial Review and Gap Analysis

Although the requirements for self-certification and third-party registration for conformance with ISO 14001 are well defined, methods for planning and implementing an EMS vary depending on the extent to which a business already has management system protocols in place and the level of documentation that exists.

Because this varies with each business, a *gap analysis* is traditionally conducted to identify the current status of each required EMS characteristic and the steps necessary to achieve conformance with the standard. After this is conducted, a plan can be orchestrated to develop and implement an effective system.

A school of thought of late is that extensive gap analyses should be the first order of business in establishing an EMS. However, because this is a new standard and, as such, company operators are uncertain as to how it will affect their business, dedicated efforts and funding for gap analyses are not commonplace. Therefore, a more conservative approach is to invest adequate time up front educating key management representatives on the issues of ISO 14001. This should work in tandem with an initial review of a company's environmental practices. This initial review is an assessment of where the company *is*. This is in contrast to a gap analysis that evaluates the *distance* that will need to be traveled to meet ISO standards. In addition, a gap analysis may include an environmental aspects review, whereas the initial review will stick to management systems and organizational functions.

Before an initial review is begun, a checklist of the basic elements of the 14001 EMS specification should be compiled. These items can then be compared to the following checklist of suggested review items:

- Identify statutory and regulatory requirements;
- Identify internal environmental policies, procedures, and management practices;

- Assess compliance and effectiveness of the existing EMS compared to the afore-mentioned criteria;
- Identify and assess other management systems and how they affect EMS elements;
- Identify benefits of ISO 14001 certification and viewpoints of internal organizations; and
- Estimate level of effort, costs, and time line associated with ISO 14001 certification.

It is important to remember that this is merely a checklist for quick review. As with any project, background study and evaluation will ensure steady progress while providing the time to orient and educate management and foster team member support.

The aforementioned elements are also part of a detailed gap analysis that is typically conducted when there is a sufficient level of management buy-in. However, the gap analysis will be more detailed in that it will evaluate the individual components of these items against the ISO 14001 specification requirements and determine the actions necessary to close this gap and remedy any discrepancies. Another significant element of a gap analysis is that a detailed study should be conducted to identify the environmental aspects and evaluate the environmental impacts. Environmental aspects are those portions of a business's activities, products, and services that have the potential for interacting with the environment. Because of the relatively high level of effort that is required during this phase, some companies may opt to use third-party consultants for a more objective analysis. The final result of a gap analysis should be a plan for the future that identifies responsibilities and roles, provides management and financial support, and outlines the implementation tools necessary to close the distance.

Self-Declaration Versus Registration

After an effort has been dedicated to pursuing the development of an EMS that meets 14001 standards, a choice should be made to self-declare conformity or to request a third-party assessment for registration.

Self-declaration is a more cost-effective approach, although the level of credibility may suffer because of a lack of discernible objectivity of the overall process. When a company "self-declares," this means that the company has implemented ISO 14001 and publicly asserts that conformity with the specifications has been achieved. The perception of the public or a customer determines the effectiveness of this approach. Furthermore, for this approach to be credible, an internal environmental management system audit and an objective analysis of the findings should be completed before implementation. Because the experience level and methods of interpretation differ from company to company, self-declaration may achieve varying levels of effectiveness.

Third-party registration is a different approach to conformity. Third-party registrars must receive domestic accreditation through the Registrar Accreditation Board. Individual registrars may focus on specific industry sectors. Therefore, proper attention should be dedicated to selecting a registrar that has a pertinent reputation, competence, auditors, and capabilities for performing the conformity assessment and responding to the organizational needs of a specific industry.

Like self-declaration, third-party registration is a process that also contains inconsistencies. Because the ISO 14001 standard is relatively new, consistency between individual registrars on a number of issues may vary. However, as the ISO movement becomes more popular, these discrepancies should be resolved. As with any new process, debugging will play a pivotal role in the evolution of the implementation of ISO 14000 standards. The benefits of implementation and registration to ISO 14001 or any of the subsequent standards published should be evaluated by each business. Figure 5.4 conceptualizes the potential effectiveness of sequential steps toward ISO 14001 registration. Many companies are unsure of the impacts or benefits the ISO 14001 standards will have on their operations. Therefore, a sensible approach is to begin an effort to educate and assess the current status of an EMS and strive to gain management support while closely watching domestic and international developments.

These standards promote a common-sense approach to managing environmental issues using time-proven methodologies stressing the commitment to continuous improvement. By incorporating environmental excellence into corporate business strategies as quality, customer service, and stakeholder concerns related to profitability and sustainability have been, the transition to ISO 14000 and the integration of business principles with environmental stewardship may be realized.

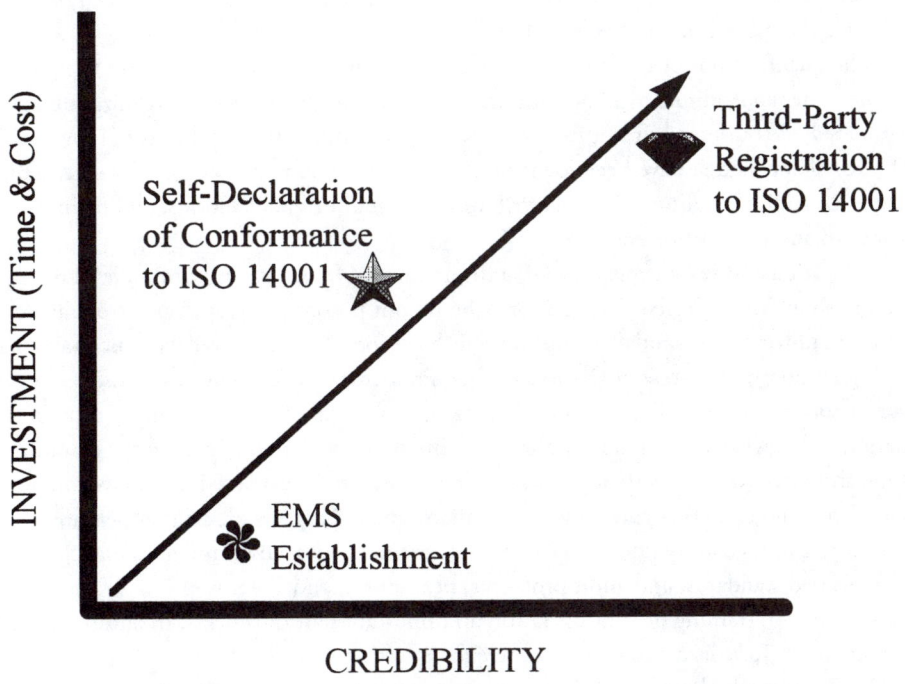

Figure 5.4   Steps toward ISO 14001 registration.

Environmental Auditing

The standards that outline the general principles, procedures, and qualifications for environmental auditing include ISO 14010, 14011, and 14012, respectively. The primary purpose of these standards is to ensure that there is consistency in the evaluation process for environmental performance. Until recently, environmental auditing in the U.S. was almost solely dedicated to determining a facility's status with respect to a set of applicable federal, state, and local regulations. The concept of auditing taken from the ISO perspective measures compliance with regulations, but also evaluates parameters as defined within ISO 14001 Environmental Management Systems, best management practices, and other elements.

Because most audits cannot be all encompassing, they are typically broken out into smaller more easily handled packages. Multimedia audits, for example, although effective for enforcement purposes, are typically not the method of choice when conducted internally. A higher level of detail is achieved when air, water, and solid waste audits are focused on individually. The principles of an environmental audit, as defined by ISO 14011, should, therefore, clearly identify the audit objectives and material to be reviewed. This is particularly effective when defined by appropriate company representatives. Once defined, it is also important that the necessary resources and personnel be allocated to achieve worthwhile results. Internal personnel may perform auditing functions or act as support staff for a third-party audit to answer questions and supply pertinent documents and materials for review. In the former case, internal auditors should be selected in a way that avoids conflicts of interest. That is, an internal auditor should not be directly responsible for the subject matter being audited.

The qualifications of an auditor, as presented in ISO 14012, Qualification Criteria for Environmental Auditors, whether internal or external to the facility or organization being evaluated, are outlined but not fully approved at the time of this publication. However, the key elements that have been identified include education and work experience, training as an auditor, auditor skill and attributes, lead auditor qualifications, and maintenance of auditing competence.

Education and work experience of auditors varies. However, a minimum level of expertise should be expected. For auditors who do not possess a 4-year degree from an accredited university or similar institution, but have completed a secondary education (that is, high school), work-relevant experience requirements should be more extensive to ensure a core competency level. In addition, formal education in areas that provide a foundation of knowledge in relation to the following topics will help to reduce the amount of time and direct work experience necessary to perform auditor-related tasks: environmental science and technology; technical and environmental aspects of facility operations; statutory and regulatory knowledge of environmental issues, environmental management systems and standards; and audit protocols, procedures, and systems.

On-the-job training of auditors is also an important part of process. In addition, the requirements for a lead auditor are more encompassing than those for an auditor. The ISO 14012 standard places requirements on auditor training by specifying a minimum number of auditing work days or a minimum number of audits performed within a specified time period, such as 3 years. The requirements of a lead auditor include a thorough understanding of the technical and regulatory aspects of the auditing process in addition to personal attributes such as leadership, management, judgment, interpersonal skills, and communication. On-the-job requirements are also more stringent for lead auditors, requiring additional workdays of experience and a greater number of audits performed, and experience as an acting lead auditor under the direct supervision and guidance of a seasoned lead auditor. In either case, core competency in the areas of education, training, work experience, and personal attributes/skills should be maintained for an auditor to be effective.

As with many types of audits, environmental audits must be held in confidence between the auditor and client, with disclosure held at the discretion of the client. In most cases, attorney–client privilege prevents "discovery" of the documents and information should legal action for various regulatory infractions exist or develop. Although this has been the traditional approach to environmental auditing, the intent of ISO 14000 is proactive public disclosure. Therefore, preventing access or refusing to make audit findings available to the public defeats the intended purpose of the standard.

The audit itself should be conducted in accordance with the standards specified by ISO 14011. These standards define the objectives of the audit, the roles and responsibilities of the auditing staff and individuals within the company being audited, and the process by which the audit is planned, executed, and summarized. Traditional objectives for an audit of this type are to evaluate a client's EMS against the criteria of ISO 14001, determine the effectiveness of its implementation, assess the continuous improvement cycle that is initiated by periodic internal management review, and identify any areas that are in need of improvement.

The responsibilities of the auditing team—guided by the experience of the lead auditor—are to

- Ensure that all auditors meet the qualifications as identified in ISO 14012 and that the activities of the audit team are performed to the guidelines of ISO 14010 and 14011;
- Conduct an adequate and thorough background investigation of the company in relation to processes, products, services, prior audits, and physical site characteristics;
- Ensure that there is not a conflict of interest between the auditors and client/auditees;
- Establish strong communication channels between the audit team and client/auditees to facilitate proper audit planning, initial and final briefings, problem resolutions, objectives restructuring, critical nonconformity notifications, and final report requirements;
- Promptly notify all critical nonconformities to the client/auditees;
- Complete the final report in a timely fashion in clear, concise, and conclusive language; and
- Provide recommendations for improvements to the EMS if it is established as one of the mutually agreed on objectives of the audit.

Proper time and effort should be taken to carefully select the most suitable and qualified auditor that matches the company's business sector and overall operations. Selection of an auditor who is not experienced in the processes of a particular organization may result in inaccurate findings, delays, and even unnecessary tension between the auditing team and any company representatives involved. After an auditor has been selected, a number of key elements should be in place before initiating the audit. These elements include the objectives and scope of the audit, the audit criteria, an established audit plan, and audit team personnel.

Before the audit is begun, several internal activities should be performed to provide a smooth transition for the process. These include the following activities: informing employees of the EMS audit, why it is occurring, and related activities that will be taking place to create an atmosphere of cooperation and understanding; establishing an internal team of individuals to assist the auditors during the process, each of whom should be familiar with the materials and processes that will be needed or evaluated during the audit; and ensuring timely access to information and areas within the facility as required for EMS auditing purposes while ensuring employee health and safety.

It is important that all audits be conducted with consistency, accuracy, and reliability. Variations in the approaches, methods, and systems used will produce differing results and inconsistencies in each facility or operation. For this reason, the specific objectives of an audit and the criteria by which the findings will be evaluated should be identified before commencing activities. This helps to ensure that the focus of the audit is well defined and the quality of the findings are competent.

The intent of ISO 14010 is to ensure that adequate thought is put into the structure of an audit before its commencement so that the client and auditor have a clear understanding of the audit's objectives. As with any audit, time and money are two prevalent constraints of EMS audits. In general, the more time taken for an auditing process, the higher the cost will be. Therefore, the more limiting these parameters, the more likely that the findings will be placed at risk and yield inaccurate or incorrect results. An example of these constraints affecting the outcome of an audit is the case wherein a sample of information collected is not representative of that which is available. This can occur when the length of time required to conduct an effective audit is too short. The definition of time and cost cannot be left open before commencing an audit. Therefore, there will always be an inherent risk. However, by understanding this potential and minimizing it where feasible, the validity of audit findings will be maximized.

The results of an audit are contained in a final written report. Most final reports are first compiled in a draft format to give the client an opportunity to review the information and findings. Although objectivity is an important element of an audit, it is standard practice for a client to review the preliminary report to ensure that the interpretation of the findings or evidence is representative of the facility's operations and that their comparison to the evaluation criteria is not understated or overstated. The final report of an audit should also contain actions to address any problems found. In some instances, recommendations are provided by the auditor to facilitate this process. The following information should be included in an EMS audit:

- Identification of the organizations involved and the individuals participating, both for the auditing team and client representatives;
- The time period during which the audit was conducted;
- The scope and objectives of the audit and the criteria against which the findings will be evaluated;
- An overview of the audit process, procedures, and systems used and an assessment of their effectiveness at the facility investigated;
- A summary of the findings in relation to the audit criteria; and
- Conclusions on the state of the facility with regard to the defined scope and evaluation criteria (recommendations for corrective action may also be provided for any discrepancies found).

LIFE-CYCLE ASSESSMENT

Life-cycle assessment (LCA) standards are continuing to evolve and may be refined in the future as the practice becomes more commonplace. The ISO 14040 standard, Life Cycle Assessment General Principles and Framework, while providing an overview of the LCA process, introduces methodologies and pertinent requirements that are further refined and presented in the following standards:

- ISO 14041—Environmental Management, Life Cycle Assessment—Life Cycle Inventory Analysis;
- ISO 14042—Environmental Management, Life Cycle Assessment—Life Cycle Impact Assessment; and
- ISO 14043—Environmental Management, Life Cycle Assessment—Life Cycle Interpretation.

Although the life-cycle inventory analysis, impact assessment, and interpretation are integral phases of the overall LCA process, these standards are in their formative stages.

Life-cycle assessments can be useful analytical tools in decision-making processes that involve a number of different issues outside of those specifically addressed by LCA efforts. The LCA is a concept that is frequently thought of as a "cradle-to-grave" approach or a systems-based analysis to managing a product or service throughout its life span. The LCA may be used for strategic planning for industrial, governmental, and nongovernmental organizations or in the design of new products or processes and the redesign of existing products or services. In many cases, LCA studies have revealed methods for cost savings associated with reduced energy use, reduced waste management, and increased use of recyclables. When taken into account, these uses yield greater cost benefits for the implementation of a wide range of pollution prevention projects.

Life-cycle analysis, which leads to life-cycle accounting, is a process useful for making business decisions associated with financial accounting, market share, inventory, overhead, and production rates.

Life-cycle assessment is a framework, or systematic approach, for examining materials and energy requirements (both incoming and outgoing) and the resulting environmental impacts caused by the construction, use, and, in many cases, disposal of products or termination of services. An LCA for an automobile, for example, takes into account its manufacture, use, service life, and disposal issues. The "life" of the product, which, in the case of the automobile, may include specific items such as fuel consumption, oil changes, emission controls, reusable thermoplastic body panels, and environmentally friendly paints, must be considered.

As previously outlined, the traditional methodology for conducting an LCA includes three phases: inventory analysis, impact assessment, and interpretation. The inventory analysis is an evaluation and quantification of the material and energy inputs and outputs over the life of a specified product or service. The impact assessment examines and attempts to understand the various types of environmental effects, their magnitudes, and their significance. Lastly, the interpretation phase draws conclusions from the compilation of information from the inventory analysis and impact assessment. These conclusions help define actions based on the technical, social, and economic issues surrounding the LCA that are executed separately from the LCA itself. The LCA is an information-gathering, analytical, decision-making process that improves the interaction of a product or service with a host of environmental issues.

Before initiating an LCA, it is important to define the goals and scope of the study so that efforts can be focused and streamlined to achieve valuable results. This is no different from conducting an environmental audit for ISO 14010, where a goal must be set and a path chosen before any data compilation or investigation is conducted. Questions that may be asked to help outline objectives and restrict the scope are as follows:

- Why is an LCA needed?
- What is the end use of the LCA and for whom is it compiled?
- What are the systems and boundaries of the LCA?
- What quality requirements for data are necessary?
- What are the review criteria?

Inventory Analysis

With initial questions such as these answered to help shape the purpose and intent of an LCA, the inventory analysis may be conducted. An LCA inventory analysis, as previously mentioned, is a measure of the inputs and outputs of materials and energy associated with a given product or service throughout its useful life. To effectively quantify these parameters, boundary conditions or specific systems should be isolated with all data kept as elemental as possible. This is not unlike mass-balance methodologies, where individual chemical constituents flowing into a process are measured against those leaving. However, conducting this exercise over the life span of a product by taking into account all the various events that may occur can become fairly onerous.

In general, the more detailed and accurate the data collected, the more reliable the results and conclusions will be. Data collected should quantify resources used and their ultimate impact on the environment via releases to air, water, and land. Multiple "balances" can be conducted on a given product in different segments and/or facets of its life, with each exhibiting varying degrees of detail.

Impact Assessment

An impact assessment is that portion of the LCA that focuses on taking the results of the inventory analysis and determining their effects on the environment. These effects may be primary or secondary. Primary effects are those that are directly apparent, or obvious. An example of this is fuel consumption of an automobile that subsequently emits various pollutants through the tail pipe. Secondary, or indirect, effects are not as apparent and require additional thought or analysis. An electric car, for example, eliminates the emission of pollutants to the atmosphere in contrast to the vehicle powered by an internal combustion engine. However, the electric utility plant producing the electricity that charges the battery that "drives" the electric car does emit pollutants. This is an example of a secondary effect. These emissions may be less and/or consist of entirely different constituents. Nonetheless, they are a factor in determining the overall environmental impact of an electrically powered vehicle.

Four potential elements exist within an impact assessment: classification, characterization, significance analysis, and valuation. Classification is an exercise that takes the inventory analysis data gathered and separates them into various well-defined categories. These categories, such as air quality, human health, recycling, and resource use, may be broad in scope. Other more specific categories include stratospheric ozone depletion, fuels combustion, and cancer risk. The selection of categories is dependent on the defined goals of the LCA. In addition, each input/output parameter from the inventory analysis may be present in several categories.

Characterization is that portion of the impact analysis that quantifies the effects of each category of classified inventory data on the environment. This exercise takes into account important scientific information, such as physical, chemical, biological, and toxicological data, as it pertains to each individual inventory parameter. The activity may quantify actual and potential impacts. Actual impacts are self-explanatory. Potential impacts are those that may be derived through various modeling methods and techniques. Examples of these include toxicological data (that is, no observable effects concentrations), air-dispersion modeling of contaminants, and the determination of ozone-depleting potential for specific chemicals and compounds. Characterization must also take into account the chain of environmental events within a particular category or elements within a category. This type of review, referred to as the stressor-effects network, pertains to the web of physical, biological, and chemical events that intertwine a specific cause with an identified environmental effect. An example of a review that takes into account the chain of events tied to a release of nitrogen oxides is illustrated in Figure 5.5.

The significance analysis and valuation portions of the impact assessment serve as technical analyses to determine the significance of the identified environmental impacts and then rank, interpret, and weigh these effects taking into account the characterization results and corresponding LCA inventory data. This procedure is somewhat subjective in nature and varies depending on prevalent socioeconomic factors. The goal of this step is to provide a sense of hierarchy, or priorities, within the overall impact assessment phase of the LCA.

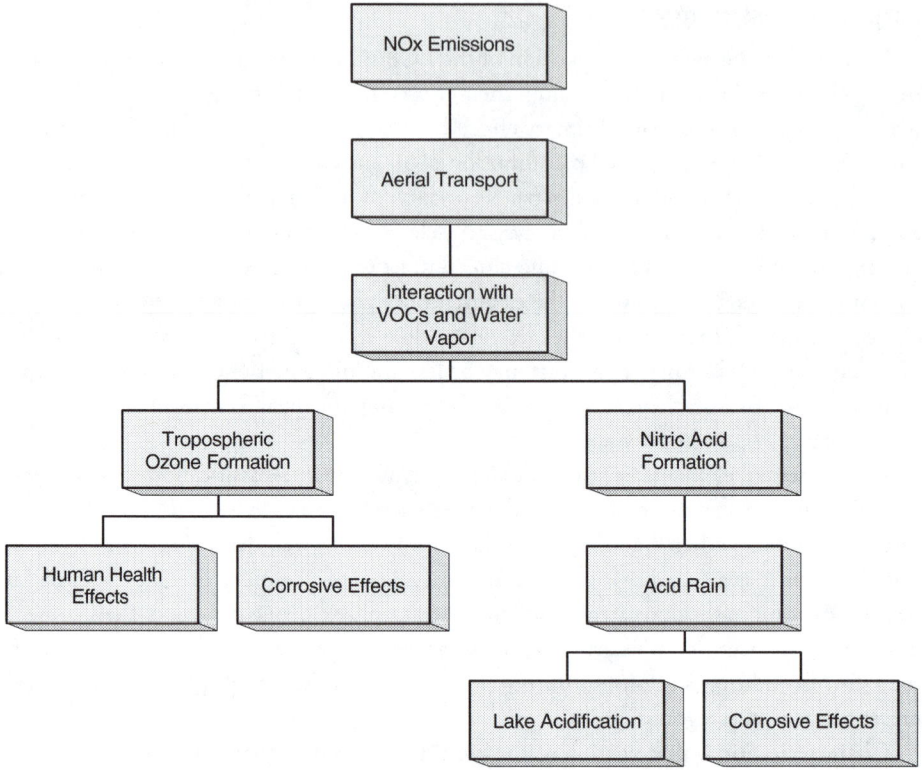

Figure 5.5   **Stressor-effects network of nitrogen oxides emissions (VOCs = volatile organic compounds).**

Interpretation

The goal of the LCA is to answer specific questions posed at the onset of the effort. By organizing the inventory analysis and/or impact assessment into comprehensible results, conclusions to these questions may be drawn. The conclusions arrived at may be iterative in nature by reviewing of the scope and data collected. As the process develops, the scope may be revised or modified, and new or additional data needs may be identified. The results of an LCA should provide an overview of the major environmental impacts/interactions of the system or product investigated, identification of the gaps in data used, and options that are pertinent to the goals and objectives of the LCA.

It is important to note that the LCA is an investigative process in which goals are set, information is gathered, assessments are made, and conclusions are drawn. No part of the LCA should be connected with actions taken as a result of the LCA effort. Such actions are governed by a decision process that takes into account a host of factors such as technical, budgetary, regulatory, and social issues and, therefore, are considered a separate initiative.

Reporting and Critical Review

Compilation of LCA investigations, data gathering exercises, and conclusions is an important element in the overall process. An LCA report should be tailored toward an intended audience and include the goals and scope of the study, methods used, detailed documentation of the results, analysis of the findings identifying critical or sensitive elements within the inventory analysis and impact assessment, and conclusions and recommendations.

A critical review of the report is a necessary process of an LCA to ensure that methods used are scientifically and technically sound, data are appropriate and reasonable, and conclusions drawn are credible based on limitations identified within the goals. This exercise verifies that an LCA study has met the methodologies, data requirements, and reporting requirements specified within the ISO LCA series of standards and has established the validity of the results.

There are three primary elements in a critical review cycle: self-review, expert review, and stakeholder review. The self-review is simply a systematic check conducted by the individual or entity that has performed the LCA. Although this may not always achieve unbiased results, it can be an effective process because of the reviewer's first-hand knowledge of the study. Expert reviews are typically conducted by a third-party independent of the given LCA process who is highly knowledgeable of the elements of the standards. This type of review, which ensures that the project is viewed from "outside the box," may either evaluate a critical review statement prepared by the entity conducting the LCA or be used to compile the statement itself. Expert review lends additional credibility to an LCA. Stakeholders may be involved in the critical review phase of an LCA. Stakeholders are parties that are directly affected by the results of the LCA and may include government agencies, environmental protection groups, industry competitors, consumers, and scientific or technical organizations. Comments from panels of these participants should be included in the critical review statement when possible.

Although the critical review process is not a required element of an LCA study, it lends additional credibility to the process and, therefore, is highly recommended. Because the LCA may incorporate and compile considerable amounts of data/information and analyze the various environmental effects of a process through numerous channels, critical review is an essential element to ensure that all pieces have been brought together in a logical fashion and that the conclusions drawn are credible and valid.

INCLUSION OF ENVIRONMENTAL ASPECTS IN PRODUCT STANDARDS

All products affect the environment to some extent during their manufacture, distribution, use, and/or disposal. The significance of these effects varies from product to product as does the time period over which they are experienced. While some environmental impacts may be limited in their geographical scope, others may be global. By integrating or providing for these impacts in the standards of a product, overall effects may be recognized and minimized significantly.

As with the standards that address life-cycle issues, ISO 14060, Guide for the Inclusion of Environmental Aspects in Product Standards, is in its formative stage. Despite this, efforts are underway to promote the following idealogies: creating a general understanding that provisions in product standards can affect the environment in beneficial and detrimental ways; defining a relationship between product standards and the environment; avoiding elements within product standards that may lead to adverse environmental impacts; promoting a general understanding that the inclusion of environmental aspects in product standards is an involved process and requires the balancing of a number of competing priorities; and instilling proactive approaches that integrate life-cycle and scientific methodologies in the development of product standards and include environmental aspects.

To achieve these goals, ISO 14060 sets forth some general considerations that should be taken into account when developing product standards to provide a balance between product performance and adverse environmental effects. Methodologies are also identified for assessing the environmental effects of product standard provisions and considering their effects throughout the various life-cycle stages of the product.

Although it may be assumed that all products affect the environment, these cause-and-effect relationships are not readily discernable. Effects can occur at local, regional, and global levels or at a combination of these levels during any portion of a product's life cycle. Trying to identify these effects at the inception of a product's life is a difficult and involved process. Nevertheless, waste minimization and pollution prevention techniques should be initiated at the forefront. As with most manufacturing process or product decisions, multiple issues come into play on any course of action selected. Therefore, consideration of parameters such as safety and health, cost, marketability, quality, product performance, and other pertinent issues should be included. The development of a product standard that takes into account environmental impacts is never truly completed or "written in stone." As such, periodic review is important because of the constant improvement of knowledge and technology in the environmental arena.

When a product is being developed, many key parameters relating to its potential effects on the environment may be known, but may not be quantifiable. Experience from the use of specific manufacturing processes and business/production practices may provide valuable input toward the specification of product standard requirements. Consideration early on may help create improved products from processes that make the most of environmental performance. Decision processes of this nature can affect

- Types of manufacturing methods and processes employed;
- Construction materials selected;
- Energy usage and forms from which it is derived;
- Types and quantities of waste streams and emissions resulting from manufacturing processes and all phases of the product's life cycle;
- Packaging methods;
- Transportation and distribution methods/channels;
- Recyclability, recovery, and reuse of the product and/or its constituents; and
- Final disposal of the product and any waste streams associated with its manufacture and use.

Environmental effects that should be considered during the formative stages of product standards are governed primarily by the inputs of material and energy and the output of material during its entire life cycle. Selection of input will therefore greatly determine output. An overview of this process is depicted in Figure 5.6.

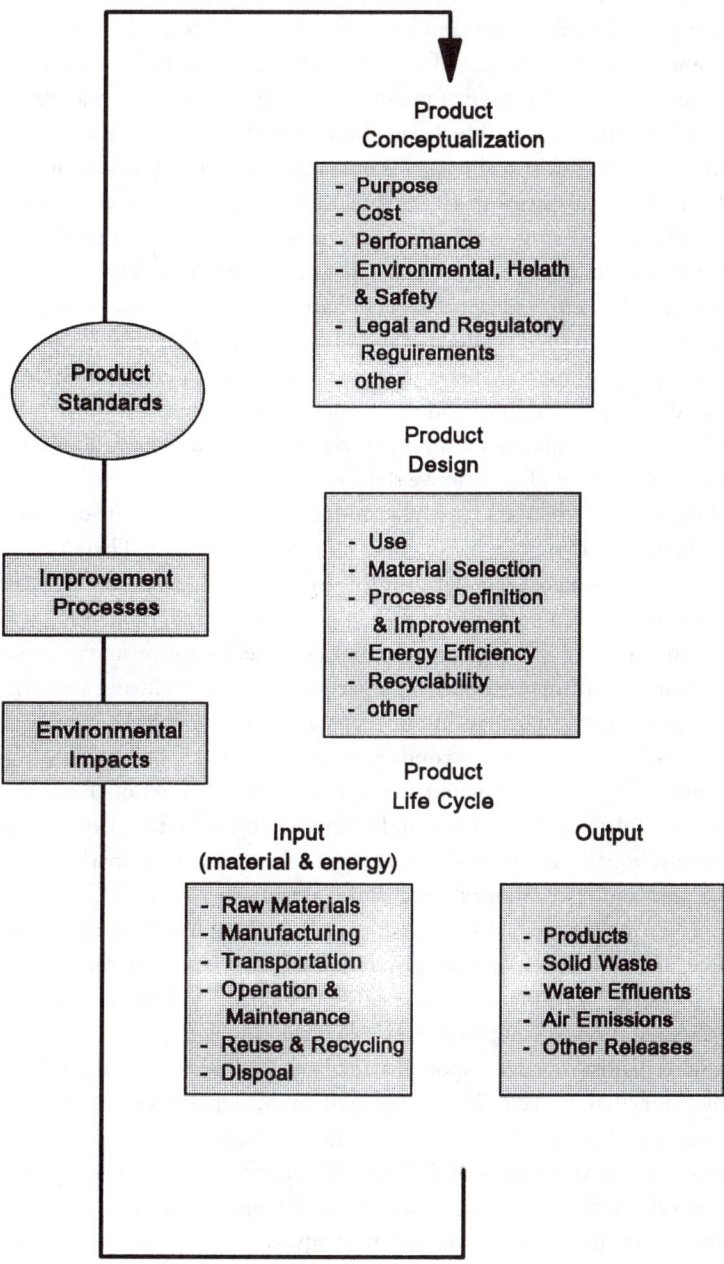

Figure 5.6   Product standards and their effects on the environment.

Inputs fall into two broad categories—energy and material. Each category poses its own set of environmental concerns and generates a wide range of output. Output, in relation to the environment, typically falls into one of three commonly defined media. These are air emissions, water effluents, and solid wastes. Here, solid wastes should not be assumed to be "solid." The definition of solid waste within the U.S. under the Resource Conservation and Recovery Act is an all-inclusive category that comprises waste streams in the form of gases, liquids, solids, and any combination thereof.

Energy input to the manufacture of a product and that required during the product's life span can directly affect the type of pollution generated and the media affected. Energy sources selected include fossil fuels, geothermal, hydroelectric, nuclear, and others. It is important to note that electrical energy is not specifically stated because it is must be generated from the use of one of these listed methods. This was cited previously in an example of an electric car. Batteries are recharged via electrical current, but electricity is generated from coal-fired, nuclear, or other types of power plants. The resulting pollution is therefore a secondary effect rather than a primary effect. Nevertheless, its generation is unavoidable and characterized by the method of the energy's derivation. Coal-fired power facilities produce environmental effects in numerous forms, the most commonly known of which is acid rain from sulfur dioxide emissions. Although nuclear power plants generate considerably less air pollutants, they create significantly more hazardous and long-term problems in the form of radioactive solid wastes.

Material inputs to the manufacture of a product, or necessary for its operation/maintenance, can have even more of a direct effect on the environment. Therefore, careful consideration should be taken into account for primary materials as well as those that are secondary, or not directly a part of the finished product. Secondary materials can include those associated with the following: creation of raw materials for primary use; consumables during manufacturing processes, not a part of the final product; transportation (that is, packaging, physical transport, and so on); operation and maintenance; reuse and recycling; and disposal of primary and secondary materials.

Environmental impacts from material inputs have their own set of effects that may be felt through the depletion of renewable and nonrenewable resources, exposure to hazardous materials, impairment to land use, and the release of contaminants, or pollutants, to air, water, and land. The latter of these issues is typically a direct effect from the output of manufacturing processes. Other than the product, emissions to air, wastewater effluents, miscellaneous releases, and the generation of solid wastes are the outputs of manufacturing processes. This topic was discussed in the Chemical Engineering and the Material Balance section in Chapter 2, and is depicted by Figure 5.7.

Air emissions consist of gases, vapors, particulates, and heat and can cause a wide range of environmental impacts. These may include the depletion of stratospheric ozone and the generation of ground level ozone, acid rain, nuisance odors, and a host of toxicological effects to humans, animals, and plants. Water effluents may include the same types of chemical constituents as contained in the aforementioned impacts, but their transport and uptake mechanisms will occur via aqueous channels rather than through suspension in air currents. Typical environmental effects from contaminated water may include toxicological effects resulting from contaminated drinking water, corrosion

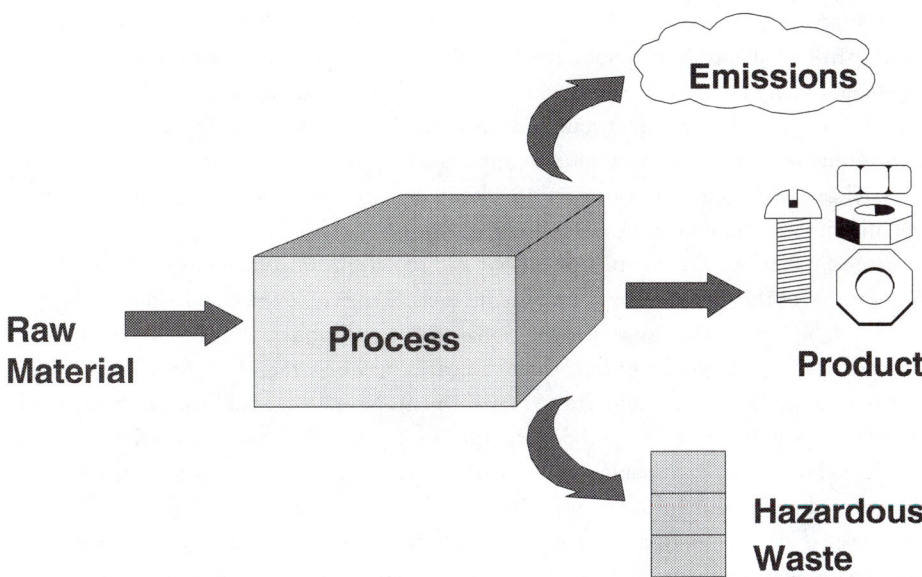

Figure 5.7   Conceptualization of the material balance.

caused by high and low pH, dispersal of radioactive contaminants, bioaccumulation of various organic and inorganic compounds, oxygen depletion caused by high biological oxygen demand, resulting in "dead" streams and bodies of water, and eutrophication caused by discharges of high levels of biological nutrients.

Solid wastes are produced concurrently with many processes. Mishandling or improper disposal of solid wastes can result in any of the aforementioned environmental effects along with many others. Waste products that are not physical in nature, such as noise, vibration, odor, and heat, should also be taken into consideration when developing product standards.

Careful consideration should be devoted to the identification and the potentiating environmental effects of provisions developed within product standards. Life-cycle analyses and related methodologies are ideal approaches for assessing the environmental impacts of a product and for identifying or formulating provisions. For an LCA to be most effective, a thorough understanding of the process and its limitations is required. Misinterpretations can lead to ineffective provisions and waste valuable personal energy, time, and money during their implementation.

Ideal provisions within product standards facilitate environmental improvement. Provisions that may hamper this process are those that specify materials to be used in a product. Product standards, or at least the provisions identified within to address environmental impacts, should not be prescriptive such that innovation and process improvement are stifled. When materials are specifically identified, new methods or materials that improve environmental performance may be difficult to incorporate. Therefore, adequate latitude should be built into the standards so that as new materials or processes are developed, they may be easily or formally integrated into the manufacture of the specified product.

Allowances for secondary or recycled materials should also be taken into consideration. This open-ended approach is not amenable in all cases, especially in areas where human health and safety are concerned; however, it should be considered wherever possible.

The compilation of product standards involves three primary strategies for instilling environmental improvement: resource conservation, pollution prevention, and design for the environment. Resource conservation, although somewhat self-explanatory, refers to the minimization of depleting various natural resources. Those resources that are nonrenewable are destined for eventual phase-out and replacement as their stocks are used up. An example of this is the current use of fossil fuels. Renewable resources, on the other hand, can be replenished. With proper management techniques, a balance between the use and replenishing modes of renewable resources may be established. Successful programs, such as timber and soil fertility management, have been established in many applications. As populations rise and the world market expands, however, this delicate balance becomes threatened. Renewable resources are subject to the effects of depletion when use rates exceed replenishment rates.

Energy is also an important resource destined for conservation. Many of the natural resources previously mentioned are those that are used in the generation of energy. The uses of some energy forms are more environmentally sound than others. As such, their effects should be weighed against each other. Regardless of the type, energy use is minimized through the maximization of efficiency. This was demonstrated during the energy crisis of the 1970s. Company operators now realize that conservation of materials and energy yield lower costs and higher profits. For these reasons, resources should be conserved regardless of the type used.

With increases in process efficiency, material and energy usage are minimized. This represents a portion of the concepts embodied by pollution prevention. Pollution prevention, as discussed in Chapter 1, introduces approaches that incorporate source reduction, reuse, recycling, material substitution, process optimization, technology introduction, and many other methods for the minimization of waste. Although these methods may not be written into product standards, recommendations on the selection of materials and processes that promote these concepts can be beneficial. In some instances, it may be appropriate to prohibit the use of certain materials such as ozone-depleting compounds, processes that accelerate climatic changes, and chemicals that can affect human health and the ecology. Proper care should be exercised before the inclusion or prohibition of any specific materials or processes until sufficient knowledge has been obtained. In some cases it may be determined that no course of action allows flexibility for the introduction of new materials and technologies.

Including environmental considerations in product standards is a relatively new initiative in designing for the environment and is embodied by the upcoming ISO 14060. Designing for the environment is a multifaceted process that can include issues that revolve around the incorporation of environmentally friendly raw materials, the use of environmentally compatible or controlled processes, the reuse of components or constituents, and the ability for disassembly and subsequent recycling.

REFERENCES

Childers, D.G. (1997) To Be or Not To Be ISO 14000 Certified. Paper presented at 33rd Annu. Aerospace/Airline Plating Met. Finish. Forum Exposition, Am. Electroplaters Surface Finish. Soc., San Francisco, Calif.

International Organization for Standardization (1993) ISO 9000 International Standards for Quality Management. 3rd Ed., Geneva, Switz.

International Organization for Standardization (1994) ISO 8402 Quality-Vocabulary, 2nd Ed., ISO 8402:1994, Geneva, Switz.

SUGGESTED READINGS

ASTM (1995) ISO 14000 Guide to Environmental Management Principles, Systems, and Supporting Techniques (Draft). ASTM PCN: 34-014000-65, U.S. Tech. Advisory Group to ISO, Philadelphia, Pa.

ASTM (1995) ISO 14001 Environmental Management Systems Specification with Guidance for Use (Draft). ASTM PCN: 34-014001-65, U.S. Tech. Advisory Group to ISO, Philadelphia, Pa.

ASTM (1995) ISO 14010 Guidelines for Environmental Auditing—General Principles of Environmental Auditing (Draft). ASTM PCN: 34-014010-65, U.S. Tech. Advisory Group to ISO, Philadelphia, Pa.

ASTM (1995) ISO 14011 Guidelines for Environmental Auditing—Auditing Procedures, Part I: Auditing of Environmental Management Systems (Draft). ASTM PCN: 34-014011-65, U.S. Tech. Advisory Group to ISO, Philadelphia, Pa.

ASTM (1995) ISO 14012 Guidelines for Environmental Auditing—Qualification Criteria for Environmental Auditors (Draft). ASTM PCN: 34-014012-65, U.S. Tech. Advisory Group to ISO, Philadelphia, Pa.

ASTM (1995) ISO 14040 Life Cycle Assessment General Principles and Practices (Draft). ASTM PCN: 34-014040-65, U.S. Tech. Advisory Group to ISO, Philadelphia, Pa.

ASTM (1995) ISO 14060 Guide for the Inclusion of Environmental Aspects in Product Standards (Draft). ASTM PCN: 34-014060-65, U.S. Tech. Advisory Group to ISO, Philadelphia, Pa.

Beaumont, L.R. (1995) *ISO 9000, The Standard Interpretation.* 2nd Ed., ISO Easy, Middletown, N.J.

Cascio, J. (1996) *The ISO 14000 Handbook.* CEEM Information Services/ASQC Quality Press, Milwaukee, Wis.

International Organization for Standardization (1996) ISO 14001 Environmental Management Systems—Specification with Guidance for Use. 1st Ed., ISO 14001:1996(E), Geneva, Switz.

International Organization for Standardization (1996) ISO 14004 Environmental Management Systems—General Guidelines on Principles, Systems, and Supporting Techniques. 1st Ed., ISO 14004:1996(E), Geneva, Switz.

A

40 CFR 261—Identification and Listing of Hazardous Waste

Subpart C—Characteristics of Hazardous Waste

40 CFR 261.20 General.

40 CFR 261.20(a)

(a) A solid waste, as defined in §261.2, which is not excluded from regulation as a hazardous waste under §261.4(b), is a hazardous waste if it exhibits any of the characteristics identified in this subpart.

40 CFR 261.20(b)

(b) A hazardous waste which is identified by a characteristic in this subpart is assigned every EPA Hazardous Waste Number that is applicable as set forth in this subpart. This number must be used in complying with the notification requirements of section 3010 of the Act and all applicable recordkeeping and reporting requirements under parts 262 through 265, 268, and 270 of this chapter.

40 CFR 261.20(c)

(c) For purposes of this subpart, the Administrator will consider a sample obtained using any of the applicable sampling methods specified in appendix I to be a representative sample within the meaning of part 260 of this chapter.

40 CFR 261.21 Characteristic of ignitability.

40 CFR 261.21(a)

(a) A solid waste exhibits the characteristic of ignitability if a representative sample of the waste has any of the following properties:

40 CFR 261.21(a)(1)

(1) It is a liquid, other than an aqueous solution containing less than 24 percent alcohol by volume and has flash point less than 60°C (140°F), as determined by a Pensky-Martens Closed Cup Tester, using the test method specified in ASTM Standard D-93-79 or D-93-80 (incorporated by reference, see §260.11), or a Setaflash Closed Cup Tester, using the test method specified in ASTM Standard D-3278-78 (incorporated by reference, see §260.11), or as determined by an equivalent test method approved by the Administrator under procedures set forth in §§260.20 and 260.21.

40 CFR 261.21(a)(2)

(2) It is not a liquid and is capable, under standard temperature and pressure, of caus-

ing fire through friction, absorption of moisture or spontaneous chemical changes and, when ignited, burns so vigorously and persistently that it creates a hazard.

40 CFR 261.21(a)(3)

(3) It is an ignitable compressed gas as defined in 49 CFR 173.300 and as determined by the test methods described in that regulation or equivalent test methods approved by the Administrator under § §260.20 and 260.21 .

40 CFR 261.21(a)(4)

(4) It is an oxidizer as defined in 49 CFR 173.151.

40 CFR 261.21(b)

(b) A solid waste that exhibits the characteristic of ignitability has the EPA Hazardous Waste Number of D001.

40 CFR 261.22 Characteristic of corrosivity.

40 CFR 261.22(a)

(a) A solid waste exhibits the characteristic of corrosivity if a representative sample of the waste has either of the following properties:

40 CFR 261.22(a)(1)

(1) It is aqueous and has a pH less than or equal to 2 or greater than or equal to 12.5, as determined by a pH meter using Method 9040 in "Test Methods for Evaluating Solid Waste, Physical/Chemical Methods," EPA Publication SW-846, as incorporated by reference in §260.11 of this chapter.

40 CFR 261.22(a)(2)

(2) It is a liquid and corrodes steel (SAE 1020) at a rate greater than 6.35 mm (0.250 inch) per year at a test temperature of 55°C (130°F) as determined by the test method specified in NACE (National Association of Corrosion Engineers) Standard TM-01-69 as standardized in "Test Methods for Evaluating Solid Waste, Physical/Chemical Methods," EPA Publication SW-846, as incorporated by reference in §260.11 of this chapter.

40 CFR 261.22(b)

(b) A solid waste that exhibits the characteristic of corrosivity has the EPA Hazardous Waste Number of D002.

40 CFR 261.23 Characteristic of reactivity.

40 CFR 261.23(a)

(a) A solid waste exhibits the characteristic of reactivity if a representative sample of the waste has *any* of the following properties:

40 CFR 261.23(a)(1)

(1) It is normally unstable and readily undergoes violent change without detonating.

40 CFR 261.23(a)(2)

(2) It reacts violently with water.

40 CFR 261.23(a)(3)

(3) It forms potentially explosive mixtures with water.

40 CFR 261.23(a)(4)

(4) When mixed with water, it generates toxic gases, vapors or fumes in a quantity suf-

ficient to present a danger to human health or the environment.

40 CFR 261.23(a)(5)

(5) It is a cyanide or sulfide bearing waste which, when exposed to pH conditions between 2 and 12.5, can generate toxic gases, vapors or fumes in a quantity sufficient to present a danger to human health or the environment.

40 CFR 261.23(a)(6)

(6) It is capable of detonation or explosive reaction if it is subjected to a strong initiating source or if heated under confinement.

40 CFR 261.23(a)(7)

(7) It is readily capable of detonation or explosive decomposition or reaction at standard temperature and pressure.

40 CFR 261.23(a)(8)

(8) It is a forbidden explosive as defined in 49 CFR 173.51, or a Class A explosive as defined in 49 CFR 173.53 or a Class B explosive as defined in 49 CFR 173.88.

40 CFR 261.23(b)

(b) A solid waste that exhibits the characteristic of reactivity has the EPA Hazardous Waste Number of D003.

40 CFR 261.24 Toxicity characteristic.

40 CFR 261.24(a)

(a) A solid waste exhibits the characteristic of toxicity if, using the Toxicity Characteristic Leaching Procedure, test Method 1311 in "Test Methods for Evaluating Solid Waste, Physical/Chemical Methods," EPA Publication SW-846, as incorporated by reference in §260.11 of this chapter, the extract from a representative sample of the waste contains any of the contaminants listed in table 1 at the concentration equal to or greater than the respective value given in that table. Where the waste contains less than 0.5 percent filterable solids, the waste itself, after filtering using the methodology outlined in Method 1311, is considered to be the extract for the purpose of this section.

40 CFR 261.24(b)

(b) A solid waste that exhibits the characteristic of toxicity has the EPA Hazardous Waste Number specified in Table I which corresponds to the toxic contaminant causing it to be hazardous.

Table 1—Maximum Concentration of Contaminants for the Toxicity
Characteristic

EPA HW No.[1]	Contaminant	CAS No.[2]	Regulatory Level (mg/l)
D004	Arsenic	7440-38-2	5.0
D005	Barium	7440-39-3	100.0
D018	Benzene	71-43-2	0.5
D006	Cadmium	7440-43-9	1.0
D019	Carbon Tetrachloride	56-23-5	0.5
D020	Chlordane	57-74-9	0.03
D021	Chlorobenzene	108-90-7	100.0
D022	Chloroform	67-66-3	6.0
D007	Chromium	7440-47-3	5.0
D023	o-Cresol	95-48-7	[4]200.0
D024	m-Cresol	108-39-4	[4]200.0
D025	p-Cresol	106-44-5	[4]200.0
D026	Cresol[4]		[4]200.0
D016	2,4-D	94-75-7	10.0
D027	1,4-Dichlorobenzene	106-46-7	7.5
D028	1,2-Dichloroethane	107-06-2	0.5
D029	1,1-Dichloroethylene	75-35-4	0.7
D030	2,4-Dinitrotoluene	121-14-2	[3]0.13
D012	Endrin	72-20-8	0.02
D031	Heptachlor (and its epoxide)	76-44-8	0.008
D032	Hexachlorobenzene	118-74-1	[3]0.13
D033	Hexachlorobutadiene	87-68-3	0.5
D034	Hexachloroethane	67-72-1	3.0
D008	Lead	7439-92-1	5.0
D013	Lindane	58-89-9	0.4
D009	Mercury	7439-97-6	0.2
D014	Methoxychlor	72-43-5	10.0
D035	Methyl ethyl ketone	78-93-3	200.0
D036	Nitrobenzene	98-95-3	2.0
D037	Pentrachlorophenol	87-86-5	100.0
D038	Pyridine	110-86-1	[3]5.0
D010	Selenium	7782-49-2	1.0
D011	Silver	7440-22-4	5.0
D039	Tetrachloroethylene	127-18-4	0.7
D015	Toxaphene	8001-35-2	0.5
D040	Trichloroethylene	79-01-6	0.5
D041	2,4,5-Trichlorophenol	95-95-4	400.0
D042	2,4,6-Trichlorophenol	88-06-2	2.0
D017	2,4,5-TP (Silvex)	93-72-1	1.0
D043	Vinyl chloride	75-01-4	0.2

[1] Hazardous waste number.

[2] Chemical abstracts service number.

[3] Quantitation limit is greater than the calculated regulatory level. The quantitation limit therefore becomes the regulatory level.

[4] If o-, m-, and p-Cresol concentrations cannot be differentiated, the total cresol (D026) concentration is used. The regulatory level of total cresol is 200 mg/l.

Subpart D—Lists of Hazardous Wastes

(Listing of wastes generated from nonspecific and specific sources only.)

40 CFR 261.30 General.

40 CFR 261.30(a)

   (a) A solid waste is a hazardous waste if it is listed in this subpart, unless it has been excluded from this list under § §260.20 and 260.22.

40 CFR 261.30(b)

   (b) The Administrator will indicate his basis for listing the classes or types of wastes listed in this subpart by employing one or more of the following Hazard Codes:

Ignitable Waste ....(I)
Corrosive Waste ....(C)
Reactive Waste...(R)
Toxicity Characteristic Waste......................(E)
Acute Hazardous Waste ....(H)
Toxic Waste..(T)

Appendix VII identifies the constituent which caused the Administrator to list the waste as a Toxicity Characteristic Waste (E) or Toxic Waste (T) in §§261.31 and 261.32.

40 CFR 261.30(c)

   (c) Each hazardous waste listed in this subpart is assigned an EPA Hazardous Waste Number which precedes the name of the waste. This number must be used in complying with the notification requirements of Section 3010 of the Act and certain recordkeeping and reporting requirements under parts 262 through 265, 268, and part 270 of this chapter.

40 CFR 261.30(d)

   (d) The following hazardous wastes listed in §261.31 or §261.32 are subject to the exclusion limits for acutely hazardous wastes established in §261.5:EPA Hazardous Wastes Nos. FO20, FO21, FO22, FO23, FO26, and FO27.

40 CFR 261.31 Hazardous wastes from non-specific sources.

40 CFR 261.31(a)

   (a) The following solid wastes are listed hazardous wastes from non-specific sources unless they are excluded under § §260.20 and 260.22 and listed in appendix IX.

Industry and EPA Hazardous Waste No. Generic:	Hazardous Waste	Hazard code
F001..............	The following spent halogenated solvents used in degreasing: Tetrachloroethylene, trichloroethylene, methylene chloride, 1,1,1-trichloroethane, carbon tetrachloride, and chlorinated fluorocarbons; all spent solvent mixtures/blends used in degreasing containing, before use, a total of ten percent or more (by volume) of one or more of the above halogenated solvents or those solvents listed in F002, F004, and F005; and still bottoms from the recovery of these spent solvents and spent solvent mixtures	(T)
F002..............	The following spent halogenated solvents: Tetrachloroethylene, methylene chloride, trichloroethylene, 1,1,1-trichloroethane, chlorobenzene, 1,1,2-trichloro-1,2,2-trifluoroethane,ortho-dichlorobenzene, trichlorofluoromethane, and 1,1,2-trichloroethane; all spent solvent mixtures/blends containing, before use, a total of ten percent or more (by volume) of one or more of the above halogenated solvents or those listed in F001, F004, or F005; and still bottoms from the recovery of these spent solvents and spent solvent mixtures	(T)
F003..............	The following spent non-halogenated solvents: Xylene, acetone, ethyl acetate, ethyl benzene, ethyl ether, methyl isobutyl ketone, n-butyl alcohol, cyclohexanone, and methanol; all spent solvent mixtures/blends containing, before use, only the above spent non-halogenated solvents; and all spent solvent mixtures/blends containing, before use, one or more of the above non-halogenated solvents, and, a total of ten percent or more (by volume) of one or more of those solvents listed in F001, F002, F004, and F005; and still bottoms from the recovery of these spent solvents and spent solvent mixtures	(I)*
F004..............	The following spent non-halogenated solvents: Cresols and cresylic acid, and nitrobenzene; all spent solvent mixtures/blends containing, before use, a total of ten percent or more (by volume) of one or more of the above non-halogenated solvents or those solvents listed in F001, F002, and F005; and still bottoms from the recovery of these spent solvents and spent solvent mixtures	(T)
F005..............	The following spent non-halogenated solvents: Toluene, methyl ethyl ketone, carbon disulfide, isobutanol, pyridine, benzene, 2-ethoxyethanol, and 2-nitropropane; all	(I,T)

Industry and EPA Hazardous Waste No. Generic:	Hazardous Waste	Hazard code
	spent solvent mixtures/blends containing, before use, a total of ten percent or more (by volume) of one or more of the above non-halogenated solvents or those solvents listed in F001, F002, or F004; and still bottoms from the recovery of these spent solvents and spent solvent mixtures	
F006..............	Wastewater treatment sludges from electroplating operations except from the following processes: (1) Sulfuric acid anodizing of aluminum; (2) tin plating on carbon steel; (3) zinc plating (segregated basis) on carbon steel; (4) aluminum or zinc-aluminum plating on carbon steel; (5) cleaning/stripping associated with tin, zinc and aluminum plating on carbon steel; and (6) chemical etching and milling of aluminum	(T)
F007..............	Spent cyanide plating bath solutions from electroplating operations	(R, T)
F008..............	Plating bath residues from the bottom of plating baths from electroplating operations where cyanides are used in the process	(R, T)
F009..............	Spent stripping and cleaning bath solutions from electroplating operations where cyanides are used in the process	(R, T)
F010..............	Quenching bath residues from oil baths from metal heat treating operations where cyanides are used in the process	(R, T)
F011..............	Spent cyanide solutions from salt bath pot cleaning from metal heat treating operations	(R, T)
F012..............	Quenching waste water treatment sludges from metal heat treating operations where cyanides are used in the process	(T)
F019..............	Wastewater treatment sludges from the chemical conversion coating of aluminum except from zirconium phosphating in aluminum can washing when such phosphating is an exclusive conversion coating process	(T)
F020..............	Wastes (except wastewater and spent carbon from hydrogen chloride purification) from the production or manufacturing use (as a reactant, chemical intermediate, or component in a formulating process) of tri- or tetra-chlorophenol, or of intermediates used to produce their pesticide derivatives. (This listing does not include wastes from the production of Hexachlorophene from highly purified 2,4,5-trichlorophenol.)	(H)

Industry and EPA Hazardous Waste No. Generic:	Hazardous Waste	Hazard code
F021..............	Wastes (except wastewater and spent carbon from hydrogen chloride purification) from the production or manufacturing use (as a reactant, chemical intermediate, or component in a formulating process) of pentachlorophenol, or of intermediates used to produce its derivatives	(H)
F022..............	Wastes (except wastewater and spent carbon from hydrogen chloride purification) from the manufacturing use (as a reactant, chemical intermediate, or component in a formulating process) of tetra-, penta-, or hexachlorobenzenes under alkaline conditions	(H)
F023..............	Wastes (except wastewater and spent carbon from hydrogen chloride purification) from the production of materials on equipment previously used for the production or manufacturing use (as a reactant, chemical intermediate, or component in a formulating process) of tri- and tetrachlorophenols. (This listing does not include wastes from equipment used only for the production or use of Hexachlorophene from highly purified 2,4,5-trichlorophenol.)	(H)
F024..............	Process wastes, including but not limited to, distillation residues, heavy ends, tars, and reactor clean-out wastes, from the production of certain chlorinated aliphatic hydrocarbons by free radical catalyzed processes. These chlorinated aliphatic hydrocarbons are those having carbon chain lengths ranging from one to and including five, with varying amounts and positions of chlorine substitution. (This listing does not include wastewaters, wastewater treatment sludges, spent catalysts, and wastes listed in §261.31 or §261.32.)	(T)
F025..............	Condensed light ends, spent filters and filter aids, and spent desiccant wastes from the production of certain chlorinated aliphatic hydrocarbons, by free radical catalyzed processes. These chlorinated aliphatic hydrocarbons are those having carbon chain lengths ranging from one to and including five, with varying amounts and positions of chlorine substitution	(T)
F026..............	Wastes (except wastewater and spent carbon from hydrogen chloride purification) from the production of materials on equipment previously used for the manufacturing use (as a reactant, chemical intermediate, or component in a formulating process) of tetra-, penta-, or hexachlorobenzene under alkaline conditions	(H)

Industry and EPA Hazardous Waste No. Generic:	Hazardous Waste	Hazard code
F027..............	Discarded unused formulations containing tri-, tetra-, or pentachlorophenol or discarded unused formulations containing compounds derived from these chlorophenols. (This listing does not include formulations containing Hexachlorophene sythesized from prepurified 2,4,5-trichlorophenol as the sole component.)	(H)
F028..............	Residues resulting from the incineration or thermal treatment of soil contaminated with EPA Hazardous Waste Nos. F020, F021, F022, F023, F026, and F027	(T)
F032[1]..........	Wastewaters (except those that have not come into contact with process contaminants), process residuals, preservative drippage, and spent formulations from wood preserving processes generated at plants that currently use or have previously used chlorophenolic formulations (except potentially cross-contaminated wastes that have had the F032 waste code deleted in accordance with §261.35 of this chapter or potentially cross-contaminated wastes that are otherwise currently regulated as hazardous wastes (i.e., F034 or F035), and where the generator does not resume or initiate use of chlorophenolic formulations). This listing does not include K001 bottom sediment sludge from the treatment of wastewater from wood preserving processes that use creosote and/or pentachlorophenol.	(T)
F034[1]..........	Wastewaters (except those that have not come into contact with process contaminants), process residuals, preservative drippage, and spent formulations from wood preserving processes generated at plants that use creosote formulations. This listing does not include K001 bottom sediment sludge from the treatment of wastewater from wood preserving processes that use creosote and/or pentachlorophenol.	(T)
F035[1]..........	Wastewaters (except those that have not come into contact with process contaminants), process residuals, preservative drippage, and spent formulations from wood preserving processes generated at plants that use inorganic preservatives containing arsenic or chromium. This listing does not include K001 bottom sediment sludge from the treatment of wastewater from wood preserving processes that use creosote and/or pentachlorophenol.	(T)

Industry and EPA Hazardous Waste No. Generic:	Hazardous Waste	Hazard code
F037..............	Petroleum refinery primary oil/water/solids separation sludge—Any sludge generated from the gravitational separation of oil/water/solids during the storage or treatment of process wastewaters and oily cooling wastewaters from petroleum refineries. Such sludges include, but are not limited to, those generated in: oil/water/solids separators; tanks and impoundments; ditches and other conveyances; sumps; and stormwater units receiving dry weather flow. Sludge generated in stormwater units that do not receive dry weather flow, sludges generated from non-contact once-through cooling waters segregated for treatment from other process or oily cooling waters, sludges generated in aggressive biological treatment units as defined in §261.31(b)(2) (including sludges generated in one or more additional units after wastewaters have been treated in aggressive biological treatment units) and K051 wastes are not included in this listing	(T)
F038..............	Petroleum refinery secondary (emulsified) oil/water/solids separation sludge—Any sludge and/or float generated from the physical and/or chemical separation of oil/water/solids in process wastewaters and oily cooling wastewaters from petroleum refineries. Such wastes include, but are not limited to, all sludges and floats generated in: induced air flotation (IAF) units, tanks and impoundments, and all sludges generated in DAF units. Sludges generated in stormwater units that do not receive dry weather flow, sludges generated from non-contact once-through cooling waters segregated for treatment from other process or oily cooling waters, sludges and floats generated in aggressive biological treatment units as defined in §261.31(b)(2) (including sludges and floats generated in one or more additional units after wastewaters have been treated in aggressive biological treatment units) and F037, K048, and K051 wastes are not included in this listing	(T)

F039..............   Leachate (liquids that have percolated through land dis   (T)
                           posed wastes) resulting from the disposal of more than
                           one restricted waste classified as hazardous under subpart
                           D of this part. (Leachate resulting from the disposal of
                           one or more of the following EPA Hazardous Wastes and
                           no other Hazardous Wastes retains its EPA Hazardous
                           Waste Number(s): F020, F021, F022, F026, F027,
                           and/or F028.)

---

*(I,T) should be used to specify mixtures containing ignitable and toxic constituents.
[§261.31(a) table revised at 57 61502, Dec. 24, 1992; amended at 60 FR 33913, June 29, 1995]

40 CFR 261.31(b)
   (b) Listing Specific Definitions:
40 CFR 261.31(b)(1)
   (1) For the purposes of the F037 and F038 listings, oil/water/solids is defined as oil and/or water and/or solids.
40 CFR 261.31(b)(2)
   (2) (i) For the purposes of the F037 and F038 listings, aggressive biological treatment units are defined as units which employ one of the following four treatment methods: activated sludge; trickling filter; rotating biological contactor for the continuous accelerated biological oxidation of wastewaters; or high-rate aeration. High-rate aeration is a system of surface impoundments or tanks, in which intense mechanical aeration is used to completely mix the wastes, enhance biological activity, and (A) the units employ a minimum of 6 hp per million gallons of treatment volume; and either (B) the hydraulic retention time of the unit is no longer than 5 days; or (C) the hydraulic retention time is no longer than 30 days and the unit does not generate a sludge that is a hazardous waste by the Toxicity Characteristic.
40 CFR 261.31(b)(2)(ii)
   (ii) Generators and treatment, storage and disposal facilities have the burden of proving that their sludges are exempt from listing as F037 and F038 wastes under this definition. Generators and treatment, storage and disposal facilities must maintain, in their operating or other onsite records, documents and data sufficient to prove that: (A) the unit is an aggressive biological treatment unit as defined in this subsection; and (B) the sludges sought to be exempted from the definitions of F037 and/or F038 were actually generated in the aggressive biological treatment unit.
40 CFR 261.31(b)(3)
   (3) (i) For the purposes of the F037 listing, sludges are considered to be generated at the moment of deposition in the unit, where deposition is defined as at least a temporary cessation of lateral particle movement.
40 CFR 261.31(b)(3)(ii)
   (ii) For the purposes of the F038 listing,
40 CFR 261.31(b)(3)(ii)(A)

(A) sludges are generated at the moment of deposition in the unit, where deposition is defined as at least a temporary cessation of lateral particle movement and
40 CFR 261.31(b)(3)(ii)(B)

(B) floats are considered to be generated at the moment they are formed in the top of the unit.

40 CFR 261.32 Hazardous wastes from specific sources.

The following solid wastes are listed hazardous wastes from specific sources unless they are excluded under §§260.20 and 260.22 and listed in appendix IX.

Industry and EPA Hazardous Waste No.	Hazardous Waste	Hazard Code
Wood preservation:		
K001..............	Bottom sediment sludge from the treatment of waste-waters from wood preserving processes that use creosote and/or pentachlorophenol	(T)
Inorganic pigments:		
K002..............	Wastewater treatment sludge from the production of chrome yellow and orange pigments	(T)
K003..............	Wastewater treatment sludge from the production of molybdate orange pigments	(T)
K004..............	Wastewater treatment sludge from the production of zinc yellow pigments	(T)
K005..............	Wastewater treatment sludge from the production of chrome green pigments	(T)
K006..............	Wastewater treatment sludge from the production of chrome oxide green pigments (anhydrous and hydrated)	(T)
K007..............	Wastewater treatment sludge from the production of iron blue pigments	(T)
K008..............	Oven residue from the production of chrome oxide green pigments	(T)
Organic chemicals:		
K009..............	Distillation bottoms from the production of acetaldehyde from ethylene	(T)
K010..............	Distillation side cuts from the production of acetaldehyde from ethylene	(T)
K011..............	Bottom stream from the wastewater stripper in the production of acrylonitrile	(R, T)
K013..............	Bottom stream from the acetonitrile column in the production of acrylonitrile	(R, T)
K014..............	Bottoms from the acetonitrile purification column in the production of acrylonitrile	(T)

Industry and EPA Hazardous Waste No.	Hazardous Waste	Hazard Code
K015..............	Still bottoms from the distillation of benzyl chloride	(T)
K016..............	Heavy ends or distillation residues from the production of carbon tetrachloride	(T)
K017..............	Heavy ends (still bottoms) from the purification column in the production of epichlorohydrin	(T)
K018..............	Heavy ends from the fractionation column in ethyl chloride production	(T)
K019..............	Heavy ends from the distillation of ethylene dichloride in ethylene dichloride production	(T)
K020..............	Heavy ends from the distillation of vinyl chloride in vinyl chloride monomer production	(T)
K021..............	Aqueous spent antimony catalyst waste from fluoromethanes production	(T)
K022..............	Distillation bottom tars from the production of phenol/acetone from cumene	(T)
K023..............	Distillation light ends from the production of phthalic anhydride from naphthalene	(T)
K024..............	Distillation bottoms from the production of phthalic anhydride from naphthalene	(T)
K025..............	Distillation bottoms from the production of nitrobenzene by the nitration of benzene	(T)
K026..............	Stripping still tails from the production of methy ethyl pyridines	(T)
K027..............	Centrifuge and distillation residues from toluene diisocyanate production	(R, T)
K028..............	Spent catalyst from the hydrochlorinator reactor in the production of 1,1,1-trichloroethane	(T)
K029..............	Waste from the product steam stripper in the production of 1,1,1-trichloroethane	(T)
K030..............	Column bottoms or heavy ends from the combined production of trichloroethylene and perchloroethylene	(T)
K083..............	Distillation bottoms from aniline production	(T)
K085..............	Distillation or fractionation column bottoms from the production of chlorobenzenes	(T)
K093..............	Distillation light ends from the production of phthalic anhydride from ortho-xylene	(T)
K094..............	Distillation bottoms from the production of phthalic anhydride from ortho-xylene	(T)
K095..............	Distillation bottoms from the production of 1,1,1-trichloroethane	(T)

Industry and EPA Hazardous Waste No.	Hazardous Waste	Hazard Code
K096..............	Heavy ends from the heavy ends column from the production of 1,1,1-trichloroethane	(T)
K103..............	Process residues from aniline extraction from the production of aniline	(T)
K104..............	Combined wastewater streams generated from nitrobenzene/aniline production	(T)
K105..............	Separated aqueous stream from the reactor product washing step in the production of chlorobenzenes	(T)
K107..............	Column bottoms from product separation from the production of 1,1-dimethyl-hydrazine (UDMH) from carboxylic acid hydrazines	(C, T)
K108..............	Condensed column overheads from product separation and condensed reactor vent gases from the production of 1,1-dimethylhydrazine (UDMH) from carboxylic acid hydrazides	(I,T)
K109..............	Spent filter cartridges from product purification from the production of 1,1-dimethylhydrazine (UDMH) from carboxylic acid hydrazides	(T)
K110..............	Condensed column overheads from intermediate separation from the production of 1,1-dimethylhydrazine (UDMH) from carboxylic acid hydrazides	(T)
K111..............	Product washwaters from the production of dinitrotoluene via nitration of toluene	(C, T)
K112..............	Reaction by-product water from the drying column in the production of toluenediamine via hydrogenation of dinitrotoluene	(T)
K113..............	Condensed liquid light ends from the purification of toluenediamine in the production of toluenediamine via hydrogenation of dinitrotoluene	(T)
K114..............	Vicinals from the purification of toluenediamine in the production of toluenediamine via hydrogenation of dinitrotoluene	(T)
K115..............	Heavy ends from the purification of toluenediamine in the production of toluenediamine via hydrogenation of dinitrotoluene	(T)
K116..............	Organic condensate from the solvent recovery column in the production of toluene diisocyanate via phosgenation of toluenediamine	(T)

Industry and EPA Hazardous Waste No.	Hazardous Waste	Hazard Code
K117..............	Wastewater from the reactor vent gas scrubber in the production of ethylene dibromide via bromination of ethene	(T)
K118..............	Spent adsorbent solids from purification of ethylene dibromide in the production of ethylene dibromide via bromination of ethene	(T)
K136..............	Still bottoms from the purification of ethylene dibromide in the production of ethylene dibromide via bromination of ethene	(T)
K149..............	Distillation bottoms from the production of alpha-(or methyl-) chlorinated toluenes, ring-chlorinated toluenes, benzoyl chlorides, and compounds with mixtures of these functional group. (This waste does not include still bottoms from the distillation of benzyl chloride.)	(T)
K150..............	Organic residuals, excluding spent carbon adsorbent, from the spent chlorine gas and hydrochloric acid recovery processes associated with the production of alpha-(or methyl-) chlorinated toluenes, ring-chlorinated toluenes, benzoyl chlorides, and compounds with mixtures of these functional groups.	(T)
K151..............	Wastewater treatment sludges, excluding neutralization and biological sludges, generated during the treatment of wastewaters from the production of alpha-(or methyl-) chlorinated toluenes, ring-chlorinated toluenes, benzoyl chlorides, and compounds with mixtures of these functional groups.	(T)
K156..............	Organic waste (including heavy ends, still bottoms, light ends, spent solvents, filtrates, and decantates) from the production of carbamates and carbamoyl oximes.	(T)
K157..............	Wastewaters (including scrubber waters, condenser waters, washwaters, and separation waters) from the production of carbamates and carbamoyl oximes.	(T)
K158..............	Bag house dusts and filter/separation solids from the production of carbamates and carbamoyl oximes.	(T)
K159..............	Organics from the treatment of thiocarbamate wastes.	(T)
K160..............	Solids (including filter wastes, separation solids, and spent catalysts) from the production of thiocarbamates and solids from the treatment of thiocarbamate wastes.	(T)

Industry and EPA Hazardous Waste No.	Hazardous Waste	Hazard Code
K161..............	Purification solids (including filtration, evaporation, and centrifugation solids), bag house dust and floor sweepings from the production of dithiocarbamate acids and their salts. (This listing does not include K125 or K126.)	(R, T)
Inorganic chemicals:		
K071..............	Brine purification muds from the mercury cell process in chlorine production, where separately prepurified brine is not used	(T)
K073..............	Chlorinated hydrocarbon waste from the purification step of the diaphragm cell process using graphite anodes in chlorine production	(T)
K106..............	Wastewater treatment sludge from the mercury cell process in chlorine production	(T)
Pesticides:		
K031..............	By-product salts generated in the production of MSMA and cacodylic acid	(T)
K032..............	Wastewater treatment sludge from the production of chlordane	(T)
K033..............	Wastewater and scrub water from the chlorination of cyclopentadiene in the production of chlordane	(T)
K034..............	Filter solids from the filtration of hexachlorocyclopentadiene in the production of chlordane	(T)
K035..............	Wastewater treatment sludges generated in the production of creosote	(T)
K036..............	Still bottoms from toluene reclamation distillation in the production of disulfoton	(T)
K037..............	Wastewater treatment sludges from the production of disulfoton	(T)
K038..............	Wastewater from the washing and stripping of phorate production	(T)
K039..............	Filter cake from the filtration of diethylphosphorodithioic acid in the production of phorate	(T)
K040..............	Wastewater treatment sludge from the production of phorate	(T)
K041..............	Wastewater treatment sludge from the production of toxaphene	(T)
K042..............	Heavy ends or distillation residues from the distillation of tetrachlorobenzene in the production of 2,4,5-T	(T)
K043..............	2,6-Dichlorophenol waste from the production of 2,4-D	(T)

Industry and EPA Hazardous Waste No.	Hazardous Waste	Hazard Code
K097..............	Vacuum stripper discharge from the chlordane chlorinator in the production of chlordane	(T)
K098..............	Untreated process wastewater from the production of toxaphene	(T)
K099..............	Untreated wastewater from the production of 2,4-D	(T)
K123..............	Process wastewater (including supernates, filtrates, and washwaters) from the production of ethylenebisdithiocarbamic acid and its salt	(T)
K124..............	Reactor vent scrubber water from the production of ethylenebisdithiocarbamic acid and its salts	(C, T)
K125..............	Filtration, evaporation, and centrifugation solids from the production of ethylenebisdithiocarbamic acid and its salts	(T)
K126..............	Baghouse dust and floor sweepings in milling and packaging operations from the production or formulation of ethylenebisdithiocarbamic acid and its salts	(T)
K131..............	Wastewater from the reactor and spent sulfuric acid from the acid dryer from the production of methyl bromide	(C, T)
K132..............	Spent absorbent and wastewater separator solids from the production of methyl bromide	(T)
Explosives:		
K044..............	Wastewater treatment sludges from the manufacturing and processing of explosives	(R)
K045..............	Spent carbon from the treatment of wastewater containing explosives	(R)
K046..............	Wastewater treatment sludges from the manufacturing, formulation and loading of lead-based initiating compounds	(T)
K047..............	Pink/red water from TNT operations	(R)
Petroleum refining:		
K048..............	Dissolved air flotation (DAF) float from the petroleum refining industry	(T)
K049..............	Slop oil emulsion solids from the petroleum refining industry	(T)
K050..............	Heat exchanger bundle cleaning sludge from the petroleum refining industry	(T)
K051..............	API separator sludge from the petroleum refining industry	(T)
K052..............	Tank bottoms (leaded) from the petroleum refining industry	(T)

Industry and EPA Hazardous Waste No.	Hazardous Waste	Hazard Code
Iron and steel:		
K061..............	Emission control dust/sludge from the primary production of steel in electric furnaces	(T)
K062..............	Spent pickle liquor generated by steel finishing operations of facilities within the iron and steel industry (SIC Codes 331 and 332)	(C, T)
Primary copper:		
K064..............	Acid plant blowdown slurry/sludge resulting from the thickening of blowdown slurry from primary copper production	(T)
Primary lead:		
K065..............	Surface impoundment solids contained in and dredged from surface impoundments at primary lead smelting facilities	(T)
Primary zinc:		
K066..............	Sludge from treatment of process wastewater and/or acid plant blowdown from primary zinc production	(T)
Primary aluminum:		
K088..............	Spent potliners from primary aluminum reduction	(T)
Ferroalloys:		
K090..............	Emission control dust or sludge from ferrochromiumsilicon production	(T)
K091..............	Emission control dust or sludge from ferrochromium production	(T)
Secondary lead:		
K069..............	Emission control dust/sludge from secondary lead smelting. (Note: This listing is stayed administratively for sludge generated from secondary acid scrubber systems. The stay will remain in effect until further administrative action is taken. If EPA takes further action effecting this stay, EPA will publish a notice of the action in the Federal Register.)	(T)
K100..............	Waste leaching solution from acid leaching of emission control dust/sludge from secondary lead smelting	(T)
Veterinary pharmaceuticals:		
K084..............	Wastewater treatment sludges generated during the production of veterinary pharmaceuticals from arsenic or organo-arsenic compounds	(T)
K101..............	Distillation tar residues from the distillation of aniline-based compounds in the production of veterinary pharmaceuticals from arsenic or organo-arsenic compounds	(T)

Industry and EPA Hazardous Waste No.	Hazardous Waste	Hazard Code
K102..............	Residue from the use of activated carbon for decoloriza-tion in the production of veterinary pharmaceuticals from arsenic or organo-arsenic compounds	(T)
Ink formulation:		
K086..............	Solvent washes and sludges, caustic washes and sludges, or water washes and sludges from cleaning tubs and equip-ment used in the formulation of ink from pigments, dri-ers, soaps, and stabilizers containing chromium and lead	(T)
Coking:		
K060..............	Ammonia still lime sludge from coking operations	(T)
K087..............	Decanter tank tar sludge from coking operations	(T)
K141..............	Process residues from the recovery of coal tar, including, but not limited to, collecting sump residues from the pro-duction of coke from coal or the recovery of coke by-products produced from coal. This listing does not in-clude K087 (decanter tank tar sludges from coking operations).	(T)
K142..............	Tar storage tank residues from the production of coke from coal or from the recovery of coke by-products from coal.	(T)
K143..............	Process residues from the recovery of light oil, including, but not limited to, those generated in stills, decanters, and wash oil recovery units from the recovery of coke by-prod-ucts produced from coal.	(T)
K144..............	Wastewater sump residues from light oil refining, includ-ing, but not limited to, intercepting or contamination sump sludges from the recovery of coke by-products pro-duced from coal.	(T)
K145..............	Residues from naphthalene collection and recovery opera-tions from the recovery of coke by-products produced from coal	(T)
K147..............	Tar storage tank residues from coal tar refining	(T)
K148..............	Residues from coal tar distillation, including but not lim-ited to, still bottoms (T)	(T)

[§261.32 table revised at 57 FR 37305, Aug. 18, 1992; 57 FR 47385, Oct. 15, 1992; amended at 60 FR 7848, Feb. 9, 1995]

B 1990 Clean Air Act Amendments—Section 112: List of 188 Hazardous Air Pollutants (HAPs)

(Note: Captrolactam was removed from the original list of 189 HAPs.)

CAA Section 112(b)

(b) List of Pollutants.—

CAA Sec. 112(b)(1)

(1) Initial List.—The Congress establishes for purposes of this section a list of hazardous air pollutants as follows:

CAS Number	Chemical Name
75070	Acetaldehyde
60355	Acetamide
75058	Acetonitrile
98862	Acetophenone
53963	2-Acetylaminofluorene
107028	Acrolein
79061	Acrylamide
79107	Acrylic acid
107131	Acrylonitrile
107051	Allyl chloride
92671	4-Aminobiphenyl
62533	Aniline
90040	o-Anisidine

CAS Number	Chemical Name
1332214	Asbestos
71432	Benzene (including benzene from gasoline)
92875	Benzidine
98077	Benzotrichloride
100447	Benzyl chloride
92524	Biphenyl
117817	Bis(2-ethylhexyl)phthalate (DEHP)
542881	Bis(chloromethyl)ether
75252	Bromoform
106990	1,3-Butadiene
156627	Calcium cyanamide
133062	Captan
63252	Carbaryl
75150	Carbon disulfide
56235	Carbon tetrachloride
463581	Carbonyl sulfide
120809	Catechol
133904	Chloramben
57749	Chlordane
7782505	Chlorine
79118	Chloroacetic acid
532274	2-Chloroacetophenone
108907	Chlorobenzene
510156	Chlorobenzilate
67663	Chloroform
107302	Chloromethyl methyl ether
126998	Chloroprene
1319773	Cresols/Cresylic acid (isomers and mixture)
95487	o-Cresol
108394	m-Cresol
106445	p-Cresol
98828	Cumene
94757	2,4-D, salts and esters

CAS Number	Chemical Name
3547044	DDE
334883	Diazomethane
132649	Dibenzofurans
96128	1,2-Dibromo-3-chloropropane
84742	Dibutylphthalate
106467	1,4-Dichlorobenzene(p)
91941	3,3-Dichlorobenzidene
111444	Dichloroethyl ether (Bis(2-chloroethyl)ether)
542756	1,3-Dichloropropene
62737	Dichlorvos
111422	Diethanolamine
121697	N,N-Diethyl aniline (N,N-Dimethylaniline)
64675	Diethyl sulfate
119904	3,3-Dimethoxybenzidine
60117	Dimethyl aminoazobenzene
119937	3,39-Dimethyl benzidine
79447	Dimethyl carbamoyl chloride
68122	Dimethyl formamide
57147	1,1-Dimethyl hydrazine
131113	Dimethyl phthalate
77781	Dimethyl sulfate
534521	4,6-Dinitro-o-cresol, and salts
51285	2,4-Dinitrophenol
121142	2,4-Dinitrotoluene
123911	1,4-Dioxane (1,4-Diethyleneoxide)
122667	1,2-Diphenylhydrazine
106898	Epichlorohydrin (1-Chloro-2,3-epoxypropane)
106887	1,2-Epoxybutane
140885	Ethyl acrylate
100414	Ethyl benzene
51796	Ethyl carbamate (Urethane)
75003	Ethyl chloride (Chloroethane)
106934	Ethylene dibromide (Dibromoethane)
107062	Ethylene dichloride (1,2-Dichloroethane)

CAS Number	Chemical Name
107211	Ethylene glycol
151564	Ethylene imine (Aziridine)
75218	Ethylene oxide
96457	Ethylene thiourea
75343	Ethylidene dichloride (1,1-Dichloroethane)
50000	Formaldehyde
76448	Heptachlor
118741	Hexachlorobenzene
87683	Hexachlorobutadiene
77474	Hexachlorocyclopentadiene
67721	Hexachloroethane
822060	Hexamethylene-1,6-diisocyanate
680319	Hexamethylphosphoramide
110543	Hexane
302012	Hydrazine
7647010	Hydrochloric acid
7664393	Hydrogen fluoride (Hydrofluoric acid)
123319	Hydroquinone
78591	Isophorone
58899	Lindane (all isomers)
108316	Maleic anhydride
67561	Methanol
72435	Methoxychlor
74839	Methyl bromide (Bromomethane)
74873	Methyl chloride (Chloromethane)
71556	Methyl chloroform (1,1,1-Trichloroethane)
78933	Methyl ethyl ketone (2-Butanone)
60344	Methyl hydrazine
74884	Methyl iodide (Iodomethane)
108101	Methyl isobutyl ketone (Hexone)
624839	Methyl isocyanate
80626	Methyl methacrylate
1634044	Methyl tert butyl ether

CAS Number	Chemical Name
101144	4,4-Methylene bis(2-chloroaniline)
75092	Methylene chloride (Dichloromethane)
101688	Methylene diphenyl diisocyanate (MDI)
101779	4,49-Methylenedianiline
91203	Naphthalene
98953	Nitrobenzene
92933	4-Nitrobiphenyl
100027	4-Nitrophenol
79469	2-Nitropropane
684935	N-Nitroso-N-methylurea
62759	N-Nitrosodimethylamine
59892	N-Nitrosomorpholine
56382	Parathion
82688	Pentachloronitrobenzene (Quintobenzene)
87865	Pentachlorophenol
108952	Phenol
106503	p-Phenylenediamine
75445	Phosgene
7803512	Phosphine
7723140	Phosphorus
85449	Phthalic anhydride
1336363	Polychlorinated biphenyls (Arochlors)
1120714	1,3-Propane sultone
57578	beta-Propiolactone
123386	Propionaldehyde
114261	Propoxur (Baygon)
78875	Propylene dichloride (1,2-Dichloropropane)
75569	Propylene oxide
75558	1,2-Propylenimine (2-Methyl aziridine)
91225	Quinoline
106514	Quinone
100425	Styrene
96093	Styrene oxide
1746016	2,3,7,8-Tetrachlorodibenzo-p-dioxin

CAS Number	Chemical Name
79345	1,1,2,2-Tetrachloroethane
127184	Tetrachloroethylene (Perchloroethylene)
7550450	Titanium tetrachloride
108883	Toluene
95807	2,4-Toluene diamine
584849	2,4-Toluene diisocyanate
95534	o-Toluidine
8001352	Toxaphene (chlorinated camphene)
120821	1,2,4-Trichlorobenzene
79005	1,1,2-Trichloroethane
79016	Trichloroethylene
95954	2,4,5-Trichlorophenol
88062	2,4,6-Trichlorophenol
121448	Triethylamine
1582098	Trifluralin
540841	2,2,4-Trimethylpentane
108054	Vinyl acetate
593602	Vinyl bromide
75014	Vinyl chloride
75354	Vinylidene chloride (1,1-Dichloroethylene)
1330207	Xylenes (isomers and mixture)
95476	o-Xylenes
108383	m-Xylenes
106423	p-Xylenes
0	Antimony Compounds
0	Arsenic Compounds (inorganic including arsine)
0	Beryllium Compounds
0	Cadmium Compounds
0	Chromium Compounds
0	Cobalt Compounds
0	Coke Oven Emissions
0	Cyanide Compounds[1]
0	Glycol ethers[2]

CAS Number	Chemical Name
0	Lead Compounds
0	Manganese Compounds
0	Mercury Compounds
0	Fine mineral fibers[3]
0	Nickel Compounds
0	Polycyclic Organic Matter[4]
0	Radionuclides (including radon)[5]
0	Selenium Compounds

NOTE: For all listings above which contain the word 'compounds' and for glycol ethers, the following applies: Unless otherwise specified, these listings are defined as including any unique chemical substance that contains the named chemical (i.e., antimony, arsenic, etc.) as part of that chemical's infrastructure.

[1] X'CN where X = H' or any other group where a formal dissociation may occur. For example KCN or $Ca(CN)_2$.

[2] Includes moni- and di-ethers of ethylene glycol, diethylene glycol, and triethylene glycol R-$(OCH2CH2)_n$-OR' where n = 1, 2, or 3 R = alkyl or aryl groups R' = R, H, or groups which, when removed, yield glycol ethers with the structure: R-$(OCH2CH)_n$-OH. Polymers are excluded from the glycol category.

[3] Includes mineral fiber emissions from facilities manufacturing or processing glass, rock, or slag fibers (or other mineral derived fibers) of average diameter 1 micrometer or less.

[4] Includes organic compounds with more than one benzene ring, and which have a boiling point greater than or equal to 100°C.

[5] A type of atom which spontaneously undergoes radioactive decay.

[§112(b)(1) amended by PL 102-187]

C Discrete Compounding Interest Factors—Single Payment and Uniform Series

2% Interest Factors for Discrete Compounding

N	P/F	P/A	F/P	F/A	A/P	A/F
1	0.9804	0.9804	1.0200	1.0000	1.0200	1.0000
2	0.9612	1.9416	1.0404	2.0200	0.5150	0.4950
3	0.9423	2.8839	1.0612	3.0604	0.3468	0.3268
4	0.9238	3.8077	1.0824	4.1216	0.2626	0.2426
5	0.9057	4.7135	1.1041	5.2040	0.2122	0.1922
6	0.8880	5.6014	1.1262	6.3081	0.1785	0.1585
7	0.8706	6.4720	1.1487	7.4343	0.1545	0.1345
8	0.8535	7.3255	1.1717	8.5830	0.1365	0.1165
9	0.8368	8.1622	1.1951	9.7546	0.1225	0.1025
10	0.8203	8.9826	1.2190	10.9497	0.1113	0.0913
11	0.8043	9.7868	1.2434	12.1687	0.1022	0.0822
12	0.7885	10.5753	1.2682	13.4121	0.0946	0.0746
13	0.7730	11.3484	1.2936	14.6803	0.0881	0.0681
14	0.7579	12.1062	1.3195	15.9739	0.0826	0.0626
15	0.7430	12.8493	1.3459	17.2934	0.0778	0.0578
16	0.7284	13.5777	1.3728	18.6393	0.0737	0.0537
17	0.7142	14.2919	1.4002	20.0121	0.0700	0.0500
18	0.7002	14.9920	1.4282	21.4123	0.0667	0.0467
19	0.6864	15.6785	1.4568	22.8406	0.0638	0.0438
20	0.6730	16.3514	1.4859	24.2974	0.0612	0.0412
25	0.6095	19.5235	1.6406	32.0303	0.0512	0.0312

30	0.5521	22.3965	1.8114	40.5681	0.0446	0.0246
35	0.5000	24.9986	1.9999	49.9945	0.0400	0.0200
40	0.4529	27.3555	2.2080	60.4020	0.0366	0.0166
45	0.4102	29.4902	2.4379	71.8927	0.0339	0.0139
50	0.3715	31.4236	2.6916	84.5794	0.0318	0.0118

4% Interest Factors for Discrete Compounding

N	P/F	P/A	F/P	F/A	A/P	A/F
1	0.9615	0.9615	1.0400	1.0000	1.0400	1.0000
2	0.9246	1.8861	1.0816	2.0400	0.5302	0.4902
3	0.8890	2.7751	1.1249	3.1216	0.3603	0.3203
4	0.8548	3.6299	1.1699	4.2465	0.2755	0.2355
5	0.8219	4.4518	1.2167	5.4163	0.2246	0.1846
6	0.7903	5.2421	1.2653	6.6330	0.1908	0.1508
7	0.7599	6.0021	1.3159	7.8983	0.1666	0.1266
8	0.7307	6.7327	1.3686	9.2142	0.1485	0.1085
9	0.7026	7.4353	1.4233	10.5828	0.1345	0.0945
10	0.6756	8.1109	1.4802	12.0061	0.1233	0.0833
11	0.6496	8.7605	1.5395	13.4864	0.1141	0.0741
12	0.6246	9.3851	1.6010	15.0258	0.1066	0.0666
13	0.6006	9.9856	1.6651	16.6268	0.1001	0.0601
14	0.5775	10.5631	1.7317	18.2919	0.0947	0.0547
15	0.5553	11.1184	1.8009	20.0236	0.0899	0.0499
16	0.5339	11.6523	1.8730	21.8245	0.0858	0.0458
17	0.5134	12.1657	1.9479	23.6975	0.0822	0.0422
18	0.4936	12.6593	2.0258	25.6454	0.0790	0.0390
19	0.4746	13.1339	2.1068	27.6712	0.0761	0.0361
20	0.4564	13.5903	2.1911	29.7781	0.0736	0.0336
25	0.3751	15.6221	2.6658	41.6459	0.0640	0.0240
30	0.3083	17.2920	3.2434	56.0849	0.0578	0.0178
35	0.2534	18.6646	3.9461	73.6522	0.0536	0.0136
40	0.2083	19.7928	4.8010	95.0255	0.0505	0.0105
45	0.1712	20.7200	5.8412	121.0294	0.0483	0.0083
50	0.1407	21.4822	7.1067	152.6671	0.0466	0.0066

6% Interest Factors for Discrete Compounding

N	P/F	P/A	F/P	F/A	A/P	A/F
1	0.9434	0.9434	1.0600	1.0000	1.0600	1.0000
2	0.8900	1.8334	1.1236	2.0600	0.5454	0.4854
3	0.8396	2.6730	1.1910	3.1836	0.3741	0.3141
4	0.7921	3.4651	1.2625	4.3746	0.2886	0.2286
5	0.7473	4.2124	1.3382	5.6371	0.2374	0.1774
6	0.7050	4.9173	1.4185	6.9753	0.2034	0.1434
7	0.6651	5.5824	1.5036	8.3938	0.1791	0.1191
8	0.6274	6.2098	1.5938	9.8975	0.1610	0.1010
9	0.5919	6.8017	1.6895	11.4913	0.1470	0.0870
10	0.5584	7.3601	1.7908	13.1808	0.1359	0.0759
11	0.5268	7.8869	1.8983	14.9716	0.1268	0.0668
12	0.4970	8.3838	2.0122	16.8699	0.1193	0.0593
13	0.4688	8.8527	2.1329	18.8821	0.1130	0.0530
14	0.4423	9.2950	2.2609	21.0151	0.1076	0.0476
15	0.4173	9.7122	2.3966	23.2760	0.1030	0.0430
16	0.3936	10.1059	2.5404	25.6725	0.0990	0.0390
17	0.3714	10.4773	2.6928	28.2129	0.0954	0.0354
18	0.3503	10.8276	2.8543	30.9057	0.0924	0.0324
19	0.3305	11.1581	3.0256	33.7600	0.0896	0.0296
20	0.3118	11.4699	3.2071	36.7856	0.0872	0.0272
25	0.2330	12.7834	4.2919	54.8645	0.0782	0.0182
30	0.1741	13.7648	5.7435	79.0582	0.0726	0.0126
35	0.1301	14.4982	7.6861	111.4348	0.0690	0.0090
40	0.0972	15.0463	10.2857	154.7620	0.0665	0.0065
45	0.0727	15.4558	13.7646	212.7435	0.0647	0.0047
50	0.0543	15.7619	18.4202	290.3359	0.0634	0.0034

8% Interest Factors for Discrete Compounding

N	P/F	P/A	F/P	F/A	A/P	A/F
1	0.9259	0.9259	1.0800	1.0000	1.0800	1.0000
2	0.8573	1.7833	1.1664	2.0800	0.5608	0.4808
3	0.7938	2.5771	1.2597	3.2464	0.3880	0.3080
4	0.7350	3.3121	1.3605	4.5061	0.3019	0.2219
5	0.6806	3.9927	1.4693	5.8666	0.2505	0.1705

6	0.6302	4.6229	1.5869	7.3359	0.2163	0.1363
7	0.5835	5.2064	1.7138	8.9228	0.1921	0.1121
8	0.5403	5.7466	1.8509	10.6366	0.1740	0.0940
9	0.5002	6.2469	1.9990	12.4876	0.1601	0.0801
10	0.4632	6.7101	2.1589	14.4866	0.1490	0.0690
11	0.4289	7.1390	2.3316	16.6455	0.1401	0.0601
12	0.3971	7.5361	2.5182	18.9771	0.1327	0.0527
13	0.3677	7.9038	2.7196	21.4953	0.1265	0.0465
14	0.3405	8.2442	2.9372	24.2149	0.1213	0.0413
15	0.3152	8.5595	3.1722	27.1521	0.1168	0.0368
16	0.2919	8.8514	3.4259	30.3243	0.1130	0.0330
17	0.2703	9.1216	3.7000	33.7502	0.1096	0.0296
18	0.2502	9.3719	3.9960	37.4502	0.1067	0.0267
19	0.2317	9.6036	4.3157	41.4463	0.1041	0.0241
20	0.2145	9.8181	4.6610	45.7620	0.1019	0.0219
25	0.1460	10.6748	6.8485	73.1059	0.0937	0.0137
30	0.0994	11.2578	10.0627	113.2832	0.0888	0.0088
35	0.0676	11.6546	14.7853	172.3168	0.0858	0.0058
40	0.0460	11.9246	21.7245	259.0565	0.0839	0.0039
45	0.0313	12.1084	31.9204	386.5056	0.0826	0.0026
50	0.0213	12.2335	46.9016	573.7702	0.0817	0.0017

10% Interest Factors for Discrete Compounding

N	P/F	P/A	F/P	F/A	A/P	A/F
1	0.9091	0.9091	1.1000	1.0000	1.1000	1.0000
2	0.8264	1.7355	1.2100	2.1000	0.5762	0.4762
3	0.7513	2.4869	1.3310	3.3100	0.4021	0.3021
4	0.6830	3.1699	1.4641	4.6410	0.3155	0.2155
5	0.6209	3.7908	1.6105	6.1051	0.2638	0.1638
6	0.5645	4.3553	1.7716	7.7156	0.2296	0.1296
7	0.5132	4.8684	1.9487	9.4872	0.2054	0.1054
8	0.4665	5.3349	2.1436	11.4359	0.1874	0.0874
9	0.4241	5.7590	2.3579	13.5795	0.1736	0.0736
10	0.3855	6.1446	2.5937	15.9374	0.1627	0.0627
11	0.3505	6.4951	2.8531	18.5312	0.1540	0.0540
12	0.3186	6.8137	3.1384	21.3843	0.1468	0.0468

13	0.2897	7.1034	3.4523	24.5227	0.1408	0.0408
14	0.2633	7.3667	3.7975	27.9750	0.1357	0.0357
15	0.2394	7.6061	4.1772	31.7725	0.1315	0.0315
16	0.2176	7.8237	4.5950	35.9497	0.1278	0.0278
17	0.1978	8.0216	5.0545	40.5447	0.1247	0.0247
18	0.1799	8.2014	5.5599	45.5992	0.1219	0.0219
19	0.1635	8.3649	6.1159	51.1591	0.1195	0.0195
20	0.1486	8.5136	6.7275	57.2750	0.1175	0.0175
25	0.0923	9.0770	10.8347	98.3471	0.1102	0.0102
30	0.0573	9.4269	17.4494	164.4940	0.1061	0.0061
35	0.0356	9.6442	28.1024	271.0244	0.1037	0.0037
40	0.0221	9.7791	45.2593	442.5926	0.1023	0.0023
45	0.0137	9.8628	72.8905	718.9048	0.1014	0.0014
50	0.0085	9.9148	117.3909	1163.9085	0.1009	0.0009

12% Interest Factors for Discrete Compounding

N	P/F	P/A	F/P	F/A	A/P	A/F
1	0.8929	0.8929	1.1200	1.0000	1.1200	1.0000
2	0.7972	1.6901	1.2544	2.1200	0.5917	0.4717
3	0.7118	2.4018	1.4049	3.3744	0.4163	0.2963
4	0.6355	3.0373	1.5735	4.7793	0.3292	0.2092
5	0.5674	3.6048	1.7623	6.3528	0.2774	0.1574
6	0.5066	4.1114	1.9738	8.1152	0.2432	0.1232
7	0.4523	4.5638	2.2107	10.0890	0.2191	0.0991
8	0.4039	4.9676	2.4760	12.2997	0.2013	0.0813
9	0.3606	5.3282	2.7731	14.7757	0.1877	0.0677
10	0.3220	5.6502	3.1058	17.5487	0.1770	0.0570
11	0.2875	5.9377	3.4785	20.6546	0.1684	0.0484
12	0.2567	6.1944	3.8960	24.1331	0.1614	0.0414
13	0.2292	6.4235	4.3635	28.0291	0.1557	0.0357
14	0.2046	6.6282	4.8871	32.3926	0.1509	0.0309
15	0.1827	6.8109	5.4736	37.2797	0.1468	0.0268
16	0.1631	6.9740	6.1304	42.7533	0.1434	0.0234
17	0.1456	7.1196	6.8660	48.8837	0.1405	0.0205
18	0.1300	7.2497	7.6900	55.7497	0.1379	0.0179
19	0.1161	7.3658	8.6128	63.4397	0.1358	0.0158

20	0.1037	7.4694	9.6463	72.0524	0.1339	0.0139
25	0.0588	7.8431	17.0001	133.3339	0.1275	0.0075
30	0.0334	8.0552	29.9599	241.3327	0.1241	0.0041
35	0.0189	8.1755	52.7996	431.6635	0.1223	0.0023
40	0.0107	8.2438	93.0510	767.0914	0.1213	0.0013
45	0.0061	8.2825	163.9876	1358.2300	0.1207	0.0007
50	0.0035	8.3045	289.0022	2400.0182	0.1204	0.0004

14% Interest Factors for Discrete Compounding

N	P/F	P/A	F/P	F/A	A/P	A/F
1	0.8772	0.8772	1.1400	1.0000	1.1400	1.0000
2	0.7695	1.6467	1.2996	2.1400	0.6073	0.4673
3	0.6750	2.3216	1.4815	3.4396	0.4307	0.2907
4	0.5921	2.9137	1.6890	4.9211	0.3432	0.2032
5	0.5194	3.4331	1.9254	6.6101	0.2913	0.1513
6	0.4556	3.8887	2.1950	8.5355	0.2572	0.1172
7	0.3996	4.2883	2.5023	10.7305	0.2332	0.0932
8	0.3506	4.6389	2.8526	13.2328	0.2156	0.0756
9	0.3075	4.9464	3.2519	16.0853	0.2022	0.0622
10	0.2697	5.2161	3.7072	19.3373	0.1917	0.0517
11	0.2366	5.4527	4.2262	23.0445	0.1834	0.0434
12	0.2076	5.6603	4.8179	27.2707	0.1767	0.0367
13	0.1821	5.8424	5.4924	32.0887	0.1712	0.0312
14	0.1597	6.0021	6.2613	37.5811	0.1666	0.0266
15	0.1401	6.1422	7.1379	43.8424	0.1628	0.0228
16	0.1229	6.2651	8.1372	50.9804	0.1596	0.0196
17	0.1078	6.3729	9.2765	59.1176	0.1569	0.0169
18	0.0946	6.4674	10.5752	68.3941	0.1546	0.0146
19	0.0829	6.5504	12.0557	78.9692	0.1527	0.0127
20	0.0728	6.6231	13.7435	91.0249	0.1510	0.0110
25	0.0378	6.8729	26.4619	181.8708	0.1455	0.0055
30	0.0196	7.0027	50.9502	356.7868	0.1428	0.0028
35	0.0102	7.0700	98.1002	693.5727	0.1414	0.0014
40	0.0053	7.1050	188.8835	1342.0251	0.1407	0.0007

N	P/F	P/A	F/P	F/A	A/P	A/F
45	0.0027	7.1232	363.6791	2590.5648	0.1404	0.0004
50	0.0014	7.1327	700.2330	4994.5213	0.1402	0.0002

16% Interest Factors for Discrete Compounding

N	P/F	P/A	F/P	F/A	A/P	A/F
1	0.8621	0.8621	1.1600	1.0000	1.1600	1.0000
2	0.7432	1.6052	1.3456	2.1600	0.6230	0.4630
3	0.6407	2.2459	1.5609	3.5056	0.4453	0.2853
4	0.5523	2.7982	1.8106	5.0665	0.3574	0.1974
5	0.4761	3.2743	2.1003	6.8771	0.3054	0.1454
6	0.4104	3.6847	2.4364	8.9775	0.2714	0.1114
7	0.3538	4.0386	2.8262	11.4139	0.2476	0.0876
8	0.3050	4.3436	3.2784	14.2401	0.2302	0.0702
9	0.2630	4.6065	3.8030	17.5185	0.2171	0.0571
10	0.2267	4.8332	4.4114	21.3215	0.2069	0.0469
11	0.1954	5.0286	5.1173	25.7329	0.1989	0.0389
12	0.1685	5.1971	5.9360	30.8502	0.1924	0.0324
13	0.1452	5.3423	6.8858	36.7862	0.1872	0.0272
14	0.1252	5.4675	7.9875	43.6720	0.1829	0.0229
15	0.1079	5.5755	9.2655	51.6595	0.1794	0.0194
16	0.0930	5.6685	10.7480	60.9250	0.1764	0.0164
17	0.0802	5.7487	12.4677	71.6730	0.1740	0.0140
18	0.0691	5.8178	14.4625	84.1407	0.1719	0.0119
19	0.0596	5.8775	16.7765	98.6032	0.1701	0.0101
20	0.0514	5.9288	19.4608	115.3797	0.1687	0.0087
25	0.0245	6.0971	40.8742	249.2140	0.1640	0.0040
30	0.0116	6.1772	85.8499	530.3117	0.1619	0.0019
35	0.0055	6.2153	180.3141	1120.7130	0.1609	0.0009
40	0.0026	6.2335	378.7212	2360.7572	0.1604	0.0004
45	0.0013	6.2421	795.4438	4965.2739	0.1602	0.0002
50	0.0006	6.2463	1670.7038	10435.6488	0.1601	0.0001

18% Interest Factors for Discrete Compounding

N	P/F	P/A	F/P	F/A	A/P	A/F
1	0.8475	0.8475	1.1800	1.0000	1.1800	1.0000
2	0.7182	1.5656	1.3924	2.1800	0.6387	0.4587

3	0.6086	2.1743	1.6430	3.5724	0.4599	0.2799
4	0.5158	2.6901	1.9388	5.2154	0.3717	0.1917
5	0.4371	3.1272	2.2878	7.1542	0.3198	0.1398
6	0.3704	3.4976	2.6996	9.4420	0.2859	0.1059
7	0.3139	3.8115	3.1855	12.1415	0.2624	0.0824
8	0.2660	4.0776	3.7589	15.3270	0.2452	0.0652
9	0.2255	4.3030	4.4355	19.0859	0.2324	0.0524
10	0.1911	4.4941	5.2338	23.5213	0.2225	0.0425
11	0.1619	4.6560	6.1759	28.7551	0.2148	0.0348
12	0.1372	4.7932	7.2876	34.9311	0.2086	0.0286
13	0.1163	4.9095	8.5994	42.2187	0.2037	0.0237
14	0.0985	5.0081	10.1472	50.8180	0.1997	0.0197
15	0.0835	5.0916	11.9737	60.9653	0.1964	0.0164
16	0.0708	5.1624	14.1290	72.9390	0.1937	0.0137
17	0.0600	5.2223	16.6722	87.0680	0.1915	0.0115
18	0.0508	5.2732	19.6733	103.7403	0.1896	0.0096
19	0.0431	5.3162	23.2144	123.4135	0.1881	0.0081
20	0.0365	5.3527	27.3930	146.6280	0.1868	0.0068
25	0.0160	5.4669	62.6686	342.6035	0.1829	0.0029
30	0.0070	5.5168	143.3706	790.9480	0.1813	0.0013
35	0.0030	5.5386	327.9973	1816.6516	0.1806	0.0006
40	0.0013	5.5482	750.3783	4163.2130	0.1802	0.0002
45	0.0006	5.5523	1716.6839	9531.5771	0.1801	0.0001
50	0.0003	5.5541	3927.3569	21813.0937	0.1800	0.0000

20% Interest Factors for Discrete Compounding

N	P/F	P/A	F/P	F/A	A/P	A/F
1	0.8333	0.8333	1.2000	1.0000	1.2000	1.0000
2	0.6944	1.5278	1.4400	2.2000	0.6545	0.4545
3	0.5787	2.1065	1.7280	3.6400	0.4747	0.2747
4	0.4823	2.5887	2.0736	5.3680	0.3863	0.1863
5	0.4019	2.9906	2.4883	7.4416	0.3344	0.1344
6	0.3349	3.3255	2.9860	9.9299	0.3007	0.1007
7	0.2791	3.6046	3.5832	12.9159	0.2774	0.0774
8	0.2326	3.8372	4.2998	16.4991	0.2606	0.0606

9	0.1938	4.0310	5.1598	20.7989	0.2481	0.0481
10	0.1615	4.1925	6.1917	25.9587	0.2385	0.0385
11	0.1346	4.3271	7.4301	32.1504	0.2311	0.0311
12	0.1122	4.4392	8.9161	39.5805	0.2253	0.0253
13	0.0935	4.5327	10.6993	48.4966	0.2206	0.0206
14	0.0779	4.6106	12.8392	59.1959	0.2169	0.0169
15	0.0649	4.6755	15.4070	72.0351	0.2139	0.0139
16	0.0541	4.7296	18.4884	87.4421	0.2114	0.0114
17	0.0451	4.7746	22.1861	105.9306	0.2094	0.0094
18	0.0376	4.8122	26.6233	128.1167	0.2078	0.0078
19	0.0313	4.8435	31.9480	154.7400	0.2065	0.0065
20	0.0261	4.8696	38.3376	186.6880	0.2054	0.0054
25	0.0105	4.9476	95.3962	471.9811	0.2021	0.0021
30	0.0042	4.9789	237.3763	1181.8816	0.2008	0.0008
35	0.0017	4.9915	590.6682	2948.3411	0.2003	0.0003
40	0.0007	4.9966	1469.7716	7343.8578	0.2001	0.0001
45	0.0003	4.9986	3657.2620	18281.3099	0.2001	0.0001
50	0.0001	4.9995	9100.4382	45497.1908	0.2000	0.0000

25% Interest Factors for Discrete Compounding

N	P/F	P/A	F/P	F/A	A/P	A/F
1	0.8000	0.8000	1.2500	1.0000	1.2500	1.0000
2	0.6400	1.4400	1.5625	2.2500	0.6944	0.4444
3	0.5120	1.9520	1.9531	3.8125	0.5123	0.2623
4	0.4096	2.3616	2.4414	5.7656	0.4234	0.1734
5	0.3277	2.6893	3.0518	8.2070	0.3718	0.1218
6	0.2621	2.9514	3.8147	11.2588	0.3388	0.0888
7	0.2097	3.1611	4.7684	15.0735	0.3163	0.0663
8	0.1678	3.3289	5.9605	19.8419	0.3004	0.0504
9	0.1342	3.4631	7.4506	25.8023	0.2888	0.0388
10	0.1074	3.5705	9.3132	33.2529	0.2801	0.0301
11	0.0859	3.6564	11.6415	42.5661	0.2735	0.0235
12	0.0687	3.7251	14.5519	54.2077	0.2684	0.0184
13	0.0550	3.7801	18.1899	68.7596	0.2645	0.0145
14	0.0440	3.8241	22.7374	86.9495	0.2615	0.0115
15	0.0352	3.8593	28.4217	109.6868	0.2591	0.0091

16	0.0281	3.8874	35.5271	138.1085	0.2572	0.0072
17	0.0225	3.9099	44.4089	173.6357	0.2558	0.0058
18	0.0180	3.9279	55.5112	218.0446	0.2546	0.0046
19	0.0144	3.9424	69.3889	273.5558	0.2537	0.0037
20	0.0115	3.9539	86.7362	342.9447	0.2529	0.0029
25	0.0038	3.9849	264.6978	1054.7912	0.2509	0.0009
30	0.0012	3.9950	807.7936	3227.1743	0.2503	0.0003
35	0.0004	3.9984	2465.1903	9856.7613	0.2501	0.0001
40	0.0001	3.9995	7523.1638	30088.6554	0.2500	0.0000
45	0.0000	3.9998	22958.8740	91831.4962	0.2500	0.0000
50	0.0000	3.9999	70064.9232	280255.6929	0.2500	0.0000

Index

A

Absorbent media cleaning, 154
Acetone, 167
Additives, 166
Adsorption, 11
Air emissions, chromium, 133
Air spraying, paint, 161
Alkalinity, solution, 143
Aluminum coating operations, 247
Aluminum, primary, 258
Analysis, inventory, 230
Anionic exchange resins, 122
Anodic electrolytic recovery, 126
Antifoaming agents, 144
Arsenic preservatives, 249
Assessment phase, waste minimization
       program, 22
Assessments
    impact, 231
    life-cycle, 229
Assignable causes, statistical process
       control, 56
Assurance, quality, 209
Atomic weight calculations, 35
Attribute control charts, 79
Attributes, statistical process control, 56
Audits
    environmental, 226
    internal quality, 218

B

Benefit/cost ratio, 199
Biosolids, recycle, 53
Blackbody emittance, 100
Blasting, stripping, 172
Bulk absorbents, 154
Business incentives, waste minimization, 4
Bypass, calculations, 52

C

C-chart, 67, 80
Capability determination, 75
Capability index, 76
Capacitance sensors, 89
Capital-recovery factor, 189
Carbon dioxide blasting, 173
Cash flow, 185
Cathodic electrolytic recovery, 124
Cationic exchange resins, 121
Characterization, impact assessment, 231
Chemical drag-out, 110
Chemical manufacturing study, 16
Chemical replacement, 132, 166
Chlorinated aliphatic hydrocarbons, 248
Chlorinated solvents, 137
Chlorobenzenes, 248
Chlorofluorocarbons, replacement, 148
Chlorophenols, 247, 249

Chromium, 133
Chromium preservatives, 249
Classification, impact assessment, 231
Clean Air Act, 108, 138, 155, 261
Clean Water Act, 108
Cleaners
    aqueous, 143
    enclosed gun, 163
    petroleum hydrocarbon, 147
    semiaqueous, 145
Cleaning
    absorbent media, 154
    solvent, 136
    spray gun, 157
    supercritical fluids, 149
    wet oxidation, 153
Closed-loop approach, 10
Coatings, surface, 155
Coking, 259
Colorants, 167
Common causes, statistical process
        control, 56
Common Sense Initiative, 4
Compound interest, 178
Compound-amount factor, 187
Computer control, 103
Conductivity, measurement, 95
Conductivity meters, 117
Conservation of mass, 40
Contaminated soil, 249
Continuous improvement, 205
Continuous level measurement, 89
Continuous-flow manufacturing, 42
Contract review, 212
Control charts, 65, 79
Control
    computer, 103
    process, 214
    quality, 209
Control limits, calculation of, 70
Control measures
    customer-supplied product, 214
    document and data, 213
Control system, pH, 94

Copper, primary, 258
Corrective actions, nonconforming
        products, 216
Corrosion inhibitors, 144
Corrosivity, hazardous waste, 242
Costs, hidden, 203
Countercurrent rinsing, 49, 113
Critical review, life-cycle assessment, 233
Critical-point constants, 150
Cumulative frequency, 60
Cumulative percentage, 60
Cyanide, 134

D

Data acquisition systems, 83
Data collection, 68
Data control, 213
Declining depreciation, 181
Defects, definition, 79
Deflocculants, 144
Degreasing
    hydrochlorofluorocarbon compounds,
        148
    solvent, 136
Delivery, product, 217
Demonstration Factory, 10
Density, 30
Depreciation, 179
Design, control processes, 212
Design for the environment, product
        standards, 238
Desorption, 12
Differential-pressure flow meters, 87
Dimensions, 30
Discrete compounding, 187
Distillates, hydrocarbon cleaners, 147
Distillation recovery, 10
Distributions, statistical process control,
        57
Document control, 213
Doppler flow meters, 86
Drag-out reduction, chemical, 110
Dry ice blasting, 173

E

Economics, engineering, 177
Effective interest rate, 179
Electrodialysis, 12, 126
Electrolysis, 12
Electrolytic recovery, 124
Electromagnetic flow meters, 86
Electroplating, bath composition, 131
Electroplating operations, 247
Electrostatic spraying, paint, 162
Electrowinning, 124
Emergency Planning and Community
       Right-to-Know Act, 3
Emission control strategies, 160
Emissivity, 100
Emulsifiers, 144
Emulsive cleaners, 145
Enclosed gun cleaners, 163
Energy inputs, product manufacture, 236
Environmental auditing, 226
Environmental impacts, product
       manufacture, 233
Environmental management systems, 221
Epoxies, 167
Equations, chemical, 35
Equipment, quality control, 215
Equipment redesign, 10
Evaporation, 130
Expert review, life-cycle assessment, 233

F

Feasibility analysis, waste minimization
       program, 24
Ferroalloys, 258
Filtration, 12
       membrane, 126
Float switch, 87
Flow measurement, 85
Flow restrictors, 117
Flumes, 86
Frequency, 60

Frequency table, 57
Future worth, calculation of, 184

G

Gap analysis, ISO 14001, 223
Grade, technical variations, 207

H

Halogenated solvents, 246
Handling, product, 217
Hazard Communication Standards, 109
Hazardous air pollutants, Clean Air Act,
       261
Hazardous and Solid Waste Amendments
       (HSWA), 2
Hazardous waste, CFR listing, 245
Hazardous waste, characteristics, 241
Heavy metals, coatings, 170
Hidden costs, 203
High-volume, low-pressure atomization,
       161
Histogram, 61
Housekeeping, 13
Hydrocarbons, petroleum, 147
Hydrochlorofluorocarbons, 148
Hydrogen-ion concentration, 92
Hyperfiltration, 128

I

Ignitability, hazardous waste, 241
Impact assessment, 231
Implementation, waste minimization
       program, 25
Improvement, quality, 209
Inflation, 183
Infrared thermometry, 100
Ink formulation, 259
Inorganic chemicals, 256
Inorganic pigments, 252
Inspections, quality control, 215

Instrumentation, 84
Integration, computer, 103
Interest, 177
Interest factors, discrete compounding,
         187, 269
International Organization for
         Standardization (ISO), 205
Interpretation, life-cycle assessment, 232
Inventory analysis, 230
Inventory control, 13
Ion exchange , 120
Iron and steel, 258
ISO 10011, 218
ISO 14000, 206, 219
ISO 14001, 221
ISO 14040, 229
ISO 840Z, 207
ISO 9000, 206, 210
ISO standards, 206

J

Just-in-Time inventory management, 13

L

Laminar flow patterns, 85
Leachate, restricted waste, 251
Lead
     primary, 258
     secondary, 258
Level change, sudden, 74
Life-cycle assessment, 229
Liquid flow measurement, 84
Liquid level measurement, 87
Lower control limits, calculation of, 70
Lower specification limit, 76

M

Management systems, 205
Manual, quality, 209

Mass flow meter, 87
Material balance, 29, 40
Material control, 13
Material inputs, product manufacture, 236
Material substitutions, 9
Mean, 63
     calculation of, 69
Media cleaning, 154
Membranes, separation, 126
Metal finishing
         process management, 109
         chromium, 133
         cyanide, 134
Metal heat treating operations, 247
Metal stripping, 134
Meters, conductivity, 117
Methane combustion, 48
Microfiltration, 127
Mixed liquor, 53
Mixing, paints and coatings, 163
Mixtures, 74
Molality, 38
Molarity, 37
Mole, 35
Molecular weight calculations, 35
Montreal protocol, 138

N

Nanofiltration, 128
National Defense Center for
         Environmental Excellence, 10
National Emission Standards for
         Hazardous Air Pollutants, 108
National Pollutant Discharge Elimination
         System, 109
Natural pattern, 72
Net present worth, 190
New process implementation, 10
Nominal interest rate, 178
Non-halogenated solvents, 246
Nonconformities, definition, 79

Normal distribution, 60
Normality, 38
NP-charts, 67, 82

O

Occupational Safety and Health
        Administration, 109
Open-channel flow measurement, 86
Operating practices, 12
Opportunity assessment, waste
        minimization program, 19
Organic chemicals, 252
Organic solvents, 151
Organization, waste minimization
        program, 20
Osmosis, 128
Outliers, 73
Oxidation
    technologies, 165
    wet, 153
Oxidation-reduction potential,
        measurement, 97
Ozone depletion, chlorinated solvents, 137
Ozone-depleting substances, 148

P

P-charts, 67, 82
Packaging, product, 217
Paints
    spray application, 160
    volatile organic compound content,
        168
    water-based, 167
Parrafin, hydrocarbon cleaners, 147
Particulates, heavy metal, 171
Patterns
    natural, 72
    unnatural, 73
Payback period, 196
Percentage, cumulative, 60
Pesticides, 256
Petroleum hydrocarbons, 147

Petroleum refinery wastes, 250
Petroleum refining, 257
pH, measurement, 91
Phased approach, opportunity assessment,
        19
Pigments, 166
Planning
    quality, 209
    waste minimization program, 20
Plastic media blasting, 172
Point level measurement, 87
Pollution prevention, product standards,
        238
Pollution Prevention Act, 3
Polyurethanes, 167
Positive-displacement flow meters, 86
Present worth, 187–190
Preservation, product, 217
Pressure conversions, 32
Pressure measurement, 101
Preventive actions, nonconforming
        products, 216
Primary materials, product manufacture,
        236
Primer operations, 157
Printed circuit board manufacturing study,
        14
Probability plot, 61
Probe constant, 95
Probes, conductivity, 95
Problem solving, 38
Process automation, 10
Process capability study, 67
Process changes, 9
Process control, 214
    statistical, 55
Process implementation, 10
Process optimization, 39
Product
    changes, 8
    identification, 214
    manufacturing and delivery, 217
    nonconforming, 216
    standards, environmental impacts,
        233

Profitability analysis, 201
Project analysis, financial, 186
Project XL, 4, 221
Proportioning, paints and coatings, 163
Proximity sensor, 87
Purchasing, 214
Purge, calculations, 52

Q

Quality
    characteristics, 79
    management responsibility, 211
    management, 207
    systems, 210, 212

R

Radiant energy principles, 101
Radio frequency sensors, 90
Range, 66, 69
Rate of return, 195
Reaction yields, 37
Reactivity, hazardous waste, 242
Reclamation, 18
Records, control of quality, 217
Recovery
    chemicals, 165
    solutions, 119
Recycle
    biosolids, 53
    calculations, 52
Recycling, 17
    solution, 119
Refrigerant recycling, 18
Regeneration, resin beds, 123
Registration, ISO 14001, 224
Reinventing Environmental Regulation
    Initiative, 4
Reporting, life-cycle assessment, 233
Resistance temperature detectors, 99
Resistivity, 95
Resource conservation, product standards,
    238

Resource Conservation and Recovery Act,
    2, 107
Resource recovery, 18
Return on investment, 195
Reverse osmosis, 12, 128
Review, life-cycle assessment, 233
Reynold's number, definition, 85
Rinsewater management, metal finishing,
    113
Risk anaysis, 201
Rotary vane meters, 86
Run, 73

S

Saponifiers, 144
Secondary materials, product manufacture,
    236
Self review, life-cycle assessment, 233
Self-declaration, ISO 14001, 224
Sensitivity analysis, 200
Sensors
    capacitance, 89
    proximity, 87
    radio frequency, 90
    ultrasonic, 90
Separation, membrane, 126
Sequestering agents, 144
Series compound-amount factor, 188
Series present-worth factor, 189
Servicing, quality , 218
Significance analysis, impact assessment,
    231
Simple interest, 177
Sinking-fund factor, 188
Sinking-fund method, depreciation, 182
Solution agitation, 117
Solvent cleaning, hydrochlorofluorocarbon
    compounds, 148
Solvents, 167
    cleaning and degreasing, 136
    miscellaneous organic, 151
Source reduction, 7
Source segregation, 12

Specific gravity, 31
Specification limits, 76
Spray application, paint, 160
Spray gun cleaning, 157
Spray rinsing, 116
Stakeholder review, life-cycle assessment, 233
Standard deviation, 63, 71
Standards,
    ISO, 206
    ISO 14000, 219
    ISO 14001, 221
    ISO 14040, 229
    product, 233
Static rinsing, 115
Statistical procedures, quality, 218
Statistical process control, 55
Steady state, 41
Stoichiometry, chemical, 35
Storage, product, 217
Straight-line depreciation, 181
Strain, definition, 101
Stratospheric Ozone Protection Public Law, 148
Stressor-effects network, 231
Stripping, paints, 171
Sudden level change, 74
Sum-of-digits depreciation, 181
Supercritical fluids, 149
Surface coating, 155
Surfactants, electroplating baths, 132
Sustainable development, 220
Switch, float, 87
Systems, quality, 209–212

T

Taxes, 183
Technology changes, 9
Technology development, 10
Temperature
    conversions, 34
    correction, pH meter, 93
    measurement, 99

Testing, quality control, 215
Thermistors, 100
Thermoelectric circuit, 99
Title VI, Clean Air Act, 138
Topcoat application operations, 157
Total quality management, 209
Toxic Release Inventory, 3
Toxicity characteristic, hazardous waste, 243
Traceability, product, 214
Training, quality, 218
Transducers, pressure, 103
Trends, 73
Turbine flow meters, 86
Turbulent flow patterns, 85

U

U.S. Environmental Protection Agency
    opportunity assessment manual, 19
    strategic waste minimization initiative user's guide, 19
    industrial pollution prevention opportunities, 20
U-charts, 67–81
Ultrafiltration, 12, 127
Ultrasonic flow meters, 86
Ultrasonic sensors, 90
Units, 30
Upper control limits, calculation of, 70
Upper specification limit, 76

V

Valuation, impact assessment, 231
Vapor degreasing, 137
Variables
    data collection, 68
    statistical process control, 56
Velocity flow meters, 86
Veterinary pharmaceuticals, 258
Volatile organic compounds, content of paints, 168
Volumetric flow measurement, 84

W

Waste minimization
  business incentives, 4
  chemical replacement, 132
  definition, 1
  operating practices, 12
  phases of, 19
  program planning and organization,
    20
Weirs, 86
Wet oxidation cleaning, 153
Wood preservation, 252
  operations, 249

X

$\bar{X}$-R charts, 66, 69

Y

Yield, definition, 79

Z

Zinc baths, cyanide replacement, 134
Zinc, primary, 258